MODERN AGRICULTURE
SCIENCE, FINANCE, PRODUCTION AND ECONOMICS

SENIOR AUTHOR AND EDITOR
DAVID P. PRICE, Ph. D.
PRIVATE CONSULTANT
Las Cruces, NM

Contributing Authors

CHARLES GLOVER, Ph.D.
NEW MEXICO STATE UNIV.

JOHN L. HAVLIN, Ph.D.
KANSAS STATE UNIV.

JOEL J. KEMPER, Ph.D.
PRIVATE CONSULTANT
MODESTO, CALIF.

E.T. KORNEGAY, Ph.D.
VIRGINIA POLYTECHNIC INST.

JOHN KUHL, Ph.D.
PRIVATE CONSULTANT
RIALTO, CALIF.

STEPHEN C. MASON, Ph.D.
UNIV. OF NEBRASKA

MARK MAYSE, Ph.D.
UNIV. OF CALIFORNIA - FRESNO

ROBERT E. MOORE, P.E.
PRIVATE CONSULTANT
PHOENIX, ARIZ.

JAMES H. LARUE, M.S.
UNIV. OF CALIFORNIA - DAVIS

VERL THOMAS, Ph.D.
MONTANA STATE UNIV.

DAN UNDERSANDER, Ph.D.
UNIV. OF WISCONSIN

Technical Consultant (aquaculture)
DAVID H. MAYER, M.S.
SAN DIEGO, CALIF.

COLLEGE OF THE SEQUOIAS
LIBRARY

Copyright © in 1989. All rights reserved. No portion of this book may be reproduced in any manner without written permission of the publisher.

Published and Distributed by
SWI Publishing
P.O. Drawer 3-A&M
University Park, New Mexico 88003
United States of America

Tel. 505-525-1370
Fax 505-525-1394

Library of Congress Catalog No. 88-64058
ISBN 0-9606246-6-X Hardcover

DEDICATION

Dr. Howard Hesby
Texas A&M University

This book is about agriculture, but more than that, it is about teaching. It is therefore dedicated to Dr. Howard Hesby, the finest example of what a teacher can be.

The university system as it exists today has one enormous weakness and flaw. Virtually all the recognition and notoriety goes to the area of research. While most campuses do give awards for teaching, the actual recognition is miniscule and provincial, compare to what can be obtained in research. For that reason, teaching is often thought of as a burden, rather than as an opportunity.

Therefore, to excel at teaching, one must have dedication, and character. Above all, they must have a selfless desire to make a contribution. Such a man is Howard Hesby.

To look at Howard Hesby, one would not think of him as a teacher. Rather, because of his enormous physical strength and muscularity, one would tend to think more on the order of a professional athlete. But professional athletes do not stay up into the pre-dawn hours tutoring students or preparing lectures. Indeed, few people in any profession make the sacrifices or put forth the raw effort that is the hallmark of Howard Hesby.

The bottom line is that Howard Hesby cares. And that is what teaching, or for that matter civilization, is all about. People who care enough to make a contribution . . . whether they are fully compensated or not.

In our society, teachers are not fully compensated. They are taken for granted. That is not the way it should be, but that is the way it is. Only when we fully recognize them for their worth, will this change. I would like to take this opportunity to recognize Howard Hesby. Howard Hesby is the finest example of what a teacher, or for that matter, a human being, should be.

TABLE OF CONTENTS

BOOK I FINANCE, BUSINESS, AND ECONOMICS 1

CHAPTER 1 FINANCE 3
    The Basic Law of Finance 3
    Levered Capital 4
    The Anatomy of Financial Failures 4
    Operating Capital 4
    Long-Term Capital Financial Failures 5
    The Motivation For Lenders 5
    Summary 7

CHAPTER 2 MARKETING (AS A MEANS OF RISK MANAGEMENT) 9
    Risk Management 9
    Forward Marketing to Avoid Risk 10
    Forward Pricing 10
    Vertical Integration 12
    The Basic Rules of Marketing 12
    Understanding Marketing Tactics When Buying Products 13
    Summary and Overview 13

CHAPTER 3 AGRICULTURAL ECONOMICS 15
    Demand 15
    Supply 16
    Costs 19
    Money and Banking 20
    Summary 21

CHAPTER 4 BASIC FINANCIAL ACCOUNTING 23
    The Balance Sheet or Financial Statement 23
    The Income Statement 24
    The Cash Flow Analysis 25
    Summary 27

CHAPTER 5 THE FUTURES MARKET 29
    What Futures Prices Are 29
    Mechanics of Hedging 29
    Figuring Basis 30
    Margin Calls 31
    Options and the Futures Market 32
    Protecting Price with a "Put" Contract 33
    Protecting Against Price Increases 33
    The "Call" Option 34
    Dangers of the Futures Market 35
    Summary 36

BOOK II AGRONOMY 37

CHAPTER 6 SOILS 39
    Productivity 39
    Development of Soils 39
    Soil Environment 41
    Soil Physical Properties 43
    Soil Chemical Properties 46

  Soil Biological Properties 47
  Soil Erosion Effects 50
  Concluding Note 52

CHAPTER 7 SOIL FERTILITY MANAGEMENT 55
  Soil Fertility and Soil Productivity 55
  Nutrient Uptake and Plant Growth 55
  Nitrogen 57
  Phosphorus 62
  Potassium 63
  Calcium, Magnesium and Sulfur 65
  Micronutrients 67
  Acid Soils and Liming 69
  Saline and Sodic Soils 70
  Soil Testing and Fertilizer Recommendations 71
  Conclusion 73

CHAPTER 8 TILLAGE PRINCIPLES AND PRACTICES 75
  Plowing 75
  Cultivation 76
  Other Purposes of Tillage 76
  Fertilization During Tillage 78
  Row Crop Spacing 78
  Herbicides 78
  Summary 79

CHAPTER 9 AN INTRODUCTION TO IRRIGATION 81
  Irrigation System 81
  Soil Type 83
  Crop Water Requirements 85
  When to Irrigate 86
  Water Source 87
  Pumping Plant 88
  Design Criteria for Irrigation Systems 88
  Summary 89

CHAPTER 10 TWO EXAMPLES OF HIGHLY SUCCESSFUL IRRIGATION PROJECTS 91
  Imperial Irrigation District 91
  System Orientation 92
  Ordering Water 92
  Maintenance Concerns 93
  Salt River Project 95
  Service to the Irrigator 95
  Crops Grown 97
  Maintenance Concerns 97
  District Direction and Politics 97
  Summary 97

CHAPTER 11 DRYLAND AGRICULTURE 99
  Dryland Agriculture: Semi-Arid Regions 102
  Fallow Systems 103
  Factors Limiting Dryland Crop Productivity 106
  Dryland Agriculture: Humid Temperate Regions 108
  Summary 111

CHAPTER 12 COMMON FIELD CROPS ... 113
- Corn ... 113
- Sorghum ... 114
- Small Grain Production ... 115
- Soybeans ... 117
- Cotton ... 118
- Summary ... 119

CHAPTER 13 SEED AND SEEDING ... 121
- Variety Development ... 122
- Seed Germination ... 122
- Seed Laws and Regulations ... 125
- Seeding the Crop ... 125
- Summary ... 126

CHAPTER 14 TREE FRUIT AND NUT PRODUCTION ... 127
- Climatic Zones ... 127
- Nursery Production ... 127
- Orchard Establishment and Training ... 128
- Pruning ... 129
- Pollination ... 129
- Fruit Growth and Development ... 130
- Nutrition and Irrigation ... 132
- Insects and Diseases ... 132
- Harvesting and Marketing ... 132
- Overview ... 135

CHAPTER 15 TROPICAL CROP PRODUCTION SYSTEMS ... 137
- Soils ... 138
- Cropping Systems ... 138
- Shifting Cultivation ... 138
- Multiple Cropping ... 139
- Rice Systems ... 140
- Pasture Systems ... 141
- Plantation Systems ... 143
- Summary ... 144

CHAPTER 16 ECONOMIC ENTOMOLOGY ... 149
- Morphology and Physiology ... 145
- Taxonomy ... 146
- Applied Insect Ecology ... 146
- Economic Injury Level and Economic Threshold ... 149
- Survey of Integrated Pest Management Tactics ... 149
- Future Challenges in Integrated Pest Management ... 153

BOOK III FEEDS AND FEEDING ... 155
(excerpted from *Modern, Practical Feeds, Feeding, and Animal Nutrition* by D. Porter Price Ph.D)

CHAPTER 17 BASIC ANIMAL FEEDS AND NUTRIENTS ... 157
- Major Nutrients Contained in Feeds ... 157
- Terms Used in Relation to Livestock Feeds ... 159
- Minerals ... 161
- Macro Minerals ... 162
- The Macro or Trace Minerals ... 163

    Vitamins 165
    Summary and Overview 167

CHAPTER 17A FEED PROCESSING FOR LIVESTOCK 169
    Pelleting 169
    Dry Grinding and Rolling of Grains 169
    Steam Rolling of Grains 170
    Steam-Flaking 170
    Pressure-Flaking 171
    Popping 171
    High-Moisture Grain 171
    Reconstituted Grain 172
    Tempering 173
    Toasting 173
    Other Forms of Heat Treatment 173
    Summary 173

CHAPTER 18 BASIC ANIMAL DIGESTION 175
    The Monogastric System 175
    The Ruminant System 176
    Summary 179

CHAPTER 19 HAY AND SILAGE MAKING (Forage Preservation) 181
    The Proper Harvesting of Forages 181
    Hay and Haymaking 181
    Silage and Silage Making 185
    Summary and Overview 188

CHAPTER 20 RATION FORMULATION 189
    Adjusting for Moisture 189
    The Pearson's Square 189
    Balancing for Two or More Nutrients 190
    Least Cost Rations 192
    Predicting Animal Performance with Calculated Rations 193
    Estimating Consumption 194
    Predicting Performance Utilizing the California Net Energy System 195
    Calculating Milk Production 196
    Overview 196

BOOK IV LIVESTOCK PRODUCTION 197

CHAPTER 21 PRINCIPLES OF FARM ANIMAL REPRODUCTION 199
    The Reproductive Tract of the Female 199
    Reproductive Anatomy of the Male 200
    Methods of Mating 201
    Special Supplement. Nutritional Aspects of Reproduction 205
    Overview 207

CHAPTER 22 PRINCIPLES OF GENETICS (as related to animal breeding) 209
    Genes 209
    Genetic Aspects of Animal Breeding 211
    Correlation Between Traits 213
    Selection Intensity 213
    Summary 213

CHAPTER 23 POULTRY PRODUCTION 215
    The Egg 215
    The Digestive System 216
    Diseases of Poultry 216
    Common Management Practices 218
    Laying Chickens 223
    Breeding Chickens 226
    Broilers 227
    Turkeys 228
    Pheasants 230
    Ducks 230
    Summary 230

CHAPTER 24 DAIRY PRODUCTION 231
    Milk Composition and Consumption 231
    Milk Pricing 231
    The Cow/The Milk 233
    Milking Systems 234
    Dairy Feeds/Nutrients 235
    Grouping Cows 237
    Challenge Feeding 238
    Feeding and Management of the Non-Milking Herd 239
    Genetic Improvement and Reproductive Efficiency 243
    Summary 244

CHAPTER 25 BEEF PRODUCTION 245
    excerpted from *Beef Production, Science & Economics, Application & Reality*
    by D. Porter Price, Ph.D.)
    Western Cow-Calf Ranching 245
    Range Management 245
    Type of Operation - Stocker Vs. Mother Cow 249
    Supplemental Feed for Breeding Cows 251
    Cow-Calf Confinement 254
    Stocker Operations 257
    Cattle Feeding and Feedlot Management 260
    Beef Cattle Breeds 261
    Summary 265

CHAPTER 26 GENERAL CHARACTERISTICS OF THE SWINE INDUSTRY 267
    Organizing and Establishing a Swine Enterprise 267
    Types and Breeds of Swine 268
    Nutrition and Feeding 273
    Reproduction and Breeding Herd Management 278
    Boar Management 280
    Sow and Pig Management 281
    Weanling Pig Management 283
    Feeding and Management of Growing-Finishing Swine 284
    Herd Health Management 285
    Marketing and Marketing Systems 287
    Waste Management 287
    Record Keeping 288
    Summary 289

CHAPTER 27 SHEEP PRODUCTION AND MANAGEMENT 291
    Products Produced 291
    Characteristics of Sheep 291

    Systems of Production 292
    Reproductive Cycle 292
    Preparations for Lambing 295
    Post-Lambing Management 297
    Ram Management 301
    Breeding Systems and Selection 301
    Types of Breeds 303
    Common Sheep Diseases and Disorders 303
    Basics of Wool Grading 304
    Breeds of Sheep 306
    Summary 311

CHAPTER 28 AQUACULTURE AND MARICULTURE 313
    Major Differences Between Fish Farming and Conventional Animal Agriculture 315
    Fish Nutrition and Feeding 317
    Fish Reproduction 319
    Summary 320
    The Aesthetic Need for Mariculture 321

APPENDICES 325
    Nutrient Requirements of Poultry 327
    Nutrient Requirements of Dairy Cattle 328
    Nutrient Requirements of Beef Cattle 329
    Nutrient Requirements of Swine 338
    Nutrient Requirements of Sheep 339
    Feed Composition Tables 340

GLOSSARY 345

INDEX 355

BOOK I   FINANCE, BUSINESS, AND ECONOMICS

CHAPTER 1 INTRODUCTION TO FINANCE
(Understanding Risk)

Finance was chosen as the first chapter for *Modern Agriculture*, because without finance, there can be no modern agriculture. Tractors, irrigation equipment, combines, hybrid seeds, etc., all require an outlay of capital.

Indeed, this is the number one problem that is holding back most underdeveloped countries. It is not the absence of technology. Technology is readily available and can be purchased very inexpensively in the form of consultants. It is the hardware of technology that is expensive, and the lack of availability of capital that keeps third world countries underdeveloped.

The Worldwide Need for Farm Credit. The U.S. recognized this problem early on and developed a government sponsored farm credit system. As early as 1916, the U.S. passed the first in a series of farm credit acts designed to ensure a readily available source of capital for agricultural producers.* Capital that would be available regardless of the borrower's social status or political ties. At that point in time, it was difficult for someone without "connections" to get financed. Agriculture was a risky business, and large city banks were not interested in loaning to someone they didn't "know". Loans went to the wealthy and famous, leaving homesteading as the only alternative to the rest of the population.

Homesteading is a form of subsistence agriculture. If a country is to prosper and grow, its people must be more productive than that. Capital must be made available to allow mechanized agriculture to come into play, thereby releasing a segment of the population to become more gainfully employed in other segments of the economy.

This is the unfortunate situation most underdeveloped nations now find themselves in. Large segments of the labor force are involved in subsistence agriculture, which not only provides very little extra food for the rest of the population, but provides very little income for the general economy. That is, in developed countries, farmers buy machinery, seeds, fertilizer, etc., which creates jobs and incomes for other people. They can do this, because they receive income from the farm products they eventually sell.

In subsistence agriculture, farmers cannot buy inputs of production, because they have no income from the sale of their products. Thus, they provide neither food, nor income, nor jobs for the rest of the population.

INTRODUCTION TO RISK

As discussed in the previous section, the availability of capital is a necessity for modern agriculture. In essence, it is a tool. But like any tool, we must learn how to use it. Each year there are thousands of farmers that find themselves in deep financial trouble because they did not manage the tool of agricultural credit properly.

The thing we must keep in mind is that the availability of capital benefits a country and its people as a whole. The availability of capital means more and cheaper food for the population at large. But for the individual farmer, the availability of capital can be either a blessing or a curse. The availability of capital can mean expanding a small farm into a much larger burgeoning operation, or even getting into farming with no previous ownership of land or equipment. But the availability of capital can also mean greatly magnified financial problems. Problems which can end in the loss of all assets and land. Problems which can lead to intense emotional and even physical distress.

THE BASIC LAW OF FINANCE

The number one rule of finance that every farmer/investor should never lose sight of . . . is that **Rate of Return is a Function of Risk**.

The obvious implication of RR = Function of Risk is that the highest yielding investments are also the riskiest. Oil drilling can pay off several thousand percent of the original investment, whereas an ordinary certificate of deposit (CD) at a bank will only pay a very low percent. The difference, of course, is that in oil drilling the chances of losing the entire investment is very great; with the CD, the investment is guaranteed.

An agricultural example of this type of thing would be planting a highly speculative and perishable crop such as lettuce, versus a staple cash crop such as alfalfa. The price of lettuce often varies as much as 1000% within a given year, whereas the price of alfalfa will typically vary by no more than about 30% per year. With lettuce, if the farmer hits the market right, he can make as much as 5 times more per acre than he could with the alfalfa. However, if the market moves the other way, he may not even make back his cost of production.

The same analogy can also be applied to a single crop. That is, as explained in Chapter 2 and 5, through the use of forward selling or contracting, there are ways a farmer can lock in a price for his crop before it is even planted. His other option, of course, is not to forward sell his crop, and hope the price goes up. Indeed, the price very well could go up; but it could also go down.

---

* The Federal Intermediary Credit Bank and the Production Credit Associations.

Thus, by leaving his crop unsold until harvest, a farmer increases his potential return, but in so doing, he also exposes himself to more risk.

PERSPECTIVE

The preceding examples of crop selection and pricing explains how farmers can increase or decrease their risk through normal farm management decisions. Some, however, may question what this has to do with finance. Finance is normally thought of as "borrowing money" or otherwise funding an investment. Indeed, some texts and training courses discuss finance primarily in terms of the financial documents and strategy needed to obtain financing. Specifically, the intent is to simply teach farmers how to obtain a loan. Unfortunately, that is precisely the attitude that gets so many farmers into trouble.

While "how to get financed" (financed to buy a farm) will be discussed in detail later on in the chapter... it is vitally important that every potential borrower first understand risk. Investment involves risk, and if we ever lose sight of that fact, then we are leaving our financial survival to chance. As discussed in Chapter 2, we must learn to manage risks if we are to survive. Chapter 2 will discuss how to manage risks, the primary purpose of this chapter is to teach the student how to understand and recognize those risks.

LEVERED CAPITAL

If there is anything that personifies risk, it is the principle of "leverage". By definition, levered capital means borrowed capital (borrowing money for the purpose of making money).

It is called "leverage" because like a lever, it helps the borrower do things he ordinarily could not do. With a lever, he can lift more weight; with levered capital he can make more money. With borrowed capital the farmer can farm more acres, buy more machinery, graze more cattle, feed more hogs, milk more cows, or otherwise expand his operation.

The ultimate purpose, of course, is (or at least should be) to make more money. Indeed, with an expanded operation the potential for increased income is there. However, one should always be cognizant of the fact that the lever can work in both directions. If things don't go well, the farmer can also lose more money.

More importantly, he can lose money he doesn't have; and this is what makes levered investments so dangerous and potentially tragic. If the farmer doesn't have the money to pay off the losses, the bank, or other financial institutions, only have two choices: 1. lend more money, or 2. foreclose.

What choice the bank makes depends primarily upon the value of the collateral (the asset the borrower pledged to ensure the loan). If the value is high enough to support further lending, or the farmer can come up with further collateral, then the bank will usually lend more money (that is, if the value of the asset, when sold, would be enough to pay off both loans). The reason for this is simple, no one wants to see a foreclosure, and the bank does not want the kind of negative publicity that accompanies a foreclosure.

But additional financing essentially puts the farmer in double jeopardy. Where only a portion of the asset or assets might have been lost previously, the farmer now stands to lose everything. The most tragic situation being where a farmer loses the land that had been in the family for generations.

THE ANATOMY OF FINANCIAL FAILURES

While it is impossible, if not presumptuous to categorize all financial failures, there are two general patterns that appear. The first involves operating capital, and occurs over a period of years. The second involves long term capital, but occurs over a relatively short period of time.

OPERATING CAPITAL

The term operating capital means just what it implies. Loans to buy the year-to-year inputs of agriculture: fertilizer, seed, feed, fuel, etc. Nearly every farmer or rancher in business requires operating capital. Even if the land was inherited free of debt, the inputs of today's agriculture are so expensive, that in almost all cases, borrowed capital is required to operate the land.

For a farmer or rancher who owns his or her land, obtaining operating capital is no problem. The amount of operating capital required is usually a small fraction of the value of the land. The farmer/rancher pledges the land as collateral, and except in unusual cases, banks or other financial institutions will lend on that basis. That is, it is a very "safe" loan for the bank.*

A farmer in this sort of a financial situation is also in a relatively "safe" position. If losses are experienced one year, it is usually fairly easy to obtain additional loans (operating capital) for the next year.

However, if the losses are substantial, and if they continue over a period of years, there can be a day of reckoning. As described in the previous section, the ultimate deciding factor will be the value of the assets versus the amount of the loans. If the amount of the loans exceeds the value of the assets, then as a general rule, there will be no loan. And, if the farmer cannot pay off the existing loan, there will be a foreclosure.

The recent farm crisis. As mentioned, in the past, this type of scenario developed over a long period of time.

---

* That is, the value of the collateral far exceeds the loan, and therefore in case of a foreclosure, the bank would be sure to be paid off.

However, just recently in the U.S., a situation developed that greatly accelerated this problem, and devastated farmers that would have otherwise been in a reasonably good financial situation.

The situation was the devaluation of land. Beginning in the early-to mid-1980's, the value of farm land began to decline. Always before, land values had appreciated. Banks and other financial institutions, therefore, could loan more and more money each year. Devaluation of land, however, broke the cycle, and suddenly reversed the financial picture of a great number of farmers. The result has been foreclosures of an unprecedented magnitude.

LONG-TERM CAPITAL FINANCIAL FAILURES

The term long-term capital is fairly self explanatory. Long-term capital refers to funds used for purchasing major assets, to be repaid over a lengthy time period. The most common asset purchased with long-term loans is land. Payment will normally be spread over 20 to 30 years, although some notes may be as short as 10 years.

The reason long-term capital is used for the purchase of major assets is due to the ability to repay. The total cost of the asset is great, but by spreading the payments over a number of years, the individual payments become more manageable. Ideally, income-producing assets should be able to produce enough income to pay for themselves. That is, ideally, income-producing assets should generate enough income each year to cover the yearly payments.

It is important to realize that the advantages of long-term capital apply almost totally to the borrower. As a general rule, financial institutions do not like to make long-term loans. Certainly there are specialized institutions set up to make home mortgage loans, and life insurance companies often make long-term land loans, but other than that, there are very few financial institutions that are willing to make long-term loans. Indeed, this is why the U.S. government set up the Federal Land Bank, as a means of making long-term credit available to farmers and ranchers.

Short- and medium-term loans, however, are much easier to come by. Therefore, there is often a temptation to finance long term assets with short-term capital. This is something that should never be done, and indeed, can be considered the "Cardinal Sin of Finance". Unfortunately, however, farmers and ranchers often do it, sometimes without ever realizing what they are doing.

Probably the best example in this area is some types of heavy farm machinery. Dealers and/or banks are often willing to lend for 3 to 5 years on machinery which is actually a long-term investment; a long-term asset both in cost and service. Another very common example is using operating capital for land improvement such as clearing, fencing, etc.

In both of these very common cases, relatively large sums of money are invested in assets that would require many years of service to justify their cost. But when financed with short or medium term loans, they create an unnecessary financial burden. That is, they must be paid for before they have made their financial contribution to the operation.

Under some circumstances, this can create what is known as a cash flow crisis (for a complete explanation of cash flow see Chap. 4). The net result can be the inability to pay off the loan with revenue generated by the business. Ultimately then, it can mean foreclosure and loss of a valuable asset to the farm, as well as the equity (payments already made) on the asset.

During the recent farm financial crisis of the late 1980's, the purchase of machinery was a significant part of this crisis. Farmers were buying unprecedented amounts of farm machinery, often on short-term type financing; the type normally associated with automobile loans. When farm incomes (commodity prices) declined, in many cases farmers were unable to meet their obligations. As a result they lost (had repossessed) the machinery they needed to work their land. On top of that, they often had already paid many, many thousands of dollars in payments, down payments, equipment trade-ins, etc., which were all lost during the foreclosures. In some cases, this was a financial blow from which many farmers could not recover. The bottom line is "never finance a long-term asset with short-term capital!".

THE MOTIVATION FOR LENDERS

In order to fully appreciate the risk involved in finance one should understand the basic motivation of lenders.

Above all, the single most important item one must remember is that if a bank lends you money **it is not doing you a favor!** Banks and bankers are not public servants or benefactors. They are businesses and business people in business to make a profit. Indeed, the term "lend" and "loan" in reality are misnomers. In reality, financial institutions do not "lend" money. What they really do is "rent" or "lease" money. Money which must be returned, along with the "rental fee" they charge for the use of that money.*

This, of course, is how banks make money. They take money that has been deposited and "lend" it out at rates greater than they pay their depositors. This, of course, does constitute a valuable service. Banks make money available and keep it in circulation. Without banks or other lending institutions savers would hoard their money and little, if any, would be available for develop-

---

* The rental fee, of course, includes interest charges, but there are also usually some "service charges" which, in actuality, simply increase the interest rate.

ment of new businesses, industry, home construction, etc.*

But even though banks perform a valuable service, never forget that they are in business to make a profit. They will "lend" money, but if, and only if, they can do so at a profit. Understanding this profit motive may seem very basic, but it is important to keep in mind when applying for a loan. Another vitally important concept to understand is how risk applies to banks.

With respect to risk, one of the most common misunderstandings concerns the profitability of the borrower. The feeling usually is that the more profitable a business, the more attractive it should be to the lender or bank. As explained in the chapter on accounting (Chap. 4), while banks and bankers certainly want their customers to be profitable, they are not necessarily concerned with profitability. The bank does not share in that profitability. The interest and other fees they charge are exactly the same, regardless of whether the business is profitable or unprofitable.

What the bank is concerned with is risk. The bank's real concern is whether the borrower will be able to repay the loan. The bottom line in case of default and foreclosure is: "is there sufficient collateral to pay off the loan?"

Now it should be pointed out that most reputable banks and bankers do not want foreclosures. As also explained in the Accounting Chapter (Chap. 4), bankers usually want to see cash flow analyses to ensure that funds will be readily available to meet bank payments. But the bottom line is collateral. Collateral, of course, is the assets a borrower pledges to a bank to "secure" the loan. Collateral is the asset or assets the bank may take possession of and sell, should a default occur.

So what does a banker's motivation mean to you? As explained in the previous section, bankers have two primary motivations: profit and avoidance of risk. The profit motive means that bankers want to make loans. Just as car rental companies want to rent the cars they have, bankers want to rent the money at their disposal; but they also want to avoid risk. They want to make sure they get their money back.

This means that banks or other lenders are going to tie up the borrowers assets. The lender wants to make sure he gets his money back, whether the borrower's business is a success or failure.

Preparing for a loan. Obviously, before attempting to get a loan, the borrower should study and analyze the feasibility of being able to repay that loan (service the debt). As discussed in Chapter 4 one of the best ways to do that is with a cash flow analysis.

Once it has been determined that the loan is feasible (that it can be repaid), then the borrower needs to consider what will be used as collateral. That is, what he or she stands to lose, should he or she be unable to repay the loan.

In some cases, the item itself is used to secure the loan. The most common example is land. In that case, the buyer is required to make a down payment which amounts to a percentage of the value of the asset. In the case of land, down payments are usually in the range of 10 to 20%. In a great number of cases, the down payments on land will come from the life savings or family inheritance of the borrower.

In cases where down payments are not required, some other asset is used as collateral. In cases where a farmer or rancher is attempting to expand the size of his or her operation, the original farm or ranch is often pledged as collateral. In that case, the original farm or ranch is often a family heirloom passed down for generations.

Is what you have to gain, worth what you have to lose? This is a question every potential borrower should ask themselves. This is especially true when attempting to expand an operation (and the original operation must be pledged as collateral).

Is the income derived from the expanded operation worth risking the loss of the original operation? Is the income to be derived from a farm or ranch worth risking the loss of a family inheritance?

Notice the use of the word "income" (income, of course, referring to dollars and cents). Put in those terms, the value of a loan is relatively easy to calculate. But when non-financial motives come into play, calculation becomes much more difficult. When emotion becomes a motivating factor, calculation not only becomes difficult, but is often overlooked. For many, the desire to enter into farming and ranching is so great that it usurps all other considerations.

When this is the situation, one must ask oneself, "is what I have to gain, worth what I have to lose?" "Is going into farming or ranching worth risking the loss of my family inheritance or life savings?" "Is expanding an existing operation, worth risking the loss of losing the original operation?"

These are tough questions, but they must be asked, and carefully considered. Once one has come to grips with these questions, then he or she can usually evaluate the financial potential for success a bit more objectively.

SHOPPING FOR A LENDER

Once one has gone through all the preceding steps of evaluating whether a loan is financially feasible or not,

---

*As discussed in the beginning of this chapter, the absence of banks (availability of capital) is the primary factor holding back many underdeveloped countries.

and has concluded that it is . . . then they are ready to begin preparation for applying for a loan.

As discussed previously, a banker is most certainly not doing us a favor by giving us a loan. It is a business transaction by which the bank intends to make money. When we walk into a bank for the purpose of taking out a loan (renting some money), we are "a customer" almost as if we had walked into an ordinary store. Therefore, it is in our best interest to bargain for the best deal possible.

Not all banks, in all situations, will bargain or negotiate their fees. But if we are a responsible person, of good reputation, and have carefully analyzed what our financial plans and capabilities are; then we are the kind of customer banks want. In many cases then, some banks will negotiate their fees, particularly as the loans get larger, and/or we deal with more senior loan officers. But even if they will not negotiate, we can always shop around.

But in shopping for a banker, there are instances in which the character, personality, and experience of the banker can be as important as the fees charged by the banker. This is especially true for operating capital loans.

In shopping for operating capital loans it is important to have a lender that understands agriculture. From time to time, special needs and situations occur that someone experienced in agriculture is in a much better position to understand.

Unfortunately, finding bank officers with outside business experience is the exception, rather than the rule. It is, therefore, that much more important to shop around. Because in agriculture, particularly with operating capital loans, unknowledgeable and/or unreasonable loan officers can make untimely and costly demands, or fail to recognize and respond to special emergency needs.

SUMMARY

To the public as a whole, the availability of borrowed capital in agriculture is extremely beneficial. In short, it allows for modern production methods which increases production while lowering cost.

For the individual farmer, however, borrowed capital is a two-way street. On the positive side, borrowed capital can allow a farmer to expand his or her operation, and potentially make more money. On the negative side, the borrowed capital can potentially cause the farmer to lose more money. Most important, borrowed capital can result in the farmer losing money he or she doesn't have.

In essence, finance involves risk. The basic law of finance is that **Rate of Return is a Function of Risk**. Before borrowing, every farmer should determine what his or her risk is. Most important, every potential borrower should determine if the potential risk is worth the potential gain.

CHAPTER 2 MARKETING
(AS A MEANS OF RISK MANAGEMENT)

If you ask just about any agricultural banker what the biggest problem facing farmers is . . . they will reply marketing. Most farmers know how to produce, but very few know how to market. Indeed, most farmers produce their commodity, and then worry about marketing it later. Seldom is an attempt made to sell their product until harvest begins.

Actually, selling or pricing the crop should begin before it is even planted. If a profitable price for that crop cannot be guaranteed, then there is little point in growing it. A large manufacturing firm wouldn't even think about manufacturing a product without a solid idea as to what price they can expect it to bring. Yet time and again, farmers plant the same crops year after year, with no idea what kind of price they will receive. As a result, agriculture is deemed to be a high risk enterprise.

RISK MANAGEMENT

Risk management is a term that has come into common usage the last few years. Basically what it implies, is managing a business to avoid potentially damaging market risk.

Farmers, of course, are used to risk. They face enormous weather risk every year. Lack of rain can keep a crop from growing, or too much can wash it away. A farmer can work all year to grow a crop, and hail can ruin it, wind lodge it, or frost damage it.

Farmers have faced these kinds of risks for generations, so it's not difficult to understand why they are willing to face market risk. But today's generation of farmers face risks unlike their predecessors.

Today's farming has become capital intensive. Each acre feeds more and more people, and therefore to get more out of it, more has to go into it. The cost of seed, fertilizer, fuel, and machinery has escalated, so the farmer has more invested in his crop. He must therefore not only worry about making a crop, but also receiving a price that will cover his production costs.

HOW MUCH RISK CAN A FARMER AFFORD?

This is a vital question, and one that must be answered by each individual farmer. Yet few farmers take the time to study it.

How much risk a farmer can afford is related to the net worth of the individual, or the farm enterprise. A wealthy farmer can afford more risk than a farmer who is not wealthy.

A farmer who does not own his land (has his farm mortgaged), and must borrow all his operating capital, cannot assume much risk. He must be able to sell his crop for at least his production costs, or he's out of business. If he cannot meet his mortgage payments, then he loses his farm. If he doesn't make his operating capital payments, then he can't work his farm.

A farmer who inherited his land, or a farmer who only has to borrow a portion of his operating capital, can afford some risk. If the price he receives doesn't meet production costs, then he can absorb the loss out of his own pocket, and hope for a better year next year.

If a better year doesn't come, then the farmer may begin to get into trouble. He may move from a position of being able to afford some risk, to a position where he cannot afford much risk.

The classic situation has been the farmer who inherited his land, and then was forced into mortgaging the land. A similar situation is putting the land up as collateral. In either case, he is essentially using his land to finance his operating expenses. This puts him in a vulnerable position.

WHY WOULD ANYONE WANT TO FACE RISK?

The only rational argument for facing risk is for extra profits. As pointed out in Chapter 1, the first law of finance is that rate of return is a function of risk.

When a crop is sold by forward contract, then the price is already set. If the cash market for the crop goes up, the farmer who forward sold his crop, doesn't make any more money. The farmer who didn't forward sell, does. He can make substantially more money than the more prudent farmer; but he can also lose money.

The question is, "How much can he afford to lose?" The classic example is the cattle feeding business from 1975 to 1985. On the average, during that period of time, the cattle feeders who didn't forward sell their cattle made substantially more money. There were periods where the market dipped, and they lost some money, but on the average they made money. After seeing that pattern for several years, a number of cattle feeders and feedlots decided that forward selling wasn't necessary.

In the spring of 1985, however, the cattle market took an extraordinary plunge. It dropped nearly 20%, and stayed down for 9 months. As a result, a number of feedlots and cattle feeders went bankrupt.

Thus, how much risk a farmer can take is not a function of the potential for extra profit. Rather, it's a matter of how much he can afford to lose and still stay in business.

FIGURING HOW MUCH YOU CAN AFFORD TO LOSE

Deciding how much you can afford to lose is both a calculation and a decision. The first step is to calculate all mortgage and operating capital payments. Then, look at your balance sheet or financial statement and look at your equity.

Equity, as explained in Chapter 4, means how much of your farming operation actually belongs to you. Compare your equity to your payments. What is very important, is to look and see where your equity is coming from. There are a lot of farmers who are very wealthy on paper . . . but most of that wealth is tied up in the ownership of their land.

Are you willing to risk your land? Is the potential for some extra profits, worth the risk of losing your land?

Is your equity tied up in machinery? If you were forced to sell some of your machinery (to cover mortgage or operating capital payments), could you still operate your farm? If not, could you lease or rent your farm for enough money to meet your mortgage payments?

These are tough questions, but they are questions that every farmer must face. All too often farmers simply accept risk as a fact of life, without really trying to quantify and identify the amount of risk they must deal with.

FORWARD MARKETING TO AVOID RISK

INTRODUCTION

Basically, there are two functions of forward marketing: 1. to ensure a profit, and 2. to avoid price risk. At first, those two goals may seem to be one and the same. That is, if you ensure a profit, then you're also guarding against price risk.

But achieving **maximum** profit can conflict with avoiding price risk. The basic reason for this is that in order to avoid price risk, you essentially must forward sell your commodity. In order to forward sell your product, you must find someone willing to forward **buy** your product.

The motivation for someone to forward buy your product is the same as your motivation for forward selling . . . avoiding price risk. In the case of the buyer, he wants to avoid the price going up, whereas you want to avoid the price going down. Thus, the price you both agree on will probably be neither very high nor very low. It will be in-between.

As mentioned previously, oftentimes a farmer who does not forward sell his crop, can get a higher price than a farmer who does. Therefore he stands to make a bigger profit. However, he also stands to suffer a loss (if the price goes down).

Therefore, the first step in setting up a marketing plan to avoid risk is to decide how much profit you want. Obviously, we want all we can get. But in this situation we have to be realistic and decide what is the least amount that is acceptable.

In some cases, with some crops, we may find that an acceptable profit is not obtainable. That is, once we decide what an acceptable price (profit) is, we may find that there is very little likelihood of finding a buyer willing to pay that price.

Good examples for many years have been grain crops.

Because of overproduction, wheat, corn, barley, and sorghum crops have often been priced below production costs.* Therefore, if the lowest price we feel is acceptable has not been achieved in recent years, then it is unrealistic to think it can be achieved in the crop year to come.

The time to reach these conclusions, of course, is **before** the crop is planted. That way, if we need to go to a different crop, there is time to make that decision. That may sound simple, yet it is probably the most commonly made mistake in farm management.

FORWARD PRICING

Basically there are two ways of forward pricing: 1. actual physically selling (forward selling) to an individual buyer, and 2. using the Futures Market. Both have advantages and disadvantages, depending upon the situation.

FORWARD SELLING

Forward selling is much like any other business contract in which a seller agrees to deliver a product at some point in the future for a specified price. As with other business contracts, it is highly important to know the reputation, integrity, and financial backing of the other party.

Indeed, the major disadvantage of forward selling to an individual is the potential for failure to execute contract. Failure to execute can come from a simple refusal to make good on the contract to bankruptcy.

Whenever the price falls below the price specified on the contract, there is always danger of the buyer refusing to honor the contract. In this case, often the only way to force him to honor the commitment is through a lawsuit. Lawsuits can be long and expensive. Most important, it can be months or even years before any payment is actually received.

For this reason, it is always a good policy to have a "liquidated damages" clause written into the contract. Liquidated damages is a legal term which means that the penalties for failing to execute the contact are already spelled out. This usually speeds up the legal process.

But again, it is highly important to know with whom you are dealing. It makes no sense to enter into a contract with someone who has a less than excellent reputation for honoring commitments.

Likewise, it is not wise to enter into a contract with someone who is financially weak. The question to ask yourself is, "if the market dropped substantially, could the other party afford to buy your crop at the higher price, and still stay in business?" For example, if you are selling grain and the buyer is a large breakfast cereal

---

* A recent exception was 1988 in which crop failures due to drought increased prices.

company the answer would probably be "yes". The cost of grain is only a fraction of the selling price of the breakfast cereal, and thus the cost of the grain could be easily absorbed.

If, however, the buyer were a small country elevator, the answer could very well be "no". In that case the buyer would have to buy your grain at a high price, and turn around and sell it at a lower price. If the volume were very great (it also pays to know what other contracts the other party may have out), he very well could have to go out of business.

When this happens, there is always the danger of bankruptcy being declared. What is particularly dangerous about bankruptcy, is that if the grain or other commodity has already been delivered . . . you may never be paid for it at all.

In a classic case that happened in Missouri, farmers delivered large tonnages of soybeans to an elevator that later declared bankruptcy. After the bankruptcy the soybeans legally belonged to the elevator, even though the farmers had not been paid for them (and probably would not be paid for them).

Finally, a group of the farmers illegally broke into the elevator, and got their soybeans out. The whole thing was filmed on national television, and the farmers faced criminal charges. Ultimately, the farmers were not prosecuted, and were allowed to keep the soybeans.

In another case that gained national attention, farmers were not so lucky. In this case an extremely large cattle feedlot was buying grain from local farmers. When the feedlot suddenly declared bankruptcy, the farmers lost everything. There was no grain to take back, since it had been fed to cattle, and the cattle belonged to owners other than the feedlot.

Therefore, when forward selling, it's extremely important to know with whom you are dealing. The point being that forward selling in itself is not a guarantee against price risk. What is required is forward selling to a reputable buyer who will honor his commitments, and is otherwise financially strong.

THE FUTURES MARKET

Among many farmers the Futures Market is a highly controversial topic. The Futures Market is also quite complex. An entire chapter (Chap. 5) is devoted to explaining the mechanics of the Futures Market. For the purposes of this chapter, only the potential role of the Futures Market as a forward pricing (risk management) tool will be discussed. In that regard, the Futures Market has some advantages over forward selling to an individual, but it also has some major disadvantages. Most specifically, **the Futures Market requires a lot of study. The Futures Market should therefore be used only by those who have the time and motivation to become familiar with its workings and complexities.**

The major advantage of the Futures Market over forward selling is the financial strength of the Commodities Exchange. When you sell in the Futures Market, you know the buyer will have to make good on his part of the contract. As mentioned in the previous section, this is not necessarily true of a contract with an individual. All you have in that case is the character and integrity of the individual. With the Futures Market, the entire system is designed to prevent defaults on the part of the participants (explained in detail in Chap. 5).

The disadvantages of the Futures Market are many. Complexity is the main disadvantage. As mentioned, it requires a large amount of specific knowledge and study. For those who are unwilling to acquire that knowledge, the Futures Market can be potentially disastrous.

Another problem relates to what are known as margin calls. The actual purpose of margin calls, however, is to ensure financial integrity of futures contracts. The problem is that they can create immediate demands for funds.

That is, whenever a futures contract is bought or sold, a portion of that contract (usually 5-15%) must be put up as margin. As the market changes, the margin required of the buyer or seller changes also. If the market moves adversely to the contract you have sold (or bought), then you must put up more margin money. As explained in Chap. 5 this requires special financing arrangements; otherwise it can cause severe short-term cash problems.

For the purpose of this chapter however, let it suffice to say that margin calls ensure that the buyer of a future's contract is financially responsible. However, farmers who sell futures contracts are also subject to margin calls, and should be prepared for them (again, see Chap. 5).

Also for the purposes of this chapter, let it just suffice to say that the Futures Market is a viable means of forward selling. Many large and prosperous farmers routinely use the Futures Market as a means of protecting against price risk. But before using the Futures Market, one must study and understand it.

Options Contracts. Another manner in which Futures trading can be used to avoid price risk, is through the trading of options contracts. These are contracts sold in the commodities exchanges which allow the holder the "option" of selling or buying a futures contract at a specified price. Although their use is somewhat complicated (explained in more detail in Chap. 5), they are essentially a form of insurance.

Like insurance, there is a fee (premium) to be paid. A farmer pays for the right to sell (or buy) a futures contract at a specified price. The farmer does not have to sell or buy a futures contract, but merely has the "option" or right to do so.

Therefore the only time a farmer would exercise his right or "option", would be if the market moved in a direction unfavorable to the price specified in his option contract. For example, if a farmer had an option contract to sell wheat at $3.50/bu., and the actual price dropped to $3.00/bu., then he would exercise his contract for the $3.50 price.* If, however, the price went to $4.00/bu., then the farmer would do nothing. He would let his options contract expire, and just consider the cost of the options contract like an insurance premium. These items are explained in much more detail in Chap. 5. For the purposes of this chapter, however, the student simply needs to be cognizant that options contracts are available, and function as a form of price insurance. Also, as mentioned, in the section on futures contracts, the use of the Futures Market is complicated, and requires intense study before it should be entered into.

PHYSICAL RISK IN FORWARD SELLING

As mentioned in the beginning of this chapter, farmers face a great number of physical risks; that is, wind, hail, flood, drought, frosts, as well as unusual insect and disease damage. Whenever we forward sell our crop before it is produced, the physical risks become magnified.

The reason is that if a crop is destroyed chances are good that a large area will be involved. This normally decreases production of that crop, which increases the price. If we are committed to deliver a crop, then it could be very expensive to have to buy a commodity to meet that commitment (contract). **Therefore, whenever a crop is forward sold, it should be insured.**

VERTICAL INTEGRATION

Aside from forward selling, most other forms of attempting to avoid price risk involve what is known as vertical integration. Basically, vertical integration means moving into another phase or segment of a business. In agriculture, vertical integration refers to moving a step closer to the consumer. Examples would be vegetable farmers investing in a cannery, dairy farmers investing in a milk processing plant, etc.

Consumer food products generally do not fluctuate in price as much as raw commodities; thus this has been a motivation in many of these cases; that is, price protection. Also, consumer food products tend to have a higher markup (profit) than raw commodities and this increased profit has also been a strong motivation.

But as mentioned in Chap. 1, profit is a function of risk. Therefore, while the idea of protecting commodity prices to the farm can be afforded by vertical integration, the additional investment creates additional risk. Therefore, vertical integration is not something to be entered into lightly. In most cases it should be considered as a separate business apart from the farming enterprise, rather than as a market for the commodities produced by the farming enterprise.

THE BASIC RULES OF MARKETING

There are two basic means of marketing a product: 1. price competition, and 2. product differentiation. That is, in selling a product we must either sell at a price equal to or below our competition, or we must convince the buyer that our product is superior to our competitor's.

Price competition is by far the easiest means of selling a product. Only minimum effort is required as buyers will usually seek out the seller. Most farm commodities are sold by this method. That is, most raw commodities are lumped into broad quality categories and considered as generic products (example; #2 Corn, etc.). They are then sold according to a "market" price. Certainly farmers try and get the highest price they can, but ultimately they must sell at or near the going or "market" price. That is, they must sell at or near the price at which other farmers are willing to sell.

As mentioned in the previous section, vertical integration toward the consumer can mean higher profit margins for the commodity. But in order to obtain those higher margins, we must usually also engage in product differentiation. That is, we are no longer selling sweet corn, we are selling XYZ canned corn; we are no longer selling fluid milk, we are selling ABC ice cream, yogurt, etc.

Of course, we don't have to become involved in product differentiation, but if we don't, then we are back competing on price only, and the much lower profit margins that accompany price competition. Therefore, nearly every branded product will make some attempt at advertising and promotion. The problem is that advertising and promotion can be very expensive, and if the advertising and promotion doesn't work then we are forced back into price competition (and we forfeit the cost of the promotion).

Therefore, in agriculture, vertical integration should normally not be thought of as a means of marketing a commodity (protecting against price risk). Rather, vertical integration should normally be thought of as a separate business. A business that will require additional investment and business risk to accompany that investment.

PRODUCT DIFFERENTIATION
WITHOUT VERTICAL INTEGRATION

In agriculture there have been various producers attempt and succeed at selling their commodity at higher prices through product differentiation. Probably the best examples have been purebred livestock breeders. In the purebred livestock business, promotion and advertising are relied upon heavily.

---

* Special Note: It is important to realize that the face value of a futures or option contract is not the net price a farmer will receive. Usually it will be less than the face value due to discounts, transportation and other costs and fees.

Purebred livestock breeders typically join together to form a breed association. The breed association then becomes heavily involved in trade magazine advertising, direct mail advertising, trade show and county/state fair promotion etc. The purpose being to convince livestock buyers that their particular breed is superior to other breeds.

In addition, individual producers will often conduct promotion of their own. Promotion that can also include print advertising, direct mail, trade show, and fair promotion. The purpose of this advertising is usually to convince buyers that their livestock is superior to others of the same breed.

As in marketing any product by means of product differentiation, if the program is successful, it will mean a greater sale price and hopefully, greater profit. If it is not successful, however, it will simply mean greater cost.

UNDERSTANDING MARKETING TACTICS WHEN BUYING PRODUCTS

Understanding the basics of price competition versus product differentiation can be useful when attempting to buy products as well. Indeed, nearly every product farmers buy for use in production agriculture will have a recognizable marketing effort behind it. That is, the seller will be attempting to compete by either price competition or product differentiation.

Fertilizer is normally sold by price competition. The farmer usually decides what level of nitrogen, phosphorous or potassium is needed, and then shops for the best price. Occasionally fertilizer companies will attempt to state that their product is superior, but most commonly, fertilizer is sold on a formula basis. When fertilizer companies do attempt to compete on a non-price basis, it is usually in the area of service (delivery, availability, and/or application of the product), or financing. This is not to say that fertilizer companies would not like to compete by product differentiation, but standardization of formulas and ingredients makes it difficult.

Products like animal feed lend themselves toward product differentiation much better. That is, there are large differences in the digestibility of ingredients; therefore, feed companies can claim that they use superior ingredients in their products. Also there are large numbers of ingredients contained in feed, and feed companies sometimes make the claim that they have special formulas that their competitors don't have.

While it is certainly true that there can be big differences in feeds, the purpose of all this is to get away from the idea of having a generic product. That is, the attempt is to get away from selling a particular feed formula but to try and sell a specific product. Instead of selling a particular level of protein, vitamins, minerals, etc., the feed is often given a trade name. The idea is to sell the customer a name rather than a formula, and thereby keep him or her from comparing the product with the competition.

Other products such as farm machinery are usually sold by both methods. Some machinery is meant to be price competitive, while other machinery is claimed to be superior (product differentiation). Often the parent company will attempt to compete by claiming product superiority, while the dealers for that equipment compete with each other on the basis of price.

With machinery, of course, there really can be big differences in technology and/or quality. The important thing to assess, however, is how much of the difference is real and how much is advertising or marketing strategy. One thing to keep in mind is that mechanical changes are sometimes brought about purely for marketing purposes. That is, the engineering department is directed to come up with changes whether they are needed or not. The purpose being product differentiation rather than function.

SUMMARY AND OVERVIEW

The time to sell a crop is before it is planted. If a crop cannot be sold for more than it costs to produce it, then it makes little sense to plant it.

Selling a crop before it is produced is a basic function of **risk management**. That is, it eliminates price risk which could otherwise be financially devastating.

There are other risks, of course, such as weather risk. However, crop insurance can be purchased to protect against weather risk. Likewise, there are means of purchasing price insurance as well. Known as "options," these financial contracts may be purchased through commodity exchanges, which (when used properly) essentially provide a guaranteed price. Like conventional insurance, however, there is a cost. There is a fee charged for options which would be comparable to an insurance premium (explained in detail in Chap. 5).

Commodity prices can be guaranteed without incurring such fees. The simplest method is forward contracting. That is, the producer simply contracts with a buyer to sell at a specified price at some point in the future (harvest, etc.). While such contracts are legally binding and legal help should probably be consulted in their creation, undoubtedly the biggest consideration is knowing the reputation of the buyer. If the buyer fails to live up to his part of the agreement, then the purpose of entering into the contract has been exacerbated. Indeed, entering into a forward contract with a dishonorable or financially weak buyer can increase price risk, rather than avoid it. That is, the most common situation is that buyers fail to execute contract only when the market price goes down. Therefore, entering into a contract with a disreputable buyer does not totally avoid down-side risk, but eliminates upside profits.

The biggest and most ever-present danger, however, is that a buyer could take bankruptcy. If he takes bankruptcy after taking delivery of the commodity, the producer not only does not get paid for his crop, he loses it entirely. **If a buyer is in possession of a commodity when he takes bankruptcy, it legally belongs to the buyer - even though the buyer has not paid for it.**

Use of the Futures Market eliminates the danger of nonpayment through what are termed "margin calls." Margin calls are sums of money demanded on a short term basis by the Commodity Exchange, in response to fluctuation in commodity prices. The purpose is to ensure payment of a contract. However, they can be demanded of the seller as well as the buyer, depending upon which direction the market has moved.

Because margin calls can occur very quickly (as quickly as the market moves), and because the amounts of money demanded can be quite substantial, margin calls can be psychologically upsetting. If a line of credit has not been set up in advance, or short-term funds are not otherwise readily available, margin calls can lead to serious financial problems. Indeed, nothing has created more problems for farmers using the Future's Market, than margin calls.

As stated in the text, the Futures Market is a sophisticated method of forward pricing that should be used only by those who have studied it and understand it. While the Futures Market can be very advantageous, it is not something someone should enter into without a sound background of knowledge. Likewise, a commitment to remain informed and abreast of factors affecting the Futures Market is essential. (This book does not supply adequate knowledge. It explains the basics only.)

In short, forward pricing is essential in good farm management but it carries a responsibility. If we forward contract with an individual, we must know with whom we are dealing. If we enter into the Futures Market (either as a hedger or as an options buyer), we must understand all the workings of the system.

Finally, whenever we forward contract we must be sure we can deliver. If we are dealing with a commodity subject to crop failure, then we must insure the crop.

Some farmers have used vertical integration as a means of guaranteeing prices. Vertical integration generally means moving into the processing or distribution of agricultural products; that is, moving closer to the consumer. While this does guarantee a market for raw agricultural products, it also creates a need for more investment. Investment which can lead to further risk. Therefore, vertical integration should not normally be thought of as a means of avoiding market risk. Rather, it should be considered a separate business.

CHAPTER 3   AGRICULTURAL ECONOMICS

A basic understanding of economics is vital to anyone interested in managing a farm or related agricultural business. However, this should present no problem as the concepts of economics are very logical and easy to understand. It is only the terms used in economics that are sometimes confusing. They are confusing in that they are often words used in ordinary English, yet when used in reference to economics can have very specific and different meanings.

<u>Demand</u>. Demand is probably the most confusing of all the terms used in economics. In plain English, when the word demand is used relative to goods or commodities, it refers to the desire or willingness to buy a product. It may also refer to the acceptance of a product by the

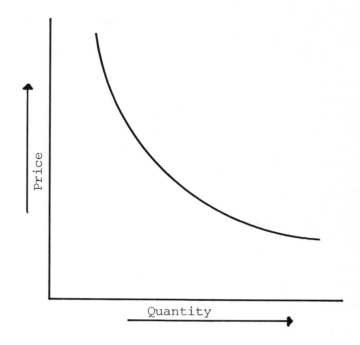

Figure 3-2. Theoretical demand curve for beef. Unlike wheat, beef has a number of competing products and therefore it is a much more price sensitive commodity. As the price changes, so also does the amount people are willing to purchase. Therefore beef (as explained in the text) would be considered a commodity with an elastic demand curve.

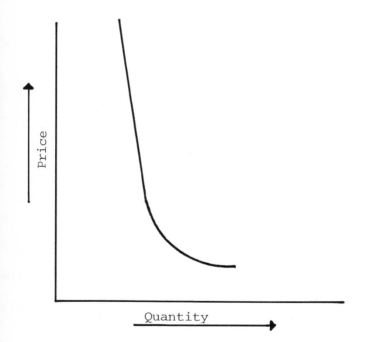

Figure 3-1. Theoretical demand curve for wheat. Wheat is a staple, and therefore the amount people are willing to purchase doesn't change very much. The only time the amount increases substantially is when wheat becomes very cheap, and begins to compete with feed grains such as corn or barley. In that case livestock feeders would begin purchasing wheat. Therefore, as a general rule it can be said that the demand curve for wheat is relatively inelastic. Inelastic (as explained in the text) means that the amount purchased does not vary greatly with price. See also figure 3-3.

public. For example, fashionable items, such as popular books, etc., are often said to be "in demand".

But in economics the term demand refers to a concept. In economics the term demand refers to the quantity of a product that the public will purchase over a wide range of prices. This concept (the concept of demand) is something that can be illustrated graphically. Figures 1 and 2 would illustrate the demand for two agricultural commodities, wheat and beef.

When illustrated in a graphic form, demand is often referred to as a demand curve. While theoretically the demand for some items might be a straight line, as a practical matter most commodities are a curve, and thus the term "demand curve".

<u>Elasticity of demand</u>. As can be seen in Figures 3-1 and 3-2, the slope of the lines representing the demand for wheat and beef are quite different. The slope of these lines is referred to as the "elasticity of demand." "Elasticity" referring to the variability of the amount of the product that will be purchased at various prices.

With wheat, the amount purchased does not fluctuate very much with respect to price. This is because bread, crackers, and other baked goods, etc. are staples in the diet; thus, the amount does not vary that much. Also, the cost of the actual wheat that goes into these products is a small fraction of the price the consumer actually pays.

It is only when wheat becomes exceedingly cheap, that we begin to see the amount purchased increase very much. In this case wheat would be competing with feed grains such as corn, barley, and sorghum, and thus an increase would be seen as use as animal feed.

With beef, however, we see very sharp differences in the amount that people are willing to purchase at various prices. This is because beef is relatively expensive compared to competing meats that can be used to satisfy this part of the diet.

Overall the differences in the amounts people are willing to purchase at various prices are said to be the "elasticity" for the product. Products in which there is much variation are said to have an elastic demand. Products in which there is little variation are said to be inelastic.

Typically, luxury items are very elastic, and staples are inelastic. Thus, in our examples beef was elastic, whereas wheat was relatively inelastic. To get out of agriculture, we could see even more personified examples in the form of Hawaiian vacations and prescription medicines. That is, Hawaiian vacations would be highly elastic, as there are a few wealthy people who would go regardless of the price, whereas there are multitudes of people who would go as the price declines. With such items as prescription medicines, however, price has relatively little effect on the amount people are willing to purchase. They will purchase what is prescribed for them, and no more.

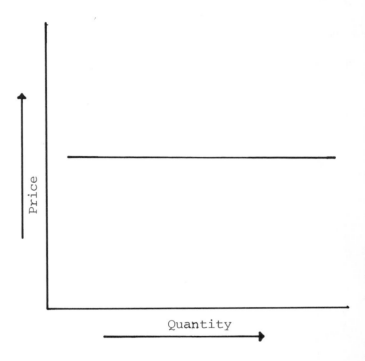

Figure 3-4. Theoretical example of a perfectly elastic demand curve. Perfect elasticity means the amount of the commodity consumers are willing to purchase is totally related to price. In actuality, however, perfect elasticity does not normally exist. There are however, some items that do have highly elastic demand curves. Usually these are luxury items, such as the luxury vacations example used in the text.

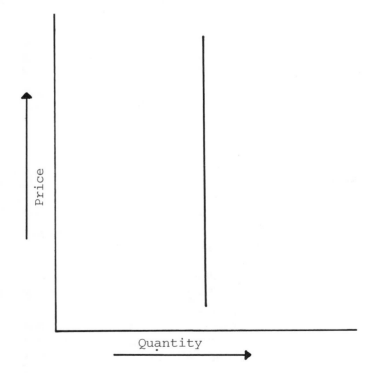

Figure 3-3. Theoretical example of a perfectly inelastic demand curve. In this case the amount consumers are willing to buy has nothing to do with price. If there is a product that comes close to being perfectly inelastic, it would be prescription medicines. In that case people usually buy whatever is prescribed for them and no more.

Supply. In economics, the term supply denotes a concept similar to the concept of demand. The difference being that supply refers to the willingness to **sell** quantities of a product at a range of prices. Obviously, the price motivation for suppliers that sell a product, is the opposite of the motivation of consumers that buy a product. As the price goes down, suppliers are willing to sell **less** product (whereas consumers are willing to buy more). Conversely, when the price goes up, suppliers are willing to sell more (and consumers to buy less).

Supply curves can be plotted on a graph just like demand curves. Also, similar to demand curves, supply curves can tend to be elastic or inelastic. The elasticity of supply depends primarily upon cost of production, and the ease upon moving into or out of production for that particular commodity.

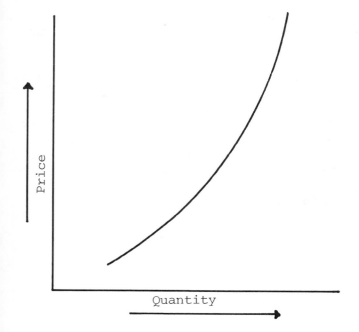

Figure 3-5. Theoretical supply curve for wheat. The slope of the supply curve is usually the opposite of the demand curve. That is, as the price of the commodity goes up, suppliers are willing to produce (supply) more, whereas consumers are willing to buy less.

As a general rule agricultural commodities have supply curves that are less elastic than manufactured goods. This is because many agricultural operations have relatively few options for producing different crops. They have heavy investment in facilities that cannot easily be moved into production of another commodity.

Probably the best example in that regard is dairy farming. There is heavy investment in milking and milk handling equipment that cannot be moved into production of anything except milk. Cattle ranches are another example. There is huge investment in land that can only be used for grazing animals. The rancher basically has only two options; he can graze the land himself, or lease it and let someone else graze it. As a result of these inflexible assets, milk and meat get produced regardless of price. Certainly, exceedingly high or low prices cause some elasticity (producers moving in or out), but the inability to produce anything else forces dairy and beef cattle operations to stay in production, even when prices are low. (Individuals may sell their operations, but the new owner continues producing the same commodity.)

In order for a commodity to have an elastic supply curve, the producer must be able to move in and out of the production of that commodity easily. Theoretically most grain crops should have elastic supply curves. That is, grain farmers have options that many other farmers do not. Fields suitable for most grain crops can be diverted to oilseed crops (soybeans, peanuts, safflower, etc.); hay crops such as alfalfa, clover, various grasses etc.; and in some areas, vegetable crops or specialty crops such as cotton. For some reason, however, even though returns to grain farming have historically been very low, production has stayed relatively high; i.e. the supply curve has been inelastic. The reason for this has probably been due to government programs on which farmers tend to rely; that is, stayed in grain farming, rather than turn to potentially more profitable crops.

As examples of crops having elastic supply curves, vegetable and many specialty crops would fall in that category. Farmers will plant these crops only if they feel they can make a profit, and will move out of production if they feel prices will be low. Another example of highly

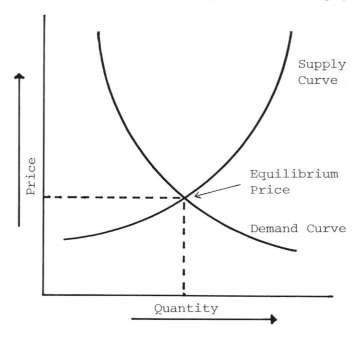

Figure 3-6. Graphic illustration of "equilibrium" price. Where supply meets demand.

elastic supply would be pigs. Because sows have the ability to litter, pig production can be increased rapidly. Farrowing operations are often relatively small sideline operations, and move in and out of production depending upon price.

Equilibrium. Theoretically, the price of a commodity or product will be where the demand and supply curves intersect. This is known as equilibrium. Figure 3-6 depicts the intersection of supply and demand curves.

A shift in demand or a change in supply. Understanding changes in supply and demand is often confusing to beginning students in economics. When a change in supply or demand is mentioned, the student usually thinks of a different position on the demand or supply curve. This is not true. **A shift in demand or supply refers to a whole new curve**.

A change in supply. A change in supply refers to something that has either increased or decreased total supply. In agriculture, most supply changes occur due to either favorable or unfavorable weather, bumper crops or crop failures.

An unexpected bumper crop would make extra commodity available, thereby lowering the price (Figure 3-7). Conversely, a crop failure would reduce the amount of commodity available, thereby increasing the price (Figure 3-8).*

---

* The effect of supply upon price is easy to understand. When large amounts are available, sellers are under pressure to sell at lower prices. This is because of storage costs, (in some cases shortage of storage facilities), interest costs of stored product, and competition among sellers. That is, for sellers to compete to sell their product to buyers, they must lower their price. Likewise, in order to induce new buyers into the market they must lower their price. Conversely, when a commodity or product is in short supply there is competition among buyers. In this case, the amount of commodity is limited and buyers must compete to obtain what amounts are available. The way they compete, of course, is to offer a higher price.

It is important to realize that increases or decreases in supply do not change demand. The new price will simply be where the new supply curve crosses or intersects with demand (demand curve).

A shift in demand. A shift in demand means a movement of the entire demand curve. Although repetitious, it should be pointed out that a shift in demand does **not** mean a movement along an existing demand curve. It is an entirely new curve. Figure 3-6 represents a shift in demand.

The causes of a shift in demand can be due to changes in the availability or price of competing products, or a change in public acceptance of the product. In short, a shift in demand is anything that will cause buyers to want to purchase more (or less) of a product along the entire range of prices.

As a means of creating more demand, commodity groups often engage in advertising campaigns. In some cases they attempt to change the image of their product. Potatoes, milk, cheese, oranges, apples, beef, and many other commodities have all been advertised in various ways as a means of increasing demand.

Changes in prices of substitutable commodities are also very much responsible for shifts in demand. For example, if soybean meal were to suddenly escalate in price, other protein meals, such as cottonseed meal, will experience an increase in demand.

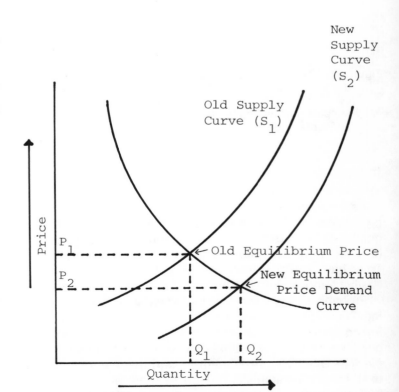

Figure 3-7. A shift in the supply curve due to a bumper crop. Greater supply means a lower price since more commodity is available for sale. (see also footnote page 18)

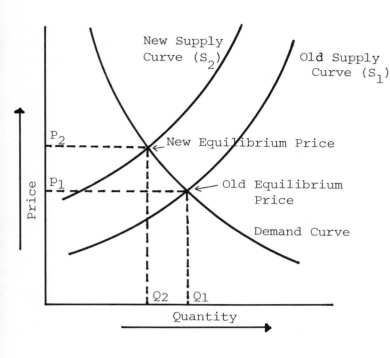

Figure 3-8. A shift in the supply curve due to a crop failure. A reduced supply means a higher price since less commodity is available for sale. (see also footnote page 18)

COSTS

Costs is another one of these economic terms that have special meanings. In general, costs are broken down into what are known as *fixed cost, variable cost*, and *total cost*. Understanding the differences between these terms is vital to management of agricultural assets.

<u>Fixed cost</u>. Fixed cost means just what it implies; costs that are locked in and unavoidable. Fixed costs remain the same, regardless of production.

The best example are land mortgages. The farmer or rancher must make mortgage payments regardless of whether he produces anything or not. Another example would be real estate taxes, they must be paid regardless.

<u>Variable costs</u>. Variable costs are the costs over which the farmer does have control. Variable costs are directly affected by production. Examples would be labor, seed, feed, fertilizer, fuel, etc.

<u>Total cost</u>. Total cost basically means fixed cost plus variable cost.

<u>Selling below cost</u>. As discussed in the section of supply, agricultural commodities often have inelastic supply curves (price has relatively little to do with how much farmers are willing to produce). This is to a great degree caused by the very high fixed costs most farmers have; that is, mortgage payments on land, machinery, and equipment. In most cases the mortgage payment on the land overshadows other fixed costs; although in the case of dairies and some other highly specialized agri-

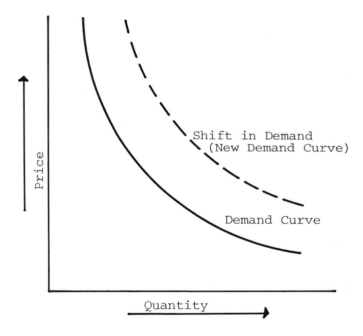

Figure 3-9. A "Shift in Demand". A shift in demand means a whole new demand curve (not just moving up or down an existing demand curve). A shift in demand occurs whenever consumers attitude toward purchasing a commodity changes. This can be either positive or negative (this graph portrays a positive change). For example, a positive change may be brought about through an advertising campaign. A negative shift in demand could occur due to negative publicity concerning the value of a commodity, etc.

culture, the equipment can also result in quite high fixed costs.

Because of these very high fixed costs, farmers are often willing to produce and sell their crops below cost. (When we say below cost, what we mean is below total cost; that is, fixed cost plus variable cost.)

Farmers, and other businesses as well, are often willing to produce and sell below total cost if they can at least recover their fixed costs. In other words, they are forced into selling at a loss in order to make their land payments. Their land or mortgage payments are a fixed cost and will occur whether or not they produce anything or not. Often the variable costs of production are less than fixed costs and, therefore, it is justifiable to sacrifice variable costs for fixed costs. Obviously this is not the kind of situation a farmer or other business wants to be in, but rather it is something they are forced into. The only alternative is not to make their mortgage payments (fixed cost) and lose their land.

OTHER TERMS USED IN ECONOMICS

Opportunity costs. "Opportunity costs" is an abstract term, that does not apply to the physical costs of production. Rather, opportunity costs refers to alternatives a farmer or business may have for investment. For example, if a farmer with a swine feeding operation feeds all his own grain (produced on his own farm), his opportunity costs would be the market value of that grain. That is, in figuring the profit of the feeding operation, the farmer should consider the market value of the grain as a cost. If that makes his feeding operation show a loss, then he knows he would be better off to sell his grain and not feed any animals.

Economies of Scale. "Economies of scale" is a term used by economists that sounds very technical and academic, but is very easy to understand. Basically, economies of scale refers to the size of an operation and the efficiency that goes along with that size.

As a general rule, as operations get bigger, they potentially become more efficient. This efficiency can be the result of a multitude of items. Larger size allows a farmer to obtain better prices for fertilizer, seed, feed; justifies the purchase of specialized machinery; and may allow bargaining for more favorable finance (interest) charges; etc. In recent years another highly significant factor has been justification of the employment of outside consultants.

Diseconomies of scale. "Diseconomies of scale" is the opposite of economies of scale. Basically, the concept of diseconomies of scale states that there are limits as to how large an operation can get and still remain efficient (compared to smaller operations). As operations grow exceedingly large, their physical needs may outstrip the advantages of size. For example, an exceedingly large dairy or feedlot could outgrow the available supply of locally grown feed, and would have to begin paying transportation costs on feed from other areas. An extremely large farm could require more labor than is available locally, and would have to begin building housing for labor solicited from other communities.

MONEY AND BANKING

When most people think of money and banking, they typically think of what is commonly known as finance. The subject of finance is covered in another chapter (Chap. 1).

Actually "money and banking" is another one of those specialized terms often used in economics. It is the study of the effect of monetary policy, the banking industry, and their effect upon the economy. It is a very specialized area of economics and volumes have been written about it. For the purposes of this book, however, only those items of practical significance will be discussed.

Banks create money. Only the Treasury Department can print money, but it is important to realize that banks actually create more money than the mint that prints currency for the government. As a result of this, banks are closely regulated and this regulation can have an enormous impact on a nation's economy.

How do banks create money? Banks create money whenever someone makes a deposit. The reason is that the bank takes the money that is deposited, and then lends it to a borrower. The borrower then takes that money and spends it. Whoever receives that money takes it and deposits it in another bank. The instant they do that, money is created.

Why? Because both depositors have bank statements showing dollars that they can take out and spend. In the beginning, however, it was the first depositor's money. At this point, however, both depositors are calling the same money their own.

Bank Failures. The scenario discussed in the previous section could go on and on. As long as a depositor puts money in a bank, the bank will, in turn, lend the money out (for an interest fee). This is how banks make money.

While the consequences of borrowing, depositing, and relending money are highly complex with respect to the economy, one effect is clear. A problem can occur if depositors and borrowers all pull their money out at the same time. Indeed this was one of the many calamities that occurred during the Great Depression of the 1930's. Depositors panicked and made "runs" on the banks. Demanding all their funds in cash, there was not enough to go around, and many banks collapsed. Many people lost their entire life savings.

To prevent that type of thing from occurring again, the U.S. government set up what is known as deposit insurance. Deposit insurance means that in case of a bank failure, each depositor will be paid off (up to $100,000). This insurance is supposed to prevent widespread panic and prevent "runs" on banks.

Reserve requirements. Deposit insurance is supposed to prevent "runs" on banks. To ensure enough funds are available for the normal course of business, however, the federal government also dictates what are known as reserve requirements. Reserve requirements mean that banks must retain a portion of all their deposits at all times. They cannot lend out the full amount of their deposits. The actual percentage they must withhold varies with the type of deposits.

Velocity of Money. The velocity of money is a term that relates to deposit requirements and activity within the economy in general. It refers to how many people or institutions will receive and spend the same money. For example, if a farmer spends a dollar on fertilizer, the dealer will in turn spend that dollar. That dollar may go to pay the wages of the truck driver that delivers the fertilizer. The truck driver may then use the dollar to buy groceries for his family. Theoretically, the dollar can eventually come back to the farmer through the purchase of those groceries (farm commodities).

Inflation. Inflation is a term that relates to the buying value of a currency. In modern times the buying value of most currencies has declined. For most developed countries the inflation rate has been low to moderate; for underdeveloped countries, inflation rates have typically been high to very high. For example, in the U.S. the annual inflation rate has varied from about 4% to 10%. In many underdeveloped countries annual inflation rates often run 100% or more. (When we say an annual inflation rate is 100%, that means it takes exactly twice as much money to buy the same amount of goods as it did one year previously.)

"Real" Interest Rates. The "real" interest rate is a term vitally important in all of business, but especially to agriculture. The reason is because of the amount of long-term financing necessary in agriculture.

The "real" interest rate refers to what is known as the "nominal interest rate" minus inflation. The nominal interest rate is the numerical rate that appears on loan contracts. For example, if the nominal interest rate is 10% and the inflation rate is 6%, then the "real" interest rate is 4%.

In the past, many U.S. farmers and ranchers increased their net worth substantially through the use of long-term financing at interest rates that ultimately came to be below the inflation rate. (That is, negative "real interest rates".) Buying their land in the 1940's and 50's at interest rates of 4% to 6%, (during the latter part of the contracts in the late 1960's and 70's, when inflation ran nearly double the nominal interest rate,) they profited greatly.

In more recent times, however, just the opposite has occurred. This is particularly true for farmers and ranchers who bought property in the 1970's and early 1980's. During that time period inflation and subsequent interest rates were relatively high. During the late 1980's, however, inflation rates declined, leaving these farmers and ranchers committed to contracts with the highest "real" interest rates in history. This was a contributory factor in the Agricultural Crisis of the late 1980's.

SUMMARY

This has been a very brief introduction to economics restricted to those topics of practical importance. While brief and elementary, it is nevertheless an important part of the education of students who would become involved in farm or ranch management.

At the very least, the student should understand how supply affects price. That is, as supply increases, price decreases. Conversely, in times of poor harvest, when a particular crop may be in short supply, price increases. The student should also understand how demand elasticity affects price. Some products have very elastic demand, meaning that the amount purchased will vary greatly with price. This also means that the price of the commodity will vary with supply to a greater degree.

Conversely, inelastic demand means that the amount consumers will purchase does not vary with price nearly as much. Changes in supply will therefore not affect price as much as products with elastic demand. Understanding the effect and interrelationship of supply and demand elasticity is important, as price volatility is affected. Price volatility means risk, and as discussed in Chaps. 1 & 2, risk is something most farmers and ranchers must avoid or minimize.

Opportunity costs is a concept many farmers overlook, that ultimately costs them a good deal of money. The best examples are farmer-feeders such as dairymen, swine operators, and/or cattle feeders. Quite often they will feed all the crops grown on their farm without considering the cash value of those crops. That, in effect, is the definition of opportunity costs. That is, examining what one's resources would be worth if used in a different manner. In the case of farmer-feeders, there are often opportunities to sell some of their crops at a higher price than other commodities of equal or higher feeding value, that they could buy back at a lower cost.

The concept of "real interest rates" is important to

anyone who would borrow or invest money. That is, we must know what the inflation rate is in order to know what the "real interest rate" is. For example, it would make little sense to put money in a fixed rate investment if the inflation rate were higher than the interest rate. In that case, we would actually lose money.

The concepts involved in money and banking are not necessarily requisites for farm management, but they do provide appreciation for the complexities of the world we live in. Everything in economics is interrelated, and the concepts of money and banking reflect this.

CHAPTER 4   BASIC FINANCIAL ACCOUNTING

The following chapter is written to provide the student with a basic familiarity of the two accounting analyses typically requested by banks; 1. the Balance Sheet (or financial statement); and 2. the Income Statement. In addition, the student will be instructed in the basics of the most important accounting analysis for the farm business itself, the Cash Flow.

THE BALANCE SHEET OR FINANCIAL STATEMENT

The Balance Sheet or Financial Statement is an accounting analysis that nearly everyone has had some kind of association with. Its purpose it to show the "balance" of assets versus liabilities and thus the name Balance Sheet. By noting the value of assets versus liabilities, the financial strength of the business can be seen and thus the synonym, Financial Statement.

Assets refer to factors that are positive for the business. Anything that has a dollar value to the business would be included as an asset. Examples would be cash, land, machinery or other items that can be sold, accounts receivable, etc.

Liabilities refer to debts that the business may have. Examples would be unpaid bills (accounts payable), loans that must be repaid, etc. Although confusing, the owner's equity is also placed with the liabilities. This is because that technically speaking, the owner's equity is a debt that the business owes to the owners.

Owner's equity is essentially the liabilities subtracted from the assets. Owner's equity therefore is an indicator of the financial strength of the business. A typical balance sheet is shown in Table 4-1.

In this balance sheet or financial statement, it can be seen that the owner has $297,449.10 in equity or about 29% of the value of the business. When borrowing money, this is one of the first things a banker will look at as it gives him some idea of the financial strength of a business ... the ability of the owner to repay loans. The greater the owner's equity, the greater the financial strength. Probably the second thing most bankers look at is what is known as "liquidity", which is defined as assets that can be turned into cash rather quickly.

Table 4-1.   EXAMPLE BALANCE SHEET.

Spring Daisy Pig Farm
Quagmire, Ohio
March 1, 1988

Assets

Current Assets

Cash on hand	$7.11
Hogs	499,800.00
Grain inventory	233,920.00
Protein supp. inventory	3,315.00
Accounts receivable	24,410.09
Total current assets	$779,452.20

Fixed Assets

Land	$24,000.00
Pens & facilities	42,050.00
Feedmill	92,850.00
Feed delivery equip.	23,248.17
Office bldg. & equip.	24,872.16
Water wells & rights	30,000.00
Total fixed assets	$236,020.33
Total Assets	$1,015,472.50

Liabilities

Current Liabilities

Bank loan on hogs	$397,853.02
Accounts payable	121,473.21
Wages payable	7,312.87
Income taxes payable	576.18
Interest payable	78,418.12
Total current liabilities	$565,633.40

Long Term Liabilities

Note on land	17,500.00
Note on constructed facilities	134,900.00
Total long term liabilities	$152,400.00
Owner's equity	$297,439.10
Total Liabilities	$1,015,472.50

When evaluating liquidity, the first thing that is looked at, of course, is cash. In this case, the cash on hand is extremely low. While exaggeratedly low in this example, agricultural enterprises do typically have little cash on hand. Usually what funds they have are tied up in the business.

The second thing the banker will look at is assets that can be sold quickly; i.e. liquid assets. In this example, nearly $500,000 in livestock are available. Hogs, of course, can be sold very quickly. The problem is that at certain times it may be unprofitable to the owner to sell them; i.e. when the hog market is down, or the livestock are halfway through a feeding period, etc. **It should be remembered, however, that bankers are not really concerned with whether their borrowers make a profit or not.** All they are actually interested in is the ability of the borrower to repay. Certainly bankers are desirous of their borrowers making a profit so that they will be able to stay in business and continue borrowing money (paying interest)... but their primary concern, first and foremost, is only to be assured that the borrower can repay the loan. As discussed in the Finance Chapter, bankers do not share in the profits. Interest charges remain the same whether the borrower makes money or loses money.

The expression "Looks good on paper," probably began with balance sheets, as they can be very misleading. The biggest area of potential misrepresentation comes in placing value on the assets. The most common values used are the "Book" values. Book values are usually the actual prices paid for the assets, minus depreciation. Due to appreciation and/or depreciation in market values, book values can be a long way from the actual sale value of an asset. With equipment, due to obsolescence or unusual wear and tear, the depreciation* charged against it may inadequate. In the case of custom built equipment, such as a feed mill or fence, much of the cost incurred may be due to installation. It would therefore have a much lower market value (unless the entire operation were sold as a unit). In the case of land, because of appreciation, the actual market value may be much higher than the original book value. For these reasons, bankers may call for a balance sheet with assets reported as appraised values, rather than book values.

While balance sheets give some idea of the financial strength of a business, **they give no indication of profitability!** Owner's equity gives an indication of ability to repay loans, but in no way says that loan repayment can come from operating revenues. Owner's equity simply gives an indication of what could be derived (toward repaying a loan) upon foreclosure. For an indication of profitability, a second analysis is required. This analysis is known as the Income Statement.

THE INCOME STATEMENT

The purpose of the income statement is to show the profit or loss of a business for a given time period. Income statements are prepared on a yearly basis for tax purposes, and often on a quarterly or monthly basis for use in management planning.

The purpose of the income statement is to show whatever the profit or loss was for the period covered by the statement. For that reason it is sometimes also called the Profit and Loss Statement.

There are two basic ways to prepare an income statement in agriculture: (1) using cash accounting and, (2) using accrual accounting. Normally, income statements using cash accounting are prepared only for income tax purposes. This is because adjustments are not made for inventory. For example, if fertilizer is used for a crop, but the crop is not sold during the accounting period, the cost of the fertilizer is deducted as an expense but the crop is not accounted for since it has not been sold. Obviously, an income statement using cash accounting does not give an accurate estimation of what income really is. For management planning, income statements using accrual accounting are considered more accurate, since inventory allowances are made. Even then, however, one must be careful in determining incomes with unsold farm commodities (inventories) included, since market prices fluctuate constantly.

Another area in income statements that can be misleading is the charging off of what is known as depreciation. Depreciation is an accounting term that refers to the decline in value of an asset. Each year a set percentage of the original cost of an asset is subtracted from its original value. That is capital items which are used over a period of years must be charged off as expense over that same number of years. Ideally, the depreciation expense used should represent the true cost to the business. However, in actual practice, obtaining a true representative value can be quite difficult. In many cases, due to price escalation, the original purchase price may be a fraction of the replacement cost of the item.

Probably the most accurate method of computing depreciation costs might be by figuring the difference in market value of the asset between the start and the end of the accounting period. Using this type of method will be inconsistent with computing depreciation for income tax purposes, since the IRS demands that a mathematical formula be used. By this time the reader is probably confused. The point of the whole discussion, however, is that income statements can be misleading (confus-

---

*Depreciation is a periodic (usually yearly) deduction from the value of an asset which theoretically represents the decline in value of the asset.

ing); e.g. fluctuating prices in farm commodities (inventories) and unrealistic depreciation charges can make a business look much better or worse than it really is. For this reason, most large corporations utilize auditing firms to examine their income and financial statements before they are presented to the stockholders. What the auditing firm does is state that the figures presented are honest and realistic. In essence, the reader has the reputation of the auditing firm as an assurance that the statement he is reading is realistic. An example income statement is shown in Table 4-2.

Table 4-2. INCOME STATEMENT EXAMPLE.

Income Statement (Jan. 1, 1987-Jan. 1, 1988)
Spring Daisy Pig Farm
Quagmire, Ohio

Receipts	
Sale of Misc. crops produced on farm	$1,825.013.06
Sale of hogs	2,012,850.11
Accounts receivable	24,410.09
Total receipts	$3,862,273.26
Adjustments to Inventory	
Swine (5553 hd @ $90 hd (150 lb @ $.60/lb)	499,800
Feed	
fertilizer	29,670
grain (2285 tons @ $85/ton)	194,250
supplement (15 tons @ $221/ton)	3,315
Total Inventory	$727,035
Expenses	
Feed puchases	$2,254,161.30
Pig purchase costs	1,916,250.00
Labor	74,318.00
Interest	194,000.00
Depreciation expense	11,000.00
Vet. medicines, supplies & services	53,504.12
Bad debts	5,114.04
Office supplies & postage	4,812.16
Consulting fees	6,000.00
Misc. expense	9,183.01
Total Expenses	$4,528.342.61
Total Receipts plus Inventory	$4,589,308.26
Less Expenses $4,428,342.61	
Net Income	$60,965.65

Learning to evaluate financial data is certainly beyond the scope of this discussion. The author would like to point out, however, that while the $60,965 income reported in the example may seem like a relatively large sum in dollars, in comparison with the amount of money handled (over 4½ million dollars) it is a very small margin ... a return of only 1.35% ($60,965 divided by $4,528,342), which is very poor indeed.

It should also be noted that nearly $500,000 of inventory is tied up in livestock. a 12% reduction in the hog market (which could easily occur) would wipe out the entire profit for the year. A 20% drop in the hog market (which could also occur), would create a $40,000 loss.

Something else the reader might want to look at is the amount of depreciation charged off; in this example $11,000. Looking back at the balance sheet, fixed assets were reported to be $236,020.33. Subtracting the land gives a remainder of $212,020.33 of depreciable assets. The amount of depreciation charged against those assets ($11,000), amounts to only about 5.2%, which might be unrealistically low, especially since the feed mill and other equipment make up the majority of the depreciable assets ... items that are subject to considerable wear and tear. No indication is given as to how depreciation was computed, and so this may be something the reader would want to check into further.

Again, the purpose of this section has in no way been intended to teach financial analysis as that is a discipline all its own, and far beyond the scope of this text. Rather, the purpose has been to simply explain some of the variables involved in making up financial reports, and to make the point that the reader should look farther than just the bottom line.

THE CASH FLOW ANALYSIS

The purpose of the cash flow analysis is to lay out the month-by-month, week-by-week, or even day-by-day flow of funds to and from the business. While usually done as a projection, the purpose is to insure that the business will always have enough cash on hand to meet its obligations.

More and more bankers are demanding to see projected cash flow analyses before lending, since it will point out directly whether the loan can be repaid out of normal operating funds. This is important, since the balance sheet may show the ability to pay, but not whether payment could be made out of operating funds; i.e. all it shows is if payment could be made upon a foreclosure.

Regardless of whether bankers want to see cash flow projections or not, it is a good business practice to prepare them. This is because it is entirely possible for businesses of substantial financial strength to get in the position of not being able to meet their short term financial obligations ... and when that happens, it can have grave consequences. In essence, it is entirely possible for a profitable business to go bankrupt. If the reader gets nothing more out of this section, it should be that profit is not a guarantee of success ... **that even profitable businesses can go broke if they cannot meet short term budgets.**

Cattle ranches typically have the greatest cash flow problem since money is expensed all year long, but income generally occurs only once a year (when their cattle are sold). For this reason, the author has chosen a ranch example to demonstrate the need for cash flow budgeting. Table 4-3 is a projected income statement for a ranch with a reasonably bright profit picture.

Table 4-3. PROJECTED INCOME STATEMENT EXAMPLE.

Rexall Ranger Ranch
No Pesos, Nevada
Projected Income Statement 1988

Receipts

Sale of Calves
128 steers, 51,2000 lb @ $.82/lb.	$41,984
102 heifers, 38,760 lb @ $.74/lb.	28,682

Sale of cull cows and bulls
26 cows, 23,400 lb @ $.52/lb.	$12,168
2 bulls, 2,250 lb @ $.62/lb.	1,395
Total Receipts	$84,229

Expenses

Hired labor	$9,600
Machinery repair (windmills & pickup trucks)	650
Payments to custom hay operator ($25/ton x 320 tons)	8,000
Purchase of bulls (2)	1,800
Protein & mineral supplements	2,920
Fence repairs	3,500
Vet expense	1,500
Property taxes	1,650
Interest on working capital	1,425
Interest on mortgage	12,375
Miscellaneous	1,200
Principle payment on land	10,000
Total expenses	$41,325
Net Income	$29,609

Table 4-4 represents a cash flow projection for the same ranch for which net income is reported in Table 4-3. As can be seen, preparing a cash flow budget is a simple matter of putting down when cash will be received and when it will be spent. The great importance of this can be seen in the months of July, August, and September. Even though the ranch is projected to have a net profit of over $29,000* ... During the months of July, August, and September the ranch will have a deficit of funds. Knowing about the deficit in advance (through the use of the cash flow projection), allows time for arranging in advance for a short term loan.

*For clarity of explanation, the net cash flow figure and the net income from the Income Statement have been shown to be an equal amount ($29,609). In a real situation the figures would actually be different since Income Statement would include a depreciation expense (decline in value of capital assets). Depreciation expenses are not included in cash flow projections, since funds do not actually flow out (for depreciation).

Table 4-4. EXAMPLE CASH FLOW PROJECTION.

Receipts	Jan.	Feb.	March	April
Operating capital loan	15,000			
Sale of steer calves				
Sale of heifer calves				
Sale of cull cows				
Sale of cull bulls				
Total	15,000			
Expenses				
Hired labor	800	800	800	800
Machinery repair			250	100
Custom hay paymt.				
Bull purchase				
Protein & Min. supplement	600	800		
Fence repairs			1,000	1,500
Vet expense			300	300
Property taxes				
Interest, working capital				
Interest, mortgage				
Miscellaneous	100	100	100	100
Operating capital				
Principle paymt. on land				
Total	1,500	1,700	2,450	2,800
Balance	13,500	11,800	9,350	6,550

The danger in not setting up a cash flow projection, of course, is coming up in the deficit months unexpectedly, and having to obtain emergency loans. If emergency loans cannot be obtained, the situation can become serious. In this example the rancher would unable to pay the custom hay operator for his services. If the hay operator refused to work, and the hay was needed for winter feed, it could have obviously grave consequences for the ranch.

Similar situations exist for farms that harvest only one crop per year. They must budget their cash flow to ensure that they have sufficient liquidity (cash) to meet their short term obligations all year (in-between harvests).

Rexall Ranger Ranch
Cash Flow

May	June	July	Aug.	Sept.	Oct.	Nov.	Dec.	Total for year
								$15,000
					41,984			$41,984
					28,682			$28,682
					12,168			$12,168
				1,395				$1,395
				1,395	82,834			$99,229
800	800	800	800	800	800	800	800	$9,600
							300	$650
		4,000	4,000					$8,000
	1,800							$1,800
	620					400	500	$2,920
1,000								$3,500
200	200	200			300			$1,500
	825						825	$1,650
							1,425	$1,425
							12,375	$12,375
100	100	100	100	100	100	100	100	$1,200
							15,000	$15,000
							10,000	$10,000
2,100	4,345	5,100	4,900	900	1,200	1,300	41,325	$69,620
4,450	105	(-4,995)	(-9,895)	(-9,400)	72,234	70,934	29,609	$29,609

SUMMARY

The purpose of this chapter has not been to instill any significant degree of expertise in the subjects discussed. Indeed, each of the subjects discussed are quite complex and would require an entire text to cover them thoroughly. Accordingly, the purpose of this chapter has been only to point out the need for obtaining either more knowledge or professional consultation when attempting to analyze accounting statements.

CHAPTER 5   THE FUTURES MARKET

The purpose of this chapter is to provide basic information concerning the Futures Market. Anyone wishing to actually engage in Futures transactions would need to seek out additional information. This is particularly true with respect to the individual commodities to be traded. The purpose of this chapter is to provide the reader with an understanding of the mechanics of the Futures Market. A basis that can be used as a background for further detailed study.

As discussed in chapter 3, the Futures Market is quite complex. Because it is complex, and has some major pitfalls for those who are not familiar with it, among many farmers the Futures Market is a highly controversial subject. Indeed, there have been a number of farm groups who have attempted to get the Futures Market completely banned. The major complaints have been that the Futures Market allows non-farm entities (speculators) to control farm prices and that it has allowed large corporate businesses to move into farming.

It is certainly true that large businesses involved in farming have used the Futures Market. The Futures Market is an effective method of risk management, and large business-oriented farms and farming enterprises use it routinely.

As with any kind of market, there is always potential for manipulation and abuse. Whether it be the Stock Market, or a country livestock auction, there are always individuals who would seek to manipulate prices. The Futures Market is no different and, from time to time, apparently there have been individuals attempt to manipulate it. But just like the Stock Market, the Futures Market has a Board of Governors (overseen by the Federal Government) and an investigative body whose job it is to head-off and prevent abuse. This doesn't mean that they have been 100% successful, but the rules of trading are strict and the regulations are rigidly enforced. There are civil and criminal penalties for violators.

While the record may or may not be perfect, the Futures Market is here to stay. Like it or not, the Futures Market provides an important service and it behooves everyone in agriculture to learn how it functions.

WHAT FUTURES PRICES ARE

Futures prices are prices quoted for most major agricultural crops and livestock for delivery within a specified month (sometime in the future). For most crops (corn, soybeans, wheat, etc.), the prices are quoted in bushels, and for most livestock, the prices are quoted on a hundred weight basis. For example, live hogs might be quoted at $65.00/cwt., which would be the same as $.65 per pound.

Futures contracts for crops usually deal in rail-car load quantities. For example, corn contracts consist of 5,000 bushels. Most futures contracts for livestock deal in truck load quantities (about 40,000-44,000 lbs.) For example, a live cattle contract for steers would be for about 38 to 40 steers (40,000 lbs.) Each commodity has subtle differences, and it's important to know all the specifics of the commodity you deal in.

WHAT FUTURES PRICES ARE NOT

Futures prices are not indicators of what the actual market price will be! On the contrary, in some cases the futures prices have had an **inverse relationship** with what the actual cash prices turned out to be. Farmers saw high futures prices quoted several months ahead and geared up to produce extra quantities of that commodity. As a result, at the time of delivery there was a surplus and the price came down. This is particularly true for livestock, and has been documented with feedlot cattle.

Futures prices are nothing more than a combined estimate of what a large number of people believe the price might be. The farther away the delivery time, the greater the opportunity for variation between the futures price and the actual cash price. The actual price is highly dependent upon weather (both in the U.S. and the rest of the world), government programs, import-export regulations, the world economy, and a myriad of other factors.

In short, **never consider the futures price as an indication of what the actual market price will be**. The futures price is simply a target price that can be used in a "hedging" program. In addition, the futures price should never be considered as the actual price that could be received (in a hedging program). The actual price received may be lower due to what is known as "basis". In addition, if delivery is required, there may be delivery costs and discounts. These reductions in price received are explained in the following sections.

THE MECHANICS OF HEDGING

Hedging is defined as the use of the Futures Market by a farmer or producer to ensure a price for his crop or livestock. To do that, he sells one or more futures contracts to coincide with the amount of crop or livestock he wishes to protect (from the price falling). In essence, he sells a contract in which he promises to deliver the type of commodity indicated during a specified delivery month at a set price. The contract is bought by another individual, usually a speculator.* The commodity ex-

---

* The speculator is often maligned by producers. In actuality, the Futures Market could not function without speculators as they provide liquidity. Without the extra capital they provide, the Futures Market would be mostly sellers, and very few buyers.

change doesn't buy or sell any contacts itself, but acts as an agent, providing a place for transactions and rules to govern those transactions.

When the delivery month arrives, the usual procedure is for the farmer to buy an equal number of offsetting contracts. That is, if he has contracted to deliver a specific commodity such as corn or soybeans, he simply buys back an equal number of contracts (for corn or soybeans) for that same month. Having sold and bought the same number of contracts, they nullify each other.

If he wishes, however, with some commodities the seller has the option of actually delivering the commodity. Instead of buying back the contracts, he can actually deliver the commodity to whoever it was that bought the contracts he sold. The commodity exchange specifies certain delivery points (cities in the U.S.), and the farmer can deliver to any one of the delivery points he or she desires.

If the farmer delivers, however, there will be additional costs involved. First of all, many delivery points carry discounts and the farmer will not receive the full face value of the contract. Then too, there will be the usual transportation costs. There will also be a grading fee. All commodities delivered must be graded to ensure they meet the contract specifications. For grain and soybeans, the grading fee is minimal as the product simply has to meet routine USDA standards. Livestock grading, however, is much more involved and costly. Each animal has to be inspected individually and individuals can be docked (discounted) or disqualified entirely. If too many animals don't meet the specifications the entire shipment can be rejected.

As a practical matter, very few futures contracts are terminated with actual delivery; most are simply terminated with a buy-back of like contracts. The option of delivery is simply to ensure that cash prices and futures prices during the month of delivery do not become too far apart. If they do, then the farmer can physically deliver his commodity.

A TRUE HEDGE

For a true hedge, the farmer must sell his commodity on the cash market and buy back his futures contracts simultaneously. The difference between the actual price of the commodity in his area and the price of the futures contracts he must pay during the delivery month, is known as the **basis**. Usually the difference between the futures price and the local cash price is the transportation and delivery costs. However, if a disparity arises and the current futures price is significantly higher than the local price, then the producer can deliver. As mentioned previously, it is the potential for delivery that keeps the futures and local price in line during the delivery month.

FIGURING BASIS

"Basis" is the difference between the futures price, during the delivery month, and the actual cash price that can be obtained locally. Normally futures prices will be slightly higher than local cash prices due to delivery costs. This is what is known as a negative basis. Most livestock have a negative basis, due to the relatively high delivery costs.

In order to place an intelligent hedge, it is imperative that the producer have a good idea what the basis or price spread will be. Usually this is arrived at by looking at historical data.

For example, let's assume that a corn farmer carefully studies historical price records and finds that a negative -15¢ per bushel is an average basis for his area. If he figures he needs $3.20 per bushel for an acceptable profit, then he will have to hedge his corn at $3.35 per bushel.

LIQUIDATING A HEDGE

As mentioned, in a true hedge the producer sells his commodity on the open market and "lifts" his hedge (buys back an equal number of contracts) at the same time. Assume in the previous example that the producer sold his corn on the Futures Market for $3.35 per bushel. This means that, if necessary, he may deliver the corn and receive exactly that price as long as the corn meets the contract specification (#2 or better).

More often, the hedger buys back an equal number of contracts for that same delivery month he has his contracts sold for. The Futures Market will have fluctuated from whatever the price was when he placed his hedge several months back. But regardless of whether the future's price is higher or lower, he will net about the same amount of money . . . if he has figured his basis accurately.

For example, if he sold his corn at planting time in April for October delivery at $3.35 per bushel, and when October comes around the futures price is only $2.80, then he will make $.55 in the Futures Market. That is, he will make $.55 per bushel when he buys back his contracts. He sold at $3.35 and now he is buying at $2.80.

The local cash market, however, will be close to the current futures price. If the -$.15 basis is accurate, then the local cash price should be $2.65 per bushel. If his target price was $3.20, then he will in fact receive that figure. The transactions will look like this:

Transaction #1 (April-planting time) sell Oct. contracts @	$3.35/bu.
Transaction #2 (Oct.-harvest time) buy Oct. contracts @	2.80/bu.
Result of Transaction #1 & 2	+.55/bu.
Transaction #3 (Oct.-harvest time) Sell corn on cash market @	$2.65/bu.
Net sale price of corn	$3.20/bu.

Now, assume that October corn futures went up instead of down. Instead of being $2.80/bu., suppose they went to $3.45 per bushel. If basis price has been calculated accurately (15¢ below futures month of delivery), the producer will still receive a net of $3.20 per bushel:

Transaction #1 (April-planting time)	
sell Oct. contracts @	$3.35/bu.
Transaction #2 (Oct. harvest)	
buy Oct. contracts @	3.45/bu.
Result of Transactions #1 & 2	-.10/bu.
Transaction #3 (Oct.-harvest)	
sell corn on cash market @	3.30/bu.
Net sale price of corn	$3.20/bu.

Note that obtaining the target price ($3.20/bu.) was dependent upon calculating the correct basis. For this reason it is wise to be conservative when calculating the basis.

BASIS RISK

Basis risk is a common term used in hedging. A literal definition would be the actual basis being greater than the anticipated basis.

A simple example of basis risk can be shown by using the previous example in which $3.20/bu. was deemed to be the target price that would yield a reasonable profit. Anticipating a -$.15/bu. basis, the corn was hedged at $3.35/bu. There was a risk however, that the actual price spread between the futures price (during the month of delivery) and the local cash price could have been more. If it had been -.25 or -$.35/bu. then the producer would have only netted $3.05 or 2.95/bu., which could conceivably force him into a break-even or loss situation.

As mentioned earlier, if the basis becomes too high, the producer has the option of delivering the commodity and collecting the face value of the futures contract he originally sold. In order to do that, however, his commodity must meet the contract specifications. If it does not, he may receive a discount or have his product disqualified for delivery. For that reason, it is extremely important that the producer be thoroughly familiar with the specifications for the actual commodity he desires to hedge.

MARGIN CALLS

Whenever a futures contract is sold or purchased the entire value of the contracts is not paid for, but the seller or purchaser is required to put up a margin. The actual amount will vary with the brokerage house, but 5% to 15% of the value of the contract is the usual minimum. As the price fluctuates from the original contract price the contract holder will either get part of his margin back, or will have to put up additional margin money, depending upon which way the price fluctuates. For example, if a producer sells (hedges) a corn contract for $3.35/bu., the contract would be worth $16,750 ($3.35 x 5000 bushels). At a 10% margin rate, he would have to put up $1,675. If the futures price went down to $2.90/bu., then he would receive some of his margin back. This is because if he went to liquidate his contract, he could do it cheaper (buy a contract for less money than the one he sold). If the price goes up, say to $3.45/bu., then he must put up extra margin money to cover the price spread between what it would cost to buy a contract to cover (liquidate) the cheaper contract he sold. In this case, an additional $500 would be needed since a $3.35 contract is worth $16,750, and a $3.45 contract is worth $17,250.

When extra margin money is required of a contract holder, he receives what is known as a "margin call". The purpose of margins and margin calls are to maintain the financial integrity of the futures market, but there is something inherently demoralizing about receiving margin calls. When margin calls come, they must be attended to quickly, and they can require substantial amounts of money. It is not uncommon to see 5% to 10% price swings in just one week; and when this happens, it can create a short-term financial emergency.

For example, if 100 acres of corn are hedged, at 100 bu./acre, then 2 contracts would be required (10,000 bushels). At 3.35/bu., hedge price, a 10% increase in price would generate a margin call of $3,350 (10,000 bu. x $.335). This is a cash requirement that must be attended to quickly. If an entire farm of 500 acres were hedged, the farmer would have to come up with $16,750 . . . just as quickly. **There is nothing in the Futures Market that has caused so many problems for farmers as margin calls.**

THE CLASSIC "WRECK"

In the U.S. there are individual farm groups who are vehemently opposed to the use of the Futures Market and are replete with stories of financial disasters caused by hedging. Usually these "wrecks" have been caused by unexpected margin calls (failure to prepare for margin calls).

The typical situation has been that producers unfamiliar with the Futures Market would sell several contracts as a hedge. The futures price would then rise rather quickly, creating margin calls for a substantial amount of money. Not having set up an account or line of credit for margin calls, the contract holder was forced to go to his banker for additional funds. The usual procedure with most financial institutions is for loan applications to be reviewed by a loan committee which meets periodically. But margin calls must be met quickly . . . at some point then, the contract holder was either not able to obtain the necessary funds without going through the usual channels, or was cut off completely.

Not being able to meet margin calls, he was forced to liquidate his contracts. This meant he had to buy contracts at the current higher price to offset the contracts he originally sold as a hedge. In so doing, he was forced into taking a loss in the Futures Market, while lifting his hedge and leaving his crop unprotected. As mentioned earlier, whenever the Futures Market goes up substantially, oftentimes a larger than normal amount of that respective commodity will go into production. At harvest time the oversupply drives the price down. In many situations then, the ill advised "hedger" also took a loss in the cash market. In other words, he took a double loss (a loss in both the cash market and the Futures Market).

PLACING A "GOOD" HEDGE

The most commonly cited rules for good hedging are:
1. Know the cost of production.
2. Figure basis objectively and conservatively.
3. Set up a line of credit for margin calls before hedging.
4. **Be a hedger - not a speculator.**

A producer obviously cannot place an intelligent hedge unless he knows what his cost of production is. The idea behind placing a hedge is to lock in a profit. If the cost of production is not known, then the producer really doesn't know what price constitutes a profit, and he can conceivably lock in a loss. This may sound childishly basic, but there are still large numbers of farmers that do not keep detailed records and do not accurately know what their costs are.

Once the cost of production can be estimated with reasonable precision, the producer can determine what price (target price) will yield him an acceptable profit. To the target price, the basis must be added. Estimating the basis must be given at least the same consideration as estimating cost of production, because it is part of the producer's break-even figure. That is, if break-even sale price is determined to be $3.20/bu. and basis is estimated at $.15/bu., then the producer would anticipate $3.35/bu. to be a break-even hedge (futures) price. He knows (or thinks he knows), that he must wait until the futures market moves to $3.35/bu. or more, before he can hedge. Obviously then, if basis is actually $.25/bu., the producer would lock in a loss at $3.35/bu.

Whenever going into a hedging program always establish an account or line of credit for margin call purposes **before** any contracts are sold. Most financial institutions will establish a line of credit without charging any interest until the money is actually used. As mentioned in the previous section, failure to prepare for margin calls can place the hedger in an extremely unfavorable position.

As a final note, most authorities stress that a hedger should not be a speculator. When using the futures market as a hedging tool, stay with it. There is always the temptation to lift the hedge early (before the commodity is sold) and take a quick profit (leaving the commodity unprotected). Then too, after following the futures market for a period of time, one often gets a feeling of competence in predicting the market and there is a temptation to openly speculate (buy and sell contracts without the intention of hedging a given commodity). Speculation is, of course, the exact opposite of hedging ... the object of hedging is to reduce risk, whereas speculation increases risk. (According to the Chicago Mercantile Exchange, 90% of all speculators lose money.)

OPTIONS AND THE FUTURES MARKET

A relatively new concept to come into the area of risk management is what is known as options trading. Through the use of options two major disadvantages to conventional hedging with the Futures Market can be overcome. Through the use of options, a producer can: 1. avoid margin calls, and 2. avoid limiting his profit potential. This second advantage to options is undoubtedly the most important.

With conventional hedging, a producer knows exactly what he will receive for his commodity (if basis has been figured accurately). If the price goes up substantially, he does not share in the higher price. He will receive only the amount for which he originally hedged his commodity.

Using options as price protection, however, the producer does not limit potential profits. He is guaranteed a minimum price, but if the price goes up he will experience a higher profit margin.

THE MOST COMMON OPTIONS STRATEGY

The most common price risk strategy with options is what is known as buying a "put". A "put" is a contract that gives you the right, but not the obligation, to sell a futures contract at a specified price.

In essence, it's like an insurance policy. If the price goes down, the producer is protected. But if the price goes up, the producer sells his commodity at whatever the prevailing cash price is and reaps the extra profit. He is not committed to a particular price (as in conventional hedging). That is, he does not have to exercise the options contract. He exercises it only if it is to his advantage.

Thus, with unlimited profit potential (and no margin calls), it would seem that options would be vastly superior to conventional hedging. And this would be true, if it were not for the cost.

Options trading carries a significant cost. The cost is the price that is paid for the options contract. A good analogy would be insurance premiums.

With conventional insurance, the insurance company assumes the policyholder's risk in exchange for a fee (the insurance premium). Options contracts are price insurance. Only instead of a large company issuing the policy (contract), it is an individual speculator. In options trading, a speculator assumes the price risk, but for that service he charges a fee. The fee is the price that is paid for the options contract.

The price (or fee) for a particular options contract is open to bidding and will vary with the commodities market. The price will be a function of where speculators believe the market is going, versus how much producers are willing to pay for price protection (where they think the market is going).

If the speculators believe the price of the commodity in question is going down, then they will want more money for assuming that risk. Likewise, if producers also believe the market is going down, then they will be willing to pay more for price protection. Conversely, if the indications are that the market will go up, then speculators will be willing to accept a lower fee for assuming downward price risk (and farmers will not be willing to pay as much as before). The whole thing takes place on the floors of the commodity exchanges, and the ultimate prices are determined by the bid and ask system that is used throughout the commodities market.

PROTECTING PRICE WITH A "PUT" CONTRACT

Figuring price protection with a put contract is similar to conventional hedging. The most important similarity is figuring basis. As explained previously, basis is the difference between your local cash price and the futures price.

Again, as in conventional future's hedging, the producer must decide what price will give him a reasonable profit. In the hedging example we used $3.20/bu. for corn. Assuming a $.15 basis, the producer would have to sell a $3.35/bu. contract to net his $3.20/bu. price.

But in options trading, we also have to include the cost of the option. For example, options (put) costs for corn might look like this:

Futures Price	Put Option Cost
$3.30/bu.	$.03/bu.
3.40/bu.	.12/bu.
3.50/bu.	.19/bu.
3.60/bu.	.25/bu.
3.70/bu.	.30/bu.
3.80/bu.	.34/bu.

In this case, the producer will have to buy the $3.60 put option, in order to guarantee his $3.20/bu. cash price. That is, $3.60 minus a $.15 basis, minus a $.25 option cost, equals $3.20.

STRATEGY AT HARVEST

When the grain is harvested, the prevailing cash prices will determine what the farmer does with his put option. If cash corn is $3.45/bu. or more (his $3.20 minimum price plus the $.25 he paid for his options contract), he will do nothing. He will receive the $3.20/bu. price he originally wanted ($3.45 - .25 = $3.20).

If however, the cash price is less than $3.45/bu., then he will have to exercise his option. That is, because of the $.25 he paid for his option, less than a $3.45/bu. cash price will give him less than the net $3.20/bu. he needs to make a reasonable profit. For example, if the cash price is $3.25/bu., then he will only net $3.00/bu. ($3.25 - .25 = $3.00).

By exercising the options contract, he can force the speculator who sold it to him to buy a corn futures contract from him at the $3.60/bu. price. If the $.15 basis was figured accurately, the real futures price will be the cash price ($3.25/bu.), plus the basis ($.15/bu.) . . . $3.40/bu.

Thus the farmer will make $.20/bu. by exercising his option. That is, he can sell futures for $3.60 but buy them back simultaneously at $3.40.

OTHER MEANS OF LIQUIDATION

In the previous example the farmer does not necessarily have to buy back his futures. If he likes, he can leave them in effect and eventually deliver his grain. If he does this however, he may be faced with margin calls. As explained previously, the only real reason he might want to do this, is if the cash and Futures Market got out of alignment.

The farmer could also sell his options to someone else. Since, in this example, the options are for $3.60/bu. and the futures price is assumed to be $3.40/bu., then the options should be worth close to $.20/bu.

PROTECTING AGAINST PRICE INCREASES

The previous examples showed how a farmer involved in grain production can protect himself against prices going down (and forcing him into a possible loss situation). But some farmers, livestock producers in particular, may need protection against commodities going up in price. That is, they may need to guarantee feed prices in order to hedge or otherwise guarantee a profit in their livestock.

Using the futures market to hedge against a price increase is essentially the same as hedging against a decrease . . . the only difference is that you buy futures rather than sell them. There is, however, one big potential danger with buying that does not exist with selling . . . and that is the potential for delivery.

When buying futures contracts, there is always the danger that the other party will decide to physically deliver the commodity. What makes this so dangerous is that the seller can deliver to whatever delivery point he wants.

During the expiration month of the contract, you can get a phone call (with no prior notice) informing you that however many truckloads of commodity are being delivered in your name . . . several hundred or even thousands of miles away. In the case of commodities such as corn or soybeans, unanticipated delivery can be only moderately inconvenient and expensive. With livestock, particularly fat or fed livestock, it can be extremely expensive.

Granted, very little physical delivery ever occurs, but if it does, it can be very awkward and expensive. To avoid potential delivery, the futures buyer can liquidate his contracts just prior to the month of delivery. For example, if a contract for June is purchased, the contract could be liquidated in May, to avoid any opportunity for delivery.

The problem with that, however, is that it leaves the price unprotected during the specific month in which protection is needed. Also, many commodities have contract months spread far apart during the year. For example, the contract months for corn are March, May, July, September, and December.

In actual practice, there is no one simple answer to the threat of delivery, and each type of operation deals with it as best they can. In almost all situations, however, the danger of market risk far outweighs the threat of delivery. (Remember, only about 2-3% of all contracts are settled by delivery.)

PROTECTING AGAINST PRICE INCREASES WITH CONVENTIONAL HEDGING

Let's assume a soybean processing plant decides that it must be able to buy soybeans for no more than $6.60/bu. in order to make a reasonable profit.

As in all use of the Futures Market, the hedger must determine what his basis is. In this case let's assume the basis is -$.30/bu. That is, the local price for soybeans is usually $.30/bu. less than the futures' price. This means that he must be able to buy futures contracts for no more than $6.90/bu.

Now let's say that during the month of delivery (termination month of the contract), soybeans have gone to $7.10/bu. This means that when the hedger liquidates his contract he will make $.20/bu. That is, he bought soybean contracts at $6.90/bu., but is now able to sell them for $7.10/bu.

However, when he actually buys his soybeans on the cash market, he will have to pay more than the $6.60 target price he originally decided was the maximum he could afford to pay. If the -$.30/bu. basis is correct, the local cash price will be $6.80/bu. ($7.10-$.30 basis). However, when he subtracts the $.20/bu. profit he made in the Futures Market, he achieves his $6.60/bu. target.

As in the other hedging example given, even if the price went in the opposite direction, the hedger will receive the same net hedge price. For example, instead of going up, let's assume the bottom fell out of the market, and the price dropped substantially . . . say to $5.80/bu. The soybean processor would pay the same net price of $6.60/bu.

That is, when he went to buy his soybeans, if he figured his basis correctly, he would only pay $5.50/bu. ($5.80-$.30 basis = $5.50). Thus, on the cash market, he would be able to buy soybeans substantially cheaper than his target price. However, when he liquidates his futures contracts, he will lose money on the trade. In this case he lost $1.10/bu. on the trade. That is, he bought futures contracts for $6.90/bu. but in liquidation he was only able to sell them for $5.80.

Therefore, if we add the $1.10/bu. he lost in the Futures Market to the cash price of $5.50/bu., we get his target price of $6.60/bu.

USING OPTIONS TO PROTECT AGAINST PRICE INCREASES

Options purchased for futures contracts can be used to protect against price increases, just as they can be used to protect against price decreases. The main advantages to options, . . . providing adverse price protection while also allowing unlimited favorable price movement, are just as true for upward price protection as for downward price protection.

In the previous example of conventional hedging, options would have allowed the soybean processor to actually keep some of the $1.10 savings he experienced when soybeans decreased in price (the actual savings would be $1.10 minus the cost of the options). Also, the use of options would have eliminated the margin calls the soybean processor would have experienced. (Remember, in conventional hedging whenever the futures market moves in the opposite direction of your position, you must make up the difference in margin calls.)

But, in protection against price increases, the use of options have another advantage . . . and that is protection against delivery. If liquidated correctly (discussed later), the threat of physical delivery can be avoided.

THE "CALL" OPTION

To protect against price increases, a "call" type option is purchased. A call option means that at any time during the life of the option, the holder of the option can "call" the seller, and force him to sell a futures contract at the specified price. Thus, the holder of the call option is guaranteed the ability to buy at a certain price.

So once again, options can be considered as insurance against adverse price movement. As explained previously, like any other insurance there is a price, or premium, that must be paid.

The price or premium is the fee charged by whoever issues or sells the call option. That is, the call option is sold by someone (usually a speculator) who is willing to assume the risk of adverse price movement. In essence, he believes that the price will not move in that direction.

The premiums or prices for call options for soybeans

might look like this:

Futures Price Call Option	Premium
$7.20/bu.	$.15/bu.
7.00/bu.	.28/bu.
6.80/bu.	.38/bu.
6.60/bu.	.45/bu.
6.40/bu.	50/bu.
6.20/bu.	.56/bu.
6.00/bu.	.60/bu.

Just as with put options, the premium cost as well as the basis, must be figured into the futures contract price. What may be confusing, in this example, is that while the basis is added, the premium cost is subtracted. This is because it is a low price, not a high price, that is desired. In order to meet the target price ($6.60), the soybean processor must be able to buy at a lower price, to offset the cost of the call option.

The basis is added, since it represents the difference between the futures and the local cash price. In this case it was assumed that the futures price is normally $.30/bu. higher than the local price. Therefore, the soybean processor can consider a futures contract (or an option to buy a future's contract) $.30/bu. higher than his target price. That is, he can consider a contract for $6.90/bu. (If soybeans are $6.90/bu. on the futures market, then he should be able to buy cash soybeans for $6.60/bu. at home.)

The price he pays for the call option is an expense, and therefore, he must be able to buy his future's contract a little cheaper to offset the cost of the option contract.

Looking at the Call Option, we can see that he will have to buy the $6.40/bu. option. That is, if we add the premium price ($.50) to the futures price ($6.40), we get the target futures price ($6.90). Likewise, when we subtract the basis ($.30), we get the target cash price ($6.60).

LIQUIDATING A CALL OPTION

As mentioned earlier, a call option has an added advantage over conventional hedging in that potential for delivery can be avoided.

If the commodity price does not move adversely (to where the holder of the call option does not have to exercise his option), then everything is OK. He just lets the call option expire and nothing happens.

In the previous example, as long as the soybean futures price stayed below $6.40/bu. (the contract price), then everything would be OK . . . the soybean processor would not have to exercise his option. If, however, the price rose above $6.40/bu., then he would have to take some action.

Basically, he has two choices. He can "call" and force the writer of the option (the speculator) to sell him futures contracts at the $6.40/bu. price. Or, he can sell the option itself, and use the price he receives to offset the higher soybean price he will have to pay. For example, if soybean futures are currently at $7.00, then his call options should be worth about $.60. That is, if they can be used to buy futures at $6.40. If this is the mode of liquidation he uses, then there is no chance for delivery.

If his choice for liquidation is to exercise the call and receive a futures contract, then he has two ways of disposing of it. He can turn around and sell a futures contract to offset the one he has received. In other words, if current futures are at $7.00, then he can make $.60 on the trade, since the ones he called are at $6.40.

His other choice is to simply hold the futures contract until expiration (last day of the month specified in the contract). At that time, he will receive the margin difference as settlement from the commodity exchange. Using the price in the previous example, he will receive about $.60/bu. (times however many bushels in the contract or contracts). This is money the commodity exchange has received from whomever wrote the call (the speculator), in the form of margin calls.*

However, if this choice is taken, there is the possibility that physical delivery will be made. Again, the probability is quite low. However, anytime a futures contract is bought and held into the expiration month, physical delivery is always a possibility.

DANGERS OF THE FUTURES MARKET

The preceding examples were used to explain the Futures Market as a means of risk management. There is, of course, another side to the Futures Market . . . and that is the area of speculation.

Speculation is the exact opposite of risk management. Speculation is the assumption of risk. When a farmer takes a position in the Futures Market to protect the price of his crop, it is usually a speculator that takes the opposite position. Likewise, when a farmer buys an option contract, it's a speculator that sells it to him.

The speculator assumes the risk that the farmer wants to avoid. This is important to realize. Because the instant a farmer enters the Futures Market without the intention of hedging or protecting the price on a specific crop . . . he is speculating.

Using the Futures Market as a means of risk management requires a lot of study. It is complicated and no one should attempt to use it without a thorough knowledge

* The speculators who sell option contracts must pay margin calls, just as if they had sold futures contracts directly.

of it.* But as the saying goes, "familiarity breeds contempt". In studying the Futures Market as intensely and thoroughly as should be done, there is always the danger of beginning to believe to know which way the market is headed. There is always the temptation to step in and buy a contract, purely for the purpose of making money. That is, take a position in the futures market without the intention of covering a crop.

Whenever a farmer enters the Futures Market without the intention of covering a crop, he has stepped over the line. He is no longer a risk manager, he is a risk taker. Instead of averting risk, he is assuming risk.

Ninety per cent of all speculators lose money. There is nothing in the way of organized business markets as potentially disastrous as the Futures Market. The stock or bond market can't even begin to compare to the roller coaster of the commodities market.

In the stock market (or the bond market), if the market is headed down, you can get out. Stocks or bonds can always be sold and some salvage of their value obtained. Not so, in the commodities market. All commodities have a price limit in which the price is allowed to fluctuate (for example, for corn it is $.15). Once that limit is reached, trading is halted.**

Thus, if the market starts down, trading can be halted before your contracts are sold. Stories of unfortunate speculators unable to sell their contracts day after day are legend. As the stories go, they are forced to ride their contracts all the way to the bottom.

While some of these stories might be exaggerated... the principle is still true. It is very possible to get in a position and not be able to get out for a period of time.

The volatility of the commodities market itself is something that cannot be over-emphasized. Indeed, the volatility of agricultural prices was the very reason the Futures Market was created. Likewise, it is the volatility of prices that poses so much risk to farmers.

Therefore, when a farmer decides to move from a position of protecting from risk, to one of assuming extra and unnecessary risk... it is a very serious matter indeed.

SUMMARY

The Futures Market is complex, and it is difficult to summarize the mechanics. The concept to remember is that the Futures Market is indeed complex and no one should attempt to utilize it without a full and complete understanding of it. It should also be realized that this chapter (and text) is merely an introduction only, and in no way supplies enough information to justify actual futures trading.

---

* The information contained within this book should not be considered an adequate explanation. It is only an introduction. The mechanics should be studied in more detail, and specifics concerning the actual commodities must be studied.

** Occasionally trading may be halted in the stock market on specific issues for a short time.

BOOK II AGRONOMY

CHAPTER 6 SOILS

John L. Havlin, Ph.D.
Dept. of Agronomy
Kansas State University

INTRODUCTION

Soil is a very thin layer covering the earth, varying from several centimeters to several meters in depth. Because it is so thin, it is an extremely fragile resource. Erosion, through wind and water forces can quickly remove topsoil developed over millions of years. Soil pollution through improper industrial waste disposal, and soil productivity degradation through excessive and improper cultivation or management, can render soils useless for food production. In addition, soil losses have significant detrimental effects on the environment, especially water quality. Therefore, understanding soils, their properties, and their management, is crucial for efficient use and preservation of soil resources.

This fact in itself is very powerful, and should instill a deep respect for this valuable resource. History provides numerous examples of civilizations which failed because of improper soil and water management (e.g. Nile, Euphrates, and Tigris River valleys). Although the Roman, Syrian, Babylonian, and other civilizations recognized the importance of water management, their water canals and aqueduct systems eventually filled with silt and rendered them useless. The extensive soil erosion resulted from excessive tree removal which left an unstable and erodible soil surface.

Present day cultures are faced with similar soil and water conservation problems. Agricultural productivity in the United States, Europe, Russia, Brazil, China, and other countries is threatened by excessive soil erosion.

PRODUCTIVITY

A well managed, productive soil is necessary for efficient crop production. Some soils are less productive than others in a given environment, regardless of management. For example, on one soil, the average winter wheat yield may be 3000 lbs./acre, while 5 miles away on another soil, average wheat yields may be only 2000 lbs./acre. With similar crop management history, only differences in soil can account for productivity differences. Table 1 gives an example of several soils (Larimer County, Colorado, USA) with their respective productivity ratings for several crops. Despite similarities in environment and crop management, wide differences in agricultural productivity exist between soils.

Because of inherent variability in soil properties, some crops cannot be grown on some soils, whereas they are very productive on others. For example, many berry crops (blueberries, raspberries, blackberries, etc.) cannot grow or are marginally productive on calcareous (high pH), high clay soils, whereas on acid, sandy soils (low pH) these crops are very productive. Therefore, in favorable climates for these crops, one could predict areas for successful berry production by knowing the different soils in the area. Even when differences between soils seem minor, management differences will exist for a given crop. Understanding the variation in soil properties is very important for efficient and profitable crop production.

DEVELOPMENT OF SOILS

Soil development is a continual process, utilizing both physical and chemical factors. Essentially it is the breakdown of rocks to form the sand, silt, and clay that we collectively know as soil.

Table 6-1. Crop productivity ratings for several soil series in Larimer County, Colorado (USA).

Soil Series (Type)	Wheat	Yield Corn**	Barley*
		lbs./ac	
Ft. Collins	1320	7744	1232
Kim	1056	7480	1056
Renohill	1144	6336	880
Weld	1408	7920	1320
Keith	1496	8096	1408

*dryland   **irrigated

The processes involved in soil development are incredibly slow by human standards. That is, scientists discuss "old" and "new" soils, but these terms are used in reference to geologic time. In reference to geologic time, man has been on the earth for only a split second in time.

Thus while there may be substantial differences between the ages of soils, in human terms they are all incredibly old. Soils are formed over eons of time, and we must therefore conserve them carefully. If soil is lost to erosion, in human terms of time, it can never be replaced.

PHYSICAL WEATHERING

Physical weathering is the first step in the development of soil. It is the breakdown of rock to form smaller and smaller particles. Table 2 describes various kinds of physical weathering processes:

CHEMICAL WEATHERING

The actual processes involved in chemical weathering are beyond the scope of this text. Still it is valuable for the student of soil science to know that they do exist.

Although water is usually not considered a "chemical", it is the primary agent involved in the chemical weathering of soils. Water can function by combining with the carbon dioxide formed by plant roots to form carbonic acid. Water can also combine with minerals such as calcium, sodium, iron, silica, etc. to change and otherwise break down soil and rocks.

SOIL FORMATION

Technically speaking, soil formation is the layering of soil to form distinct combinations of soils. Different layers occur at different depths, and have distinct properties. These layers are known as soil **horizons**. The vertical (up and down) sequence of these layers is known as the soil profile (see Figure 1).

Three types of soil horizons (A, B, and C) are normally present (Figure 1). The upper layer is known as the A horizon. It is generally higher in organic matter, darker in color, and more weathered than the soil layers below. That is, water has usually leached some of the clay and minerals out, and transported them to the layers (horizons) below. This is known as **eluviation**.

The B horizon is known as the zone of accumulation. That is, the minerals and clay carried down by water from the A horizon accumulate in the lower B horizon. Organic matter itself is not soluble in water, and therefore the organic matter level is lower in the B horizon. The A and B horizons are collectively termed the **soil solum**.

Unconsolidated or weathered rock material lying below the B horizon represents what is known as the C horizon. This layer may also contain accumulations of carbonates and other soluble salts. Below the C horizon lies bedrock or parent material and is generally designated as the R layer.

As mentioned, eluviation is the term used to describe the movement of minerals and/or clay particles from one horizon to another. When the main feature of a particular horizon is the loss of minerals and clay, it is termed an E horizon; E, representing eluviation.

Occasionally there will be a heavy accumulation of organic matter over the A horizon. When this occurs, the overlying layer is termed the O horizon; the O, for organic matter.

Properties of two horizons are sometimes found within the same layer and are termed **transitional horizons**. These horizons may be designated as AB, EB, BC or E/B, B/C, etc. A BC designation indicates the horizon has both B and C characteristics, but is more like a true B horizon. When more characteristics are similar to the C horizon, then a CB is used. E/B indicates a horizon comprised of individual parts of E and B components, where E properties predominate. When a B/E designation is used, B horizon components are dominant.

Table 6-2. Physical weathering processes in soil formation.

Freezing and thawing	Expansion of freezing water in cracks, depressions, voids, etc. can further crack or split rock materials.
Heating and cooling	Differential expansion/contraction of different minerals within a rock cause fracturing.
Wetting and drying	Swelling (wet) and shrinking (dry) of clay and silt particles will cause abrasion between particles.
Grinding	Wind, water, or gravity can promote disintegration through particle abrasion. Reduction in vegetative cover can accelerate this process.
Organisms	Plant roots growing into cracks can split apart rocks. Animal or human digging action (e.g. plowing/cultivation) contribute to physical disintegration.

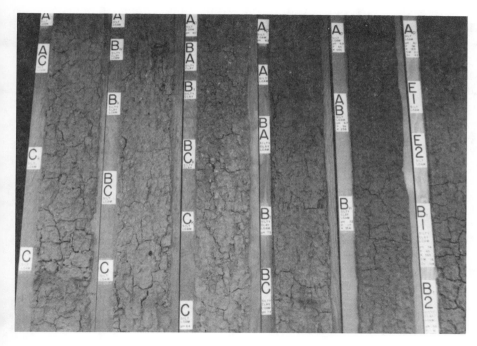

Figure 6-1. Example of several soil profiles. The depth of the dark A horizon and rainfall increases from left to right. In the bottom picture the first profile on the left has a shallow A with no B horizon, indicating semi-arid conditions. Profiles in the middle have well developed A and B horizons formed under moderate rainfall. The furthest right profile has a leached (E) horizon indicative of high rainfall climates.

SOIL ENVIRONMENT

SOIL WATER

Several very important soil processes require water and, thus, it is necessary to understand soil water relationships. Water is essential for the dissolving of minerals (weathering), and soil-forming processes such as leaching. Soil microorganisms also require water to facilitate organic matter decomposition. Their activity combined with mineral weathering provide all the soluble nutrients necessary for plant growth. In addition, the transport of nutrients from the soil to plant roots is through water.

Although the importance of water is obvious, too much water can be detrimental. Excess water will leach below the rooting zone and carry nutrients required for plant growth with it. Also, excess water displaces oxygen which can inhibit normal root activities and plant growth.

The soil water content depends on many above and below ground factors. Obviously the total quantity and frequency of precipitation (rainfall and snowfall), or irrigation water, affects soil water content.

Water is continuously moved through the soil-plant system by the **hydrologic cycle**. Evaporation from plant, water, and soil surfaces, and transpiration through plants, moves water vapor to the atmosphere, where it cycles back to the earth through precipitation.

Depending on the soil and its water content, part of the precipitation runs off the soil surface and the remainder will infiltrate into the soil. Part of the infiltrating water will remain as soil water, and the rest will percolate through the soil profile and leach to the ground water.

One of the most important factors that influence infiltration and percolation, is the amount of water already contained in the soil at the time of precipitation. When the soil is saturated and cannot store additional water, most or all of the rainfall will run off the soil surface. If surface vegetation or residues are not sufficient, then exposed surface soil can be eroded away. If the soil is dry and soil surface conditions allow the water to infiltrate, the soil profile can absorb large quantities of water.

The remaining soil water is defined by several different terms depending on the quantity. **Gravitational** or free water is the quantity of water in excess of the soil water storage capacity. The excess or "gravitational" water will percolate down through the soil profile into the ground water. After a rainfall or irrigation saturates the soil, and all the gravitational water has been drained away, the remaining water is considered **capillary water**. Essentially all the air spaces in the soil are filled with water and further drainage due to gravity cannot continue because of an electrochemical attraction between water molecules and soil particles. Capillary water can however, move from wet soil to dry soil.

As evaporation and the uptake of water by plants occurs, the soil water content is reduced. Unless further precipitation or irrigation occurs, ultimately it will reach what is known as the **hygroscopic** water level. Hygroscopic water is water held to the soil particles very tightly and does not generally move. Plants cannot extract hygroscopic water.

For practical application, it is helpful to further define soil water in relation to plant growth. Once all gravitational water is drained, the soil is said to be at **field capacity**. Generally, capillary water is considered to be water available to plants. When soil moisture content is reduced just to the point of hygroscopic water, then the soil is at **permanent wilting point**, or wilting percentage. At this point, plants become water stressed and begin to wilt. Again, soil moisture content between field capacity and wilting point is considered plant available water. Soil water contents below the wilting point is unavailable water.

The amount of water a soil will hold depends primarily upon its physical characteristics. The most important factor is soil texture. Clay soils can hold more water because particle size is smaller than in sandy soils, which allows more particle surface area and air space or voids to hold water. Fine-textured clay soils contain more soil

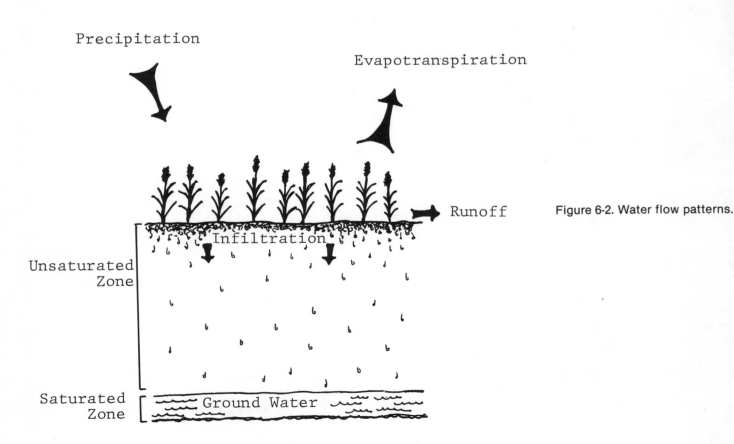

Figure 6-2. Water flow patterns.

water at field capacity than coarse-textured sandy soils, which is also true at the wilting point. Thus, even though clay soils hold more total water, they may, in some cases, have less plant available water than lighter textured soils.

Soil particle arrangement or structure also affects water holding capacity. Tightly packed soil particles (poor structure) will hold less water than a well-structured soil with loose particles. Only in very sandy soils where large air spaces exist, would compaction improve moisture storage.

Organic matter is very important for water holding capacity, especially in sandy soils. Since organic matter can hold water in excess of its own weight, then adding organic matter and/or maintaining surface residues can greatly improve the soil moisture storage in sandy soils. The influence of organic matter is much less in fine-textured soils, because organic matter generally coats soil particles. Thus, in fine-textured soils, the reduction in pore space offsets the effect of the organic matter.

SOIL AERATION

Air contains 78% nitrogen gas, 21% oxygen, and 0.03% carbon dioxide (CO_2). Soil air contains about the same nitrogen, but oxygen and CO_2 can vary greatly. Plant root growth and nutrient absorption by roots require oxygen and release CO_2. Respiration by roots will therefore increase the proportion of CO_2, and reduce the proportion of oxygen. Microbial activity also produces CO_2. Plant growth can be seriously limited by oxygen deficiency, or excess CO_2. Under low oxygen conditions, some minerals normally used as nutrients can actually become toxic. Iron (Fe), aluminum (Al), and manganese (Mn) are examples of these.

Because gases exchange between the above ground atmosphere and the soil atmosphere, the oxygen levels in the soil are replenished. Two processes, mass flow and diffusion, are responsible for gas exchange. **Mass flow** refers to the movement of wind over the soil surface. The contribution to gas exchange is much less for mass flow than diffusion. **Diffusion** refers to the movement of individual gases. As the CO_2 concentration in soil pores exceeds the concentration in the above ground atmosphere, CO_2 will migrate toward the soil surface and escape into the atmosphere. Diffusion processes are very efficient in maintaining similar CO_2 and oxygen levels above and below ground. The rate of diffusion will be greater in a sandy soil with larger soil pores, compared to clay soil. As soil water content increases, diffusion rates will decrease because diffusion is much slower in water than in air. **Therefore, in a high clay, waterlogged soil, most plants could not survive for very long because of reduced oxygen supply.**

SOIL TEMPERATURE

Soil temperature is very important for numerous biological and chemical processes. Over the long term, temperature affects mineral weathering and organic matter accumulation, which was discussed earlier. In the short term, temperature affects seed germination, plant growth rate, water and nutrient availability, and microbial activity.

The energy that warms the soil is the sun's radiation. Maximum radiation occurs at noon when the sun is at its highest point in the sky. The rate of increase in soil temperature depends primarily on soil water content, soil texture, and residue/vegetation cover.

Soil moisture influences the **heat capacity** of a soil, which is the heat required to raise the temperature of 1 square centimeter of soil 1 degree C. As soil moisture increases, the soil heat capacity increases. That is, it takes more heat to increase the temperature of the soil.

Soil moisture also affects what is known as the **thermal conductivity**. This is a term that refers to the ability to transmit heat down through the soil profile. The surface of a dry soil will warm faster than a wet soil, but the heat will not penetrate to any depth as fast. It takes longer to heat a wet soil, but the heat is conducted better. This means that when the sun is shining a wet soil will warm to a greater depth at a faster rate than a dry soil. Surface residues will absorb and reflect solar radiation and "insulate" the soil. Thus, if warm soils are needed to germinate a crop, the residue should be incorporated or removed to expose the soil surface. No-till cropping systems often have lower soil temperatures, which may reduce crop vigor and yields.

SOIL PHYSICAL PROPERTIES

Soils contain all three physical states of matter: solids, liquids, and gases. The proportion of liquids and gases is determined by the size and arrangement of the solid particles. Smaller particles have greater total surface area and thus, more voids of free space for water and air. The opposite is true for larger particles. Tightly packed soil particles will have less air space than loosely packed particles of the same size. Another important solid component in soil is organic matter. Organic matter helps bind particles together and plays an important role in plant nutrient availability. Soil physical properties are extremely important to plant growth and crop production. Rooting depth is determined by the physical makeup of the various soil horizons. Soil texture, structure, porosity, consistency, and depth are the important physical properties in soils.

SOIL TEXTURE

Soil particles represent about 50% of the volume of a soil. In an average, well-aerated soil, the remaining soil volume is air (25%) and water (25%). Obviously, these values fluctuate with soil moisture content. **Soil texture** is determined by the proportion of sand, silt, and clay-sized particles. These fractions are defined by average particle diameter where sand fractions range between 2.0 and 0.05 mm, silt between 0.05 to 0.002 mm, and clay fraction below 0.002 mm. Soil texture can be distinguished by "feeling" the wet or saturated soil with the fingers. A sandy soil will feel "gritty" while clay soils feel smooth and sticky. A crude estimate of soil texture can be determined in the field by this method. Precise determination requires laboratory analysis.

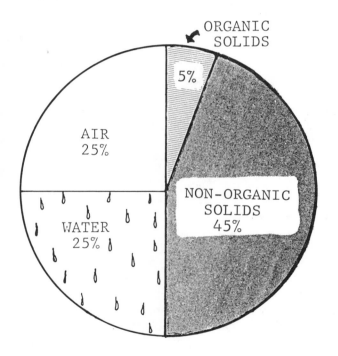

Figure 6-3. Composition of solids, air, and water by volume in a typical soil.

Soil texture names are based on the percentage of sand, silt, and clay. The term **loam** is defined as a mixture of sand, silt, and clay particles with properties of each displayed equally. Silty loam refers to a loam soil with more silt than the other two fractions, while a sandy loam soil exhibits slightly more sand than a loam soil.

Generally, soils that contain a balance of sand, silt, and clay fractions (i.e. loam soils) have few limitations for agricultural uses. These soils have a good water holding capacity but are usually well-drained. With sufficient organic matter (at least 2%), loams and silt loams can be very productive. In contrast, sandy soils have low water holding capacities and under low rainfall, are very droughty if not irrigated. Because of the large pore space in sands, water percolates rapidly through the soil, carrying away valuable plant nutrients. The large particle size of sands cannot hold or store nutrients as effectively as loam or clay soils.

Total air space can be much higher in clays than in sands, thus water contents can be greater. Heavy clay soils can often hold too much water for normal plant growth, because of poor aeration. Clay soils are slow to dry out and are also hard to till, resulting in numerous soil and crop management difficulties. However, with adequate organic matter and aeration, clay soils can be very productive if managed properly.

Generally, soil texture varies with soil depth. Therefore, knowing surface soil texture will not provide information about physical properties in the lower subsoil, or B horizons. As a soil ages, clay is carried by water into the lower horizons. In some older soils, the clay content of the B horizons can be twice that of the A horizon. Heavy clay subsoils can be a valuable source of available water, but can also prevent roots from penetrating below the clay layer.

As discussed, soil and crop management can be difficult in soils with high sand or clay contents. It is important to recognize that soil texture cannot be altered. Over long time periods erosion can remove topsoil and expose subsoils that may have a higher clay content, but for all practical purposes, man cannot change the texture of the soil. In contrast, the structure of the soil can be altered by management.

SOIL STRUCTURE

Soil structure refers to the physical order or arrangement of soil particles. These include sand, silt, clay, and the organic fractions. Wetting and drying, freezing and thawing, and biological activity are natural processes which promote the development of soil structure. Individual particles are held or cemented together into what are known as aggregates. Aggregates or **peds** are organized into specific forms which determine soil structure. An example of six common types of soil structure is shown in Figure 4. **Platy** structure exhibits a flat, layered appearance while **massive** structure is compact and almost structureless. Water infiltration is very slow in massive or platy structured soils. **Block** structured soils have aggregates that are cube-like with sharp edges **(angular Block)** or rounded edges **(subangular Block)**. **Prismatic** (flat tops) or **columnar** (rounded tops) aggregates have long vertical axis with flattened sides. Moderate water infiltration rates are exhibited in these soils. **Granular** aggregates are nearly spherical with rounded corners and are sand sized. **Single grain** structure is simply single solid particles (not aggregates) which include most sands or sandy soils. Rapid water infiltration rates occur in granular or single grain soils.

Good soil structure not only depends on the degree of aggregation or how readily aggregates form, but also on the stability of aggregates, especially in water. Weakly

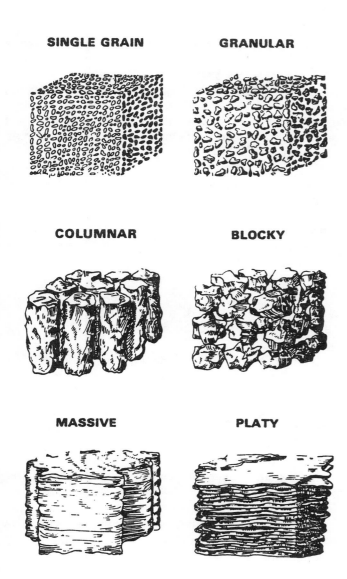

Figure 6-4. Examples of six major types of soil structure. (Source: USDA and U.S. Dept. of Interior, 1959).

developed aggregates will "melt" or disperse when the soil becomes water saturated. The dispersed clay (non-aggregated) and silt particles will move with the water and occupy pore spaces. Once pore spaces are filled with fine soil particles, water cannot readily infiltrate and will run off the surface. Weakly aggregated and dispersed soils are more susceptible to erosion than well aggregated soils.

Aggregate stability is related to: (1) soil texture, (2) organic matter content, (3) kind of clay, and (4) the nature of the minerals associated with the clays. Soils that contain some sand particles (> 30%) will enhance aeration and water infiltration because of larger pore spaces. Some clays are "stickier" than others and, therefore, the extra cementing action can greatly increase aggregate cohesion. Organic matter also acts to help bind particles together. Clays associated with calcium and magnesium exhibit better aggregate stability than sodium saturated clays.

To summarize, a granular structure promotes good aeration and water infiltration and is especially important in the topsoil or A horizon. If water cannot penetrate the topsoil because of poor structure, then erosion will occur. Poorly structured soils will be poorly aerated and will inhibit plant growth. Weak topsoil aggregates break apart under a hard pounding rain and, after drying, a hard crust forms on the surface. Crusting prevents seed germination and reduces oxygen diffusion into the soil.

Poor subsoil structure will inhibit root penetration. Weakly aggregated subsoils are also poorly aerated and can become waterlogged under high rainfall. Thus, topsoil and subsoil structure are extremely important to crop productivity. Soil structure can be improved by crop rotations, residue management, and proper cultivation, although observable improvements in soil structure can take years to accomplish.

SOIL POROSITY

Porosity refers to the portion of the total soil volume not occupied by solid particles. The pore space is either filled with air or water, but usually both, depending on soil moisture. Sandy soils have less total pore space than heavier clay soils, but the pores are larger. As a result, water readily infiltrates and partially drains. Thus, sandy soils usually are better aerated than clay soils. The disadvantage is that the water holding capacity is lower in sands, which may limit plant growth during dry periods.

Clay soils have smaller pores, but larger total pore space. Therefore, water holding capacity is high. However, infiltration and percolation are much lower, thereby reducing aeration (small pores stay filled with water longer).

For example, clay subsoils can have 60% pore space, but they can have so little air space that roots may not grow. However, good structure and aggregation can greatly improve infiltration and aeration. Cultivation can improve aeration in the short run, but over time, soil compaction will increase, and organic matter will decrease. Increasing the organic matter content through limiting tillage, and planting deep rooted legume crops or crops with a high root density like wheat, barley, or oats, will improve soil porosity.

SOIL CONSISTENCY

Soil consistency refers to the cohesive nature of the soil or how well particles stick or hold together. Consistency is dependent on water content and, thus, soil consistency is described at several moisture contents from dry to saturated. Evaluations of soil consistency are usually made for engineering purposes when the soil will be used for supporting roads and buildings.

SOIL DEPTH

Soil depth is an important physical property when evaluating crop growth potential. Generally, shallow soils will have a smaller reservoir for available water and nutrients than a deep soil. Crop productivity will be greater on a deeper soil provided there are no other physical constraints. The properties of the individual horizons are also important. Water and plant roots may not be able to penetrate a heavy clay subsoil, leaving only the shallower A horizon as the rooting zone. Some soils have heavy textures overlying sand. Although roots may penetrate to the sandy subsoil, water may not infiltrate through the surface horizons rapidly enough for optimum use by the roots.

SOIL CHEMICAL PROPERTIES

The chemical processes that occur in soils affect mineral weathering, soil formation, biological activity, and plant growth and nutrition. Weathering of parent materials releases elements or ions that react with other ions and soil constituents to form new minerals. Some of these ions are also incorporated into organic fractions or absorbed by plant roots.

SOIL pH

The most important chemical property in soils is the reaction of pH. Soil reaction indicates the degree of acidity (low pH) or alkalinity (high pH) of a soil. Soil pH determines and controls numerous chemical and biological soil processes. Designation of pH is given in a scale of 1 to 14. The lower the number the more acid the soil, and the higher the number the more basic. The most acid designation would be 1, and 14 the most basic. Neutral is 7. The numbers represent logarithms so one number difference actually means a ten fold difference. For example, a pH of 5, would be 10 times more acid that a pH of 6.

Table 6-3. Terms used to describe various pH ranges.

pH	Description
<5.5	Strongly acid
5.6-6.0	Medium acid
6.1-6.5	Slightly acid
6.6-7.0	Very slightly acid
7.1-7.5	Very slightly alkaline
7.6-8.0	Slightly alkaline
8.1-8.5	Medium alkaline
>8.5	Strongly alkaline

Soil pH strongly influences the chemical processes which determine nutrient availability. In acid soils basic minerals are leached faster and the weathering intensity is greater. Medium or strongly acid soils may not be as fertile as a result. High pH soils can tie up some micro minerals, thereby making them unavailable for plant growth. Iron and zinc deficiencies are common in high pH soils.

Generally, crops like blueberries, cranberries, pineapple, azaleas, and conifer trees prefer acid soils. Alfalfa, beans, barley and wheat do better in near neutral or slightly alkaline soils. The optimum soil pH value for most crops is 6.5 - 7.0.

Generally, acid soils are caused by one or more of the following factors or processes:

-soils developed from acid forming parent rocks.

-depletion of basic cations by leaching (generally associated with humid regions). Crop removal of calcium with crops such as alfalfa can tend to make the soil more acid.

-biological or microbial decomposition of organic residues producing organic and inorganic acids.

-application of acid producing fertilizers, primarily nitrogen sources, will increase soil acidity especially at the soil surface.

In contrast, similar but opposite factors contribute to soils which have developed an alkaline pH:

-basic parent materials, which during weathering processes, produce large quantities of Calcium and Magnesium.

-moderate weathering environments with low leaching potential.

-grassland vegetation with high nutrient recycling capacity.

BUFFERING CAPACITY

The buffering capacity is related to soil acidity or pH, and is highly important. PH indicates the acidity of a soil, whereas the buffering capacity indicates the resistance of a soil to a change in pH.

The term pH actually refers to the amount of chemical ions available to form acid; but it does not indicate how many total chemical ions are present. Some soils have large quantities of minerals which combine with acid forming ions and neutralize them. Calcium (Ca^{++}), sodium (Na^{++}), and magnesium (Mg^+) are very common minerals that tend to reduce acidity.

Formulas for calculating buffering capacity are beyond the scope of this text. What is important for the student to realize is that it can be done and, that it can be a vitally important criteria to know. For example, acid soils are often made more suitable for agriculture by applying limestone. On the other side of the scale, highly basic soils can be made more usable by adding acid such as sulphuric acid. Knowing how much lime or acid to add is determined by the buffering capacity of the soil.

SOIL BIOLOGICAL PROPERTIES

The biological activity in a soil affects both chemical and physical properties of that soil, primarily through interactions with organic matter. With all other factors being equal, organic matter content has the greatest input to soil and crop productivity. Organic matter is formed from the breakdown and decay of plant materials by soil organisms.

SOIL ORGANISMS

Many organisms are responsible for the decay of plant and animal residues. However, microorganisms are significantly more important to organic matter degradation. The microorganisms primarily consist of bacteria, fungi, actinomycetes, and algae.

Most soil organisms require oxygen to survive. They are known as aerobes. (The easiest way to remember their name is to think of the prefix **aer** as air, the Greek word for oxygen.) Other organisms cannot survive in the presence of oxygen, and are known as anaerobes (**an** meaning without ... without oxygen). Most soil organisms are aerobic since only flooded, or water-saturated, soils are oxygen deficient.

Normally anaerobic bacteria account for an average 15% of the total bacteria present in well-drained aerobic soils. The ability of anaerobic bacteria to survive in aerated soils is due to anaerobic "microsites". These microsites occur in small soil pore spaces, which are water-saturated on the inside of soil aggregates.

ENVIRONMENTAL INFLUENCES ON MICROORGANISMS

Compared to other environmental factors, temperature has the greatest affect on microbial activity. Optimum temperatures for most soil microorganisms is between 68-86 degrees F (20 to 30 degrees C).

Moisture content in soils can fluctuate greatly and cause changes in the types of bacteria and bacterial activity. As moisture content increases, and aeration decreases, anaerobic conditions are favored. This causes a shift in the breakdown of organic matter. Residue decomposition still occurs in an anaerobic environment but is carried out by anaerobic bacteria. Instead of CO_2 being the major gaseous by-product, anaerobic decomposition results in methane (CH_4), nitrogen gas (N_2) and hydrogen sulfide (H_2S). A characteristic of anaerobic decomposition is the foul odor associated with the above gases.

Soil pH can also have an effect on soil microbial populations. Generally, bacteria favor neutral (pH 7.0) conditions and are more active than other microorganisms. **Nitrogen-fixing bacteria responsible for nodulation on legume crops (alfalfa, beans, clover, etc.) cannot fix nitrogen in acid soil.** Thus, legume production in acid soils requires liming to raise soil pH to near neutral. Nitrogen deficiency will result without proper nodulation by these bacteria.

CROP RESIDUE DECOMPOSITION

Crop residues consist of numerous plant parts that contain different chemical compounds which differ in ease of decay by microorganisms. For example, leaf tissues generally decompose at a faster rate than the stem or stalk because lignin content in stem tissue is greater than in leaves. The ease of decomposition of the major organic compounds in plant materials increases according to the following order: Amino acids and Sugars > Fats or Oils > Proteins > Hemicellulose > Cellulose > Lignin.

The rate of decay not only depends on the structure and composition of the plant materials, but also on the availability of nutrients, especially nitrogen. If decomposable plant materials do not contain sufficient N, P, S, and other inorganic minerals to meet microbial requirements, then the microbes will take these nutrients from the surrounding soil. Microbial utilization of nutrients from the soil (which would otherwise be available to plants) is called "immobilization". **Once the microbial requirement for nutrients has been satisfied, nutrients remaining in decaying plant tissues will be released to the soil solution for plant utilization.** Microbial release of plant nutrients from organic to inorganic form is termed "mineralization". **Nutrients can only be mineralized if the supply in decaying plant material exceeds microbial demand.**

MINERALIZATION

Mineralization is the breakdown of rocks, soil, or organic matter to release nutrients in a form that is available to plants. The mineralization process is extremely important to nitrogen availability. The release of N in forms available to plants (NH_4^+ or NO_3^-) by microbial decay will depend on the carbon to nitrogen (C:N) ratio of the plant residue. Table 4 gives average C:N ratios for several plant residues. Microbial decay processes reduce the C:N ratio over time.

When residues low in carbon, but high in nitrogen are returned to the soil, the microorganisms have an excess of N for their needs and, thus, net N mineralization occurs (N is released to the soil). As soon as the material is added to the soil, microbial activity increases until the carbon source is consumed. At this point, activity will slowly decline to the original level. Without plant uptake or leaching, the soil N level will be higher than before the residues were added.

Table 6-4. Approximate Carbon to Nitrogen ratios of common plant residues and soil organic matter.

RESIDUE	C:N RATIO
Alfalfa/Clover	13:1
Manure	20:1
Corn Stover	60:1
Oat Leaves	70:1
Wheat Straw	80:1
Pine Needles	100:1

In contrast, when crop aftermath or other residues low in nitrogen are decomposed, the microbes consume the available soil N. If available soil N is not sufficient to meet microbial demand, then decomposition will slow down. Once the quantity of residue is nearly decomposed, the microbial activity will decrease as the carbon and energy sources decline. As the microbe population returns to the original level before residue addition, N is mineralized (released) from the dead cell tissue (microorganisms), and available soil N will increase to the initial or a slightly higher soil N level.

ORGANIC MATTER

The level of organic matter in a soil will vary with the kind of soil, and climate. Sandy soils store less water, and are usually better aerated and warmer than finer textured soils. As a result, organic matter is decomposed more rapidly, and therefore sandy soils tend to have less organic matter.

Climate, especially moisture and temperature, influences organic matter. Higher rainfall and cooler climates encourage greater organic matter accumulation, whereas warmer and drier environments tend to lower organic matter levels. These relationships are primarily true for grassland soils. Forest soils in humid climates have lower organic matter contents than less humid climates, because severe leaching usually reduces native fertility.

CULTIVATION, CROP ROTATION AND ORGANIC MATTER

Manipulation of soils and plants by man through tillage and cropping can dramatically change organic matter contents. Generally, soil organic matter decreases, but the rate of decline will depend upon tillage frequency, crop rotation, and the quantity of residues returned to the soil.

Cultivation of virgin soils stimulates microbial growth and activity, resulting in higher organic matter breakdown (mineralization). Greater mineralization results in significant losses of organic matter and total N over time. In a study of N losses at several locations in Kansas (USA) 30 to 45% of total N was lost over 40 years of cultivation. Total N losses were 15 to 20% after only 16 years in a study in western Nebraska. In the cultivated Great Plains regions of the United States, organic matter content has declined by nearly 50% over the last 50 to 75 years. This represents about 1 to 3% decline in total N per year.

Tillage results in the mixing of soil with organic residues, increased aeration, greater microbial activity, and increased organic matter loss. Depending on environment, loss of 20 to 60% can occur after only 40 to 50 years. In addition to environment (primarily moisture and temperature), the magnitude of loss depends on the type of tillage used. Organic matter losses are considerably less from minimum or no-tillage systems, compared with those utilizing a moldboard plow. The organic matter content of some soils can increase by 25% after 5 to 10 years of no-till cropping, compared to moldboard plow tillage.

Maintenance of surface plant residues will reduce the rate of residue decomposition, resulting in increased levels of organic matter over time. Plant residue decay will occur at a slower rate on the soil surface compared to burial by plowing. Slower decomposition of surface residues is related to reduced microbial access to N and moisture. When residues are buried, microbes have access to soil N as well as residue N, which increases the decay process. Therefore, as a general rule, nutrient content will be higher in surface soils of no-till fields than plowed soils.

CROPPING SYSTEM

Cropping system or crop rotation can greatly influence the loss of organic matter and total N. Although many examples can be found to document crop rotation effects on organic matter, the most famous example is the Morrow Plot. These long-term crop rotation plots were established in 1876 at the University of Illinois campus in Champaign-Urbana, Illinois (USA). The effect of several crop rotations and fertilizer treatments on organic matter loss is depicted in Figure 5. The greatest loss in organic matter occurred under continuous corn. Application of manure, lime, and phosphorus to continuous corn rotation reduced the organic matter loss from 37% to 13% between 1904 and 1973. Including oats and/or clover slowed the rate of organic matter decline. Using a legume in the rotation maintained the highest level of organic matter in this soil compared to continuous cropping.

The long-term Sanborn plots in Missouri also show changes in organic matter content after 50 years of crop

Table 6-5. Soil organic matter as influenced by crop rotation over 50 years in Missouri. (Smith, G.E. 1942. Missouri Agr. Exp. Sta. Bull 458)

Crop Rotation	Organic Matter %
Continuous Corn	1.45
Continuous Wheat	3.40
Continuous Oats	4.08
Continuous Timothy	4.68
Corn-Wheat-Clover	3.31
Corn-Oats-Wheat, Clover	3.74
Corn-Oats-Wheat, Clover-Timothy-Timothy	3.83
Virgin	5.78

rotation systems (Table 5). Continuous cropping with corn resulted in the greatest decline in organic matter. This data indicates that corn, even in rotation with other crops, can have deleterious effects on organic matter content.

The long-term Rothamsted plots in England showed that after 100 years of continuous wheat, organic matter declined only a few tenths of a percent. In contrast, annual applications of N, P and K (86 lbs. N/Ac) and farmyard manure containing nearly 200 lbs. N/acre increased organic matter to 2.4 and 5.2%, respectively (Table 8). Similar experiments at Hoosfield and Saxmundham also verify that fertilization with manure or inorganic fertilizers increased organic matter (Table 6). **Utilizing manures increased organic matter much greater than fer-**

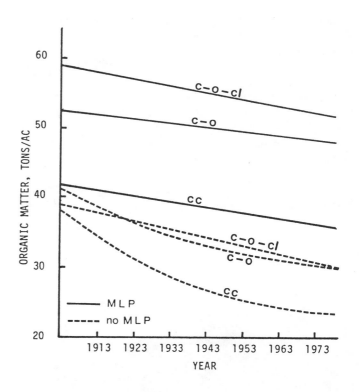

Figure 6-5. Effect of crop rotation and amendments on organic matter content of surface soil in the Morrow plots. Continuous Corn (CC), corn-oats (C-O), and corn-oat-clover (C-O-CL) rotations are illustrated with and without additions of manure, lime, and phosphorous (MLP). (Adapted from Univ. of Illinois Agric. Exp. Stat. Bull. 775.)

Table 6-6. Soil organic matter as influenced by fertilizer and farmyard manure (FYM) treatments in three long-term experiments in England. (Cooke, G.W. 1967. Control of Soil Fertility)

Treatment	Organic Matter (%)		
	Rothamsted[1]	Hoosfield[2]	Saxmundham[3]
Control	2.2	2.0	2.7
NPK Fertilizer	2.4	2.3	2.9
Farm Yard Manure	5.2	5.4	4.0
(Years)	(100)	(80)	(60)

[1] Wheat 1843-1944; [2] Barley 1862-1946; [3] Rotation 1899-1957

tilizers in these studies, indicating that the benefit of manure to maintenance of organic matter is related to organic components other than N.

The benefit of legumes, other crops, and fertilizers in a crop rotation on organic matter maintenance have been reported for many different soils and crops all over the world. **It is important to recognize that complete removal of crop residues by harvesting, grazing, or burning without returning manure will deplete organic matter and reduce soil productivity.** In many countries crop residues are utilized as livestock feed. Soils in these regions eventually will be rendered unproductive unless management practices are adapted that replace organic matter losses.

Figure 6-6. A manure spreader loaded with feedlot manure. As stated in the text, manure adds more than just nitrogen, potassium and phosphorous. Manure adds organic matter which in itself can produce an additional crop response. However, in some cases, especially with cattle manure, one must watch sodium levels. In some cases animals are fed excess salt, which can be a detriment in some soils.

SOIL EROSION EFFECTS

Soil organic matter is concentrated near the soil surface, thus, surface soils are generally more productive than subsoils. Topsoil loss by erosion reduces soil productivity by removing organic matter and plant nutrients. Soil structure can be deteriorated, resulting in poor aeration and reduced water-holding capacity. Topsoil removal may reduce soil depth available for rooting, which can seriously affect crop yields.

A recent field experiment in Indiana (USA) evaluated the effect of water erosion on various soil properties and soil productivity. Clay content increased slightly with severity of erosion because of topsoil loss and exposure of heavier textured subsoils. Both organic matter and plant-available water decreased with degree of erosion. This was directly correlated with reduced corn and soybean yield. Specifically, as clay content increased from 18 to 24%, organic matter decreased from 2.3 to 1.6%, and available water decreased from 12 to 5%. The result was a 16% reduction in corn production and a 14% reduction in soybean production. (Shertz, D.L. [et al.] 1985)

With water erosion, soils are removed from the steeper slope positions and deposited at flat positions down slope. Studies in Ontario, Canada measured corn yields and soil properties from non-eroded, eroded, and depositional areas at 7 locations. This data again indicates the relationship of yield loss to erosion severity. Topsoil loss and, specifically, reduced available water and nitrogen accounted for the corn yield losses. It is interesting to note the increases in organic matter and available water in the depositional areas resulted in no comparable increases in corn yield. (Battiston, L.A. 1985) **Unless drought conditions or nutrient deficiencies occur, added topsoil in depositional areas will not improve yields to offset yield losses in eroded areas.**

SOIL EROSION AND RESIDUES

The most effective way to reduce wind and water erosion is to maintain adequate surface residues during those times when erosion potential is the greatest. The extreme situation would be expressed in a zero or no-till management system, where all or most of the crop residues are left on the soil surface. On most soils, total surface residue maintenance may not be necessary to reduce erosion and maintain productivity. However, no-till systems will reduce wind and water erosion by 95% or greater. Generally, 20 to 30% of crop residues must be present after planting, to reduce wind and water erosion by 50%, compared to clean tilled or plowed fields. The actual amount of residue required to reduce erosion to a tolerable level will depend on the soil, the environment, and the kind of residue.

Wind erosion can occur with either sloping or flat land. In contrast, water erosion occurs only on sloping land and increases with increasing slope and rainfall.

Table 6-7. Wheat residue required to control wind erosion for a range in soil textures.*

Soil Texture	Wheat Residue (kg/ha)	Surface Cover %
Sands, loamy sands	2250	68
Sandy loams	1700	55
Clay, silty clay	1100	42
Silt loam, loams	700	33

*Estimates are for Nebraska conditions.

Tillage up and down hill will result in much greater soil loss compared to tillage across a hill or on the contour. As tillage is reduced, or surface residues increase, erosion losses will be reduced regardless of rainfall amount.

The obvious question is how many tillage operations can be performed, and with what implements, and still leave sufficient residues to control wind and water erosion. The following simple procedure can be used to determine the kind and number of tillage operations required to maintain sufficient residues. (The data presented in the following example applies to soils, crops, and conditions found in the central Great Plains region in the United States. Extrapolation of the data used here may result in serious errors for crops in other parts of the world.)

Table 7 gives the approximate quantity of wheat residue required to minimize wind erosion. The next step is to determine how much residue is left in the field after harvest. Although one could physically weigh a few square meters of surface residue or estimate percent residue cover, an estimate can be made by knowing grain yield (Table 8). Multiply final grain yield by the factor in Table 8. For example a 4000 lbs./acre wheat yield would produce approximately 4000 lbs./acre of residues.

Using Table 7, assume 1960 lbs./acre residue are needed to minimize erosion losses on a loamy sand soil. Therefore, 4000 lbs. - 1960 = only 2040 lbs. residues can be reduced by tillage to minimize erosion losses. If more than 2040 lbs./acre residue reduction occurred, then annual erosion losses would be large enough to eventually reduce yields.

The last step is to determine which tillage treatments may be used and to calculate the approximate residue reduction for all tillage treatments. Table 9 lists numerous tillage implements and the approximate percent residue reduction for each operation. Using this data, over-winter residue decomposition can account for approximately 20% loss. Thus, if 4000 lbs./acre residues are left after harvest, then 4000 x .20 = 800 lbs./acre residue will be lost over winter. If fall tillage operations are performed, then similar calculations are done for the tillage operations used. Fall plowing would eliminate most residues and thus, it is not one to use for leaving residues on the soil surface. In the spring, if a disk or field cultivator is used, then residues are further reduced by 25%. Two pre-planting field cultivation operations would reduce existing residues by approximately 50%. Examples of several tillage implements are shown in

Table 6-8. Residue estimates from grain yield for several selected crops.*

Crop	Residue (lbs./ac)
Corn/Sorghum	0.88
Wheat	1.47
Oats	1.37
Barley	1.28
Soybeans	0.66

*Values pertain to Midwest and Great Plains conditions only.

Table 6-9. Percent residue reduction for several tillage operations.*

Tillage Implement	Percent Residue Reduction After Each Tillage Operation
Moldboard Plow	95-100
Chisel Plow (straight shank)	25
Chisel Plow (curl shank)	50
Tandem Disk	25
Field Cultivation	25
Sweep Plow	10
Planter	5
Overwinter Decomposition	20

*Estimates are for Midwest and Great Plains (USA) conditions only.

Figure 6-7. Moldboard plow. As can be clearly seen, this type of implement eliminates virtually all surface residue. For this reason, moldboard plows should be used only on very flat fields ordinarily not subject to erosion. Even then, its use should be judiciously considered in each individual case. An alternative to the moldboard plow is the sweep plow (figure 11-6 shown on pg. 104). The sweep plow passes underneath the soil surface, so only about 10% of the surface residue is disturbed.

Figure 9. Using this simple procedure can help a grower maintain sufficient surface residues to reduce erosion to levels which sustain soil productivity.

CONCLUDING NOTE

Historically, agricultural crop production practices have emphasized maximum short-term returns. This approach has been fostered by short-term economic considerations, and sustained by continuing technological innovations. That is, despite less than adequate soil management, technological developments have allowed production to remain at constant or increasing levels.

The preservation of soil resources will depend upon the appreciation for, and knowledge of, soil science. Hopefully, this discussion of soil properties will provide the background and insight needed to understand the numerous processes involved in soils. Proper management of these processes will insure viable and productive soils for generations to come.

Figure 6-8. A field cultivator in use. The purpose of the cultivator is to eliminate weeds, and therefore surface residue will also be eliminated. Therefore, if contours and slope of the field are not carefully controlled, the cultivator can contribute to soil erosion.

Figure 6-9. A curled shank chisel plow. This plow is intermediate in surface residue elimination. Roughly 50% is eliminated.

Figure 6-10. A tandem disk in operation. Approximately 25% of surface residue is eliminated per pass with the tandem disk.

Figure 7-X & Y. Below, nitrogen deficiency in corn. Notice the pale green color of the deficient plants in the foreground, compared to the deep rich green color of the normal plants in the background. Notice also the greatly reduced growth in the deficient plants.
 Above, is a case of magnesium deficiency in corn. However, the browning of the leaf edges could also be indicative of potassium deficiency as well.

Figure 13-X. Most seeds are treated with fungicides to inhibit mold growth. To serve as a warning, treated seed is stained with an unnatural color, usually red or pink. **Never under any circumstances consume or allow animals to consume treated seed. Death may result. Humans have died both from eating treated seeds directly, or by consuming meat and milk from animals fed treated seed.**

CHAPTER 7 SOIL FERTILITY MANAGEMENT
John L. Havlin Ph.D.
Dept. of Agronomy
Kansas State Univ.

SOIL FERTILITY AND SOIL PRODUCTIVITY

The concepts of soil fertility and soil productivity are different but related. Adequate soil fertility is highly important, but is only one component of a productive soil. It is important to realize that a fertile soil may not be a productive soil. Plant nutrients in the soil may be at sufficient levels for maximum plant growth, but numerous other soil properties can adversely affect soil or crop productivity. Soil moisture, temperature, pH, structure, texture, topsoil depth, organic matter, tillage, and management can greatly influence soil productivity and plant growth. These factors were discussed in the previous chapter. The interaction and manipulation of all these factors to produce maximum yields is the goal and challenge of each grower.

Table 7-1. Essential Plant Nutrients.

Macronutrients	
Primary	Secondary
Nitrogen (N)	Calcium (Ca)
Phosphorous (P)	Magnesium (Mg)
Potassium (K)	Sulfur (S)

Micronutrients	
Zinc (Zn)	Molybdenum (Mo)
Iron (Fe)	Boron (B)
Copper (Cu)	Chlorine (Cl)
Manganese (Mn)	

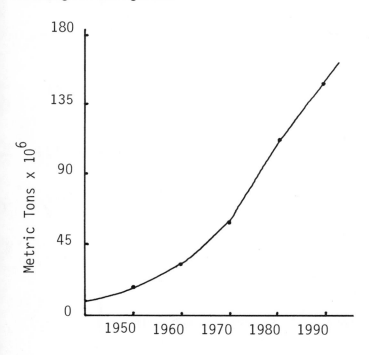

Figure 7-1. Trend in total world fertilizer consumption. (Adapted from: TVA statistics)

NUTRIENT UPTAKE AND PLANT GROWTH

There are 16 elements essential to plant growth and they are divided into primary, secondary, and micronutrients. Carbon, hydrogen, and oxygen make up over 90 percent of plant weight and are obtained from air (CO_2) and water (H_2O). The remaining 13 elements are obtained from soil minerals and organic matter (Table 7-1). These elements are considered essential because if any one is below a "critical" level in the soil, the plant will not develop normally. The plant will display a nutrient deficiency by several means, depending upon the degree of deficiency. Plant nutrient deficiency symptoms for the essential plant nutrients are described in Table 2.

If a nutrient is only marginally deficient, little or no visual symptoms may develop, and only small yield losses will occur. Moderate deficiencies usually produce distinct visual symptoms on leaf and stem tissues, and usually result in severe yield losses. Severe nutrient deficiencies will not allow the plant to develop normally, and the plant may not survive long enough to produce a harvestable yield. **Generally, once a deficiency symptom is visible, the yield potential is already reduced and correcting the deficiency by applying a fertilizer to that crop will not totally restore the lost yield potential.**

There are three mechanisms by which plant nutrients are transported to the roots: 1) mass flow, (2) diffusion, and (3) root interception. Mass flow is simply the flow of water containing nutrients from the soil to the plant roots. Nutrients that are very soluble in water like nitrate, are transported by mass flow.

Diffusion is the method by which most nutrients are transported. Diffusion refers to the movement of ions from one point to another, in response to a difference in concentration. That is, in the immediate vicinity of the root, the concentration of plant nutrients will usually be less than that at some distance from the root. This is

Table 7-2. CHARACTERISTIC EXTERNAL NUTRIENT DEFICIENCY SYMPTOMS.

Element	Leaf Coloration	Growth/Development
Nitrogen (N)	Light green or yellow on lower leaves which eventually die (necrosis)	Plants stunted and weak; early leaf loss; leaves small; plants may mature earlier
Phosphorous (P)	Purple or dark blue-green; first appears on lower, older leaves	Plants develop and mature slowly
Potassium (K)	Yellow, brown tissue between veins, 'burning' or necrosis along leaf blade; symptoms on mature, lower leaves; chlorosis or white 'specks' on clover leaves	Slow growth; short internodes, leaves can appear wilted
Sulfur (S)	Yellow-green coloration in leaf blades and veins; appears on younger plant parts	Slow growth; plants mature early
Calcium (Ca)	White strips along leaf margins in some vegetables; chlorosis on younger plant parts	Death of buds and some roots; small leaves
Magnesium (Mg)	Chlorosis beginning at the center and edges on leaf blade of mature, lower leaves	Slow growth; delayed maturity, poor tillering
Iron (Fe)	Chlorosis between leaf veins; yellow or white young leaves	Plant growth reduced or stops
Zinc (Zn)	Yellow or white coloration between midribs and leaf margin; appears on older, mature leaves	Shortened internodes; rosetling in broadleaf plants
Manganese (Mn)	Interveinal chlorosis, white or grey 'spots' in leaves of small grains; appears on younger leaves	Stunted growth; leaves in vertical position
Copper (Cu)	Chlorosis of leaves; leaf tips turn white; appears on young leaves	Poor growth, wilting
Boron (B)	Yellowing of younger leaves; deformed fruit	Buds die; flowers and fruit drop off

because the root is continually removing ions from the soil around it. In response to the nutrient depletion around the root, ions move in from areas of higher concentration.

The third mechanism, **root interception**, occurs when actively growing roots come in actual physical contact with clay particles and directly remove cations or anions absorbed on exchange sites. Of the three mechanisms,

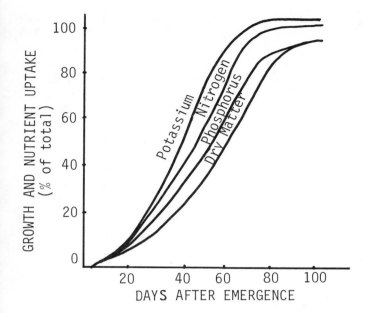

Figure 7-2. Dry Matter and nutrient accumulation patterns in grain sorghum.

Obviously, nutrient uptake must occur before plant growth can take place. Figure 2 illustrates how nitrogen, phosphorous, and potassium uptake in grain sorghum occur prior to growth (dry matter accumulation). At the half-bloom stage, only 50 percent of dry matter has been produced, while 60, 70, and 80 percent of the phosphorous, nitrogen, and potassium have accumulated.

The general pattern of plant growth and dry matter accumulation is one of small increases in size, followed by rapid growth, and finally little or no growth (Fig. 7-2). Many factors, such as variety, moisture, soil structure, etc., affect the slope of the growth curve. Any above or below ground factor that reduces growth will reduce yield. Many times the farmer cannot control some adverse effects on growth (i.e. moisture or temperature). Soil fertility factors, of course, can be controlled by the addition of fertilizers.

The challenge for the farmer is to (1) identify which nutrients in the soil are deficient; (2) select the best fertilizer source to correct the deficiency; (3) determine the "correct" fertilizer rate to give optimum crop yield and (4) determine the best method to apply the fertilizer material.

mass flow and diffusion are considered to be responsible for most of the nutrient uptake. **With either mechanism, nutrient movement to plant roots can only occur in moist soils.** Nutrient uptake cannot take place in dry (below wilting point) soils. (In mass flow nutrients move with the water; with diffusion nutrients move through the water.)

NITROGEN

Plants contain between 1 and 5 percent nitrogen by weight. The primary role of nitrogen in plants is the formation of proteins. Nitrogen content depends on the crop, soil nitrogen supply, and stage of growth.

Table 7-3. RELATIVE CONCENTRATION AND AVERAGE CONTENT OF MACRO- AND MICRONUTRIENTS IN PLANTS.

Element	Relative Concentration	Content
H	60,000,000	6 %
O	30,000,000	45 %
C	30,000,000	45 %
N	1,000,000	1.5 %
K	400,000	1.0 %
Ca	200,000	0.5 %
Mg	100,000	0.2 %
P	30,000	0.2 %
S	30,000	0.1 %
Cl	3,000	100 ppm
B	2,000	20 ppm
Fe	2,000	100 ppm
Mn	1,000	50 ppm
Zn	300	20 ppm
Cu	100	6 ppm
Mo	1	0.1 ppm

Generally, nitrogen content in leaves is higher early in the growth cycle and steadily decreases until maturity. The total nitrogen content in grain will usually increase with time as the grain develops and matures.

Plants deficient in nitrogen are stunted with yellow leaves (chlorosis). Nitrogen is mobile in the plant and thus, older leaves exhibit the deficiency symptom first. When nitrogen is severely limited, the lower leaves turn brown and die, which is called 'firing'. Early vigorous vegetative growth with dark green leaves indicates adequate nitrogen availability. Excess available nitrogen will delay crop maturity. In dryland regions where water can limit plant growth, excessive vegetative growth can reduce soil moisture. This may cause the crop to run out of water during the critical grain or fruit-filling period.

Approximately 99 percent of total soil nitrogen is contained in the organic matter. The remainder is found in inorganic forms, primarily nitrate (NO_3^-) and ammonium (NH_4^+). The transformations between organic and inorganic nitrogen occur through microbial activity. Figure 3 illustrates the nitrogen cycle in the soil and plant system. It is called a cycle because nitrogen can continually go from one form to another.

Nitrogen must be combined with hydrogen, oxygen, or carbon before it can be utilized by plants. **Nitrogen fixation** is the process of chemically combining nitrogen from the air with another element. Some soil microorganisms directly "fix" small quantities of nitrogen in the soil (Azotobacteria). Others (rhizobium) function through the indirect means of association with root

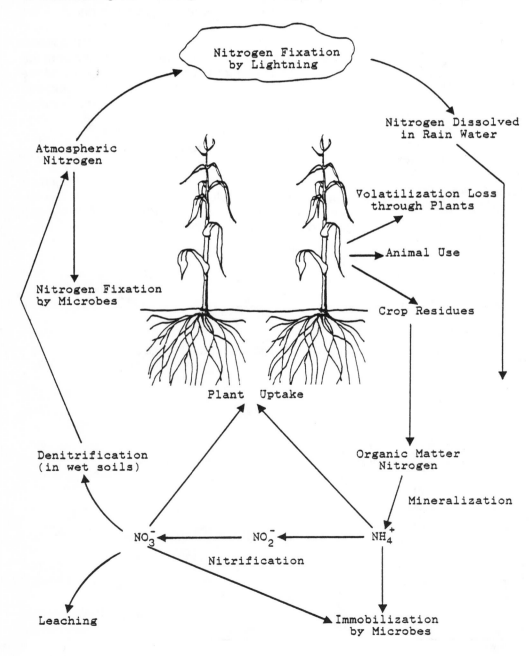

Figure 7-3. Nitrogen cycle in the soil - plant system.

nodules of legume plants and can fix much larger amounts. The value of legumes in crop rotation plans is usually related primarily to nitrogen fixation by the legume. However, the total amount of nitrogen supplied by nitrogen fixation (5 to 10 lbs. per acre per year) is not sufficient for optimum growth of most non-leguminous crops. (Lightning can also fix some nitrogen which is then dissolved in rain water and falls to the earth.)

Plants take up nitrogen in the form of nitrate (NO_3^-) and ammonium (NH_4^+). Once in the plant they are converted to proteins. When the plant dies, the residues fall to the soil surface and are slowly incorporated into the soil. Soil microorganisms break down the residues and produce or release other organic nitrogen compounds. Other soil organisms utilize these compounds for energy and can release ammonium to the soil. As discussed in the previous chapter this process is called **mineralization.**

Microorganisms will oxidize NH_4^+ to nitrite (NO_2^-) and then to nitrate (NO_3^-) which is called **nitrification.** Thus, the processes of mineralization and nitrification supply available nitrogen for direct plant uptake. In most soils with 1 to 3 percent organic matter, about 10 to 30 lbs. nitrogen per acre are mineralized per year. But nitrogen fixation and mineralization do not contribute enough plant available nitrogen for non-legume crops. Therefore, most crops will require nitrogen fertilizer for maximum production.

There are several mechanisms in the soil-plant system through which nitrogen can be lost or otherwise made unavailable for plant growth (Figure 7-3). **Immobilization** is the opposite of mineralization in that NO_3^- and NH_4^+ are taken up and used as a food source by microorganisms. That is, soil microorganisms use soil nitrogen for protein formation within their own cells. This process is set into motion whenever plant residues are added to the soil. This is because most crop residues contain very little nitrogen. The microbes will therefore utilize any free nitrogen available in the soil. Microbial activity and population will therefore increase right after crop residues are added to a soil. After several months when the energy source (crop residue) is depleted, the microbial activity slows down to its previous level, and the nitrogen is again released to the soil as the microorganisms die.

Nitrate nitrogen can be easily leached down through the soil profile and out of the root zone. The amount of nitrate lost by leaching depends on (1) the amount of rainfall or irrigation water applied, (2) the permeability of the soil, (3) the rate of nitrification by soil bacteria or the quantity of nitrate present, and (4) presence of a growing crop.

Leaching losses will be greater in sandy soils than in heavy textured soils. Leaching losses will be reduced with an actively growing crop as opposed to a bare field. Actively growing young roots take up large amounts of nitrogen and therefore reduce leaching losses. Nitrate leaching can pollute ground water. Nitrate concentration in drinking water in excess of 10 parts per million can cause health problems in infants. Nitrate contamination is primarily caused by fertilizer nitrogen applied in excess of crop needs, or application during high rainfall and/or bare surface conditions.

Another mechanism for nitrogen loss from the soil and plant systems is through **dentrification.** Dentrification is the gaseous loss of nitrogen (N_2, N_2O, NO) from soils which are usually very wet or water-logged. In well-drained soils, dentrification losses are usually very small. However, losses can be very large when these soils are flooded, such as under rice production.

Volatilization is also a mechanism for gaseous nitrogen loss and is usually associated with the application of ammonium fertilizers. Volatilization losses are greatest when: (1) ammonium containing fertilizers are broadcast on the soil surface, (2) soil surface is calcareous and/or soil pH is greater than 7.2, and (3) surface residues are present. Volatilization of urea-based fertilizers often occur when light rains or humid conditions occur immediately after application (water causes urea to break down). When the environment and soil surface condition are optimum, ammonia losses can be as high as 30 percent of the applied nitrogen fertilizer.

Currently most nitrogen fertilizers are inorganic, although organic sources such as animal manures are sometimes used. Anhydrous ammonia and urea ammonium nitrate are the most common nitrogen sources. Anhydrous ammonia is generally the least expensive nitrogen source, however, it is the most difficult to apply and most dangerous to use. A typical anhydrous applicator is shown in Figure 7-4. The ammonia is a liquid under pressure in the tank but changes to a gas when injected into the soil. Because of the high pressure, weak or faulty hoses, couplers, or connectors can be very dangerous. **Contact with the skin, eyes, or lungs can kill or blind the applicator.** Extreme care must be exercised when using anhydrous ammonia.

Anhydrous ammonia should be 'knifed' into the soil deep enough to prevent the gas from escaping (usually 6 to 8 inches). Heavy clay soils are difficult to work with, because the knife opening sometimes will not seal back and some gas will escape. Dry sandy soil can also be a problem. Without sufficient water to react with the gaseous ammonia, it can diffuse upward and escape. Wet, easily tilled or friable soils are best for ammonia application.

Urea ammonium nitrate (UAN) may be slightly more expensive than anhydrous ammonia, but it is safer to use and easier to apply. Liquid fertilizers like UAN can be broadcast or sprayed on the surface, banded or dribbled on the surface, or banded below the surface.

The most common solid or dry nitrogen fertilizers are urea, ammonium nitrate, and ammonium sulfate. Dry fer-

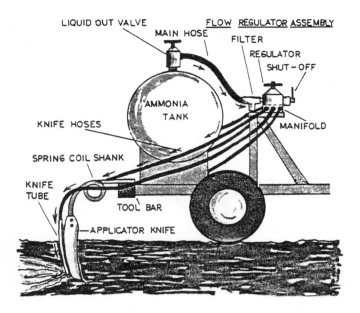

Figure 7-4. A typical anhydrous ammonia applicator. Ammonium is usually the most inexpensive source of nitrogen, but it is also very dangerous to apply. Ammonia is extremely caustic. Since anhydrous ammonia is under high pressure, one must be exceptionally cautious and cognizant of the danger. A ruptured hose, leaking valve or other fitting can mean severe burns, blindness and even death. (Drawing by the Fertilizer Institute.)

tilizer nitrogen is usually broadcast or banded on the soil surface. They can be banded below the surface but it's much easier to subsurface band liquids. Very little dry fertilizers are deep-banded.

As indicated earlier, loss of N by ammonia volatilization can occur with some nitrogen fertilizers. Surface, or broadcast applied, urea-based fertilizers (UAN, urea, sulfur-coated urea) usually exhibit the greatest ammonia

Figure 7-5. Occasionally fertilizer is applied through sprinkler irrigation systems as in this experimental solar powered unit (univ. of Calif. Frenso). Whenever fertilizer is applied through sprinklers, however, special consideration must be given to the potential for leaf burning, particularly with nitrogen. For that reason, more than one application is used. That is, the fertilizer is applied at very dilute rates. (Note the low pressure dropper tubes.)

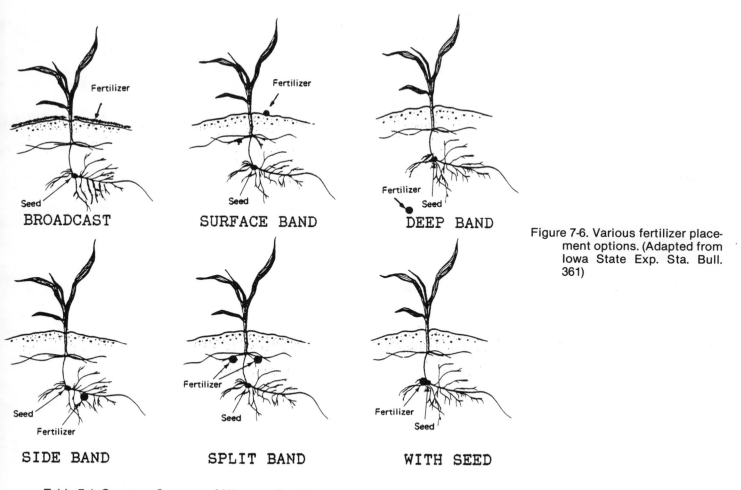

Figure 7-6. Various fertilizer placement options. (Adapted from Iowa State Exp. Sta. Bull. 361)

Table 7-4. Common Sources of Nitrogen Fertilizer.

Fertilizer	Analysis (N-P_2O_5-K_2O)	Physical Form	Method of Application
Anhydrous ammonia (NH_3)	82-0-0	high pressure liquid (gas at atmospheric pressure)	must be injected 6-8 inches deep in friable,* moist soil.
Urea Ammonium Nitrate (UNA) (NH_4NO_3 + urea + H_2O)	28-0-0 to 32-0-0	liquid	spray on surface or banded below or on the surface
Ammonium nitrate (NH_4NO_3)	34-0-0	dry solid (prills)	broadcast or banded below surface
Ammonium sulfate ($(NH_4)_2SO_4$)	21-0-0	dry granules	broadcast or sidedress
Urea ($NH_2 \cdot CO \cdot NH_2$)	45-0-0	dry solid (prills)	broadcast or banded below or on the surface
Sulfur Coated Urea	35-0-0	dry solid (prills)	broadcast or banded below or on the surface

* Friable soils are those which are easily crumbled or pulverized.

loss. Ammonium sulfate and ammonium nitrate can also volatilize. Losses can usually be avoided by placing fertilizers below the soil surface.

Broadcast fertilizers can also be tied up or immobilized if applied to surface crop residues. The quantity of nitrogen immobilized will depend on the kind and amount of residue present. Legume residues have a low carbon to nitrogen ratio (are high in nitrogen), thus, very little fertilizer nitrogen would be immobilized. Wheat straw has a high carbon to nitrogen ratio (very little nitrogen) and, thus, would immobilize about 20 to 30 lbs. nitrogen per ton of residue. However, once incorporated and degraded in the soil, part of the immobilized nitrogen in the residues will be mineralized and made available to plants. Unfortunately this cycle may take 2 or more years to complete.

PHOSPHORUS

Most plants contain 0.1 to 0.3 percent phosphorus. Phosphorus is very important in seed development, so seeds generally contain very high levels of phosphorus. Phosphorus also functions in the storage, transfer, and release of energy necessary to most metabolic processes in plant cells. When plants are phosphorus deficient plant cells cannot divide. Therefore symptoms of P deficiency are stunted growth, delayed maturity, and shriveled seed. However, the most common **early** sign is the classic purple streaking on the leaves.

Phosphorus is present in the soil in both inorganic and organic forms. The phosphorus cycle shown in Figure 7-7 illustrates several ways the soil can replenish phosphorus in the soil solution (P dissolved in the soil water and available for plants). The quantity of available or dissolved phosphate ions in the soil solution is usually very small, less than 1 kg phosphate per hectare. Crops normally require 20 to 30 lbs. of phosphate per acre. Therefore, the soil must re-supply the soil solution many times during the growing season.

Soils that contain phosphorus minerals with very low solubility are often phosphorus deficient. Soil pH strongly influences phosphate mineral solubility. Maximum solubility occurs between pH 6.5 to 7.0. Therefore, putting lime on acid soil to raise pH can often increase phosphorus availability.

Most of the available phosphorus reaches the plant

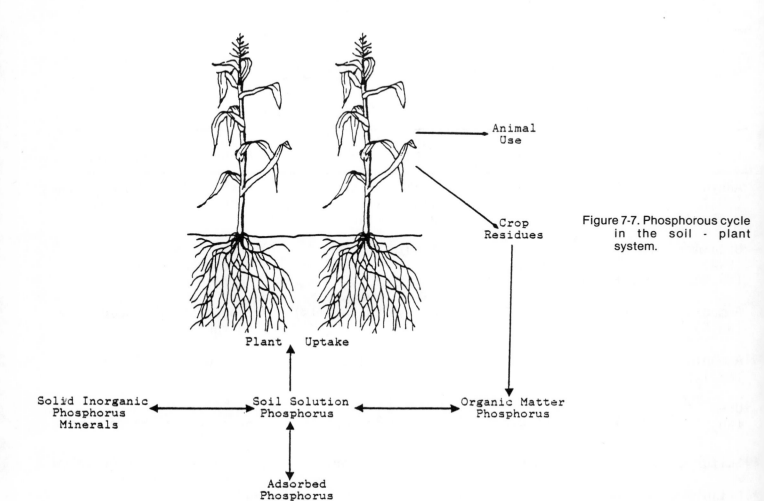

Figure 7-7. Phosphorous cycle in the soil - plant system.

root by diffusion. Thus, dry soils will not supply as much phosphorus as wet soils. Light-textured soils have more pore space or less surface area than loam or clay soils. Therefore, sandy soils, also tend to be phosphorus deficient. Increasing organic matter content by adding manure can improve phosphorus availability in sandy soils (by increasing water holding capacity).

Inorganic phosphorus fertilizers are many times more soluble than soil phosphorus minerals. Thus, when phosphorus fertilizers are added to soils, they readily dissolve and are available for plant uptake. The most common commercially available phosphorus sources are listed in Table 5. Normal, triple or concentrated superphosphates are solid granular materials. The phosphorus concentration in the fertilizer material is expressed as phosphorus oxide (P_2O_5) instead of simply phosphorus (P). The P_2O_5 content is actually only 43.47% the P content. Therefore, normal superphosphate containing 20% P_2O_5 would be the equivalent of about 8.7% elemental P.

Mono and di-ammonium phosphate contain nitrogen in addition to P. Ammonium polyphosphate is a liquid material that also contains nitrogen. Ammonium nitrogen can be toxic to seedlings if concentrated with or near the seed. Therefore, if ammonium phosphate fertilizers are applied with the seed, the amount of nitrogen should not exceed 5 to 7 lbs. per acre, depending on the crop.

Unlike nitrogen, phosphorus is not mobile in the soil therefore, P will stay within several inches of the point of application. For this reason, crops grown two and three years later may utilize some of the original phosphorus fertilizer. Some soils with high clay and/or low pH can "fix" or tie-up phosphorus fertilizer and reduce phosphorus availability. On these soils, broadcast phosphorus will not be the most effective method of application. Seed or band applications near the seed, will improve the plant uptake.

POTASSIUM

Plants require relatively large amounts of potassium. Depending on the crop, the plant potassium content ranges from about 1.5 to 3.0 percent. The primary function of potassium is in maintaining cell turgor pressure. Potassium is also involved in metabolism and the formation of some proteins and carbohydrates. Potassium is important in drought resistance, stalk strength, and disease resistance.

Potassium is very mobile in the plant so deficiency symptoms appear on older leaves first. Symptoms are difficult to detect unless the deficiency is severe. In corn, small grains, and grasses, leaf "burning" or necrosis begins at the leaf top and edges and moves toward the midrib. In some cases the veins can remain green. Lower leaves completely die in severe deficiencies. With clover or alfalfa crops, potassium symptoms appear as white spots on the leaves.

Table 7-5. Common Phosphorous Fertilizer Materials.

Fertilizer	Analysis $N-P_2O_5-K_2O$	Physical Form	Method of Application
Normal Superphosphate $Ca(H_2PO_4)_2 + CaSO_4$	0-20-0	dry granules	broadcast, with the seed, banded
Triple Superphosphate $Ca(H_2PO_4)_2$	0-46-0	dry granules	broadcast, with the seed, banded
Monoammonium Phosphate $NH_4H_2PO_4$	11-52-0	dry granules	broadcast, with the seed, banded
Diammonium Phosphate $(NH_4)_2HPO_4$	18-46-0	dry granules	broadcast, with the seed, banded
Ammonium Polyphosphate $(NH_4)_3HP_2O_7$	10-34-0	liquid	broadcast, with the seed, banded

1/Rates with the seed in excess 50 to 70 kg of fertilizer material per hectare will severely reduce stands.

Like phosphorus, most of the potassium in soils is not readily available to plants. Potassium leaches out of plant residues very rapidly, and therefore is not a major component of soil organic matter. Therefore, soil potassium is in inorganic forms associated with soil minerals. Figure 7-8 illustrates the relationship between the important potassium components in soils. Over 90% of soil potassium is found in secondary soil minerals. All of the potassium in these minerals is unavailable to plants until the very slow weathering pro-

Figure 7-8. Potassium deficiency in alfalfa. Potassium is very mobile in the plant, and symptoms will appear in the older leaves first. The classic sign being leaf "burning" which is clearly indicated in this photo.

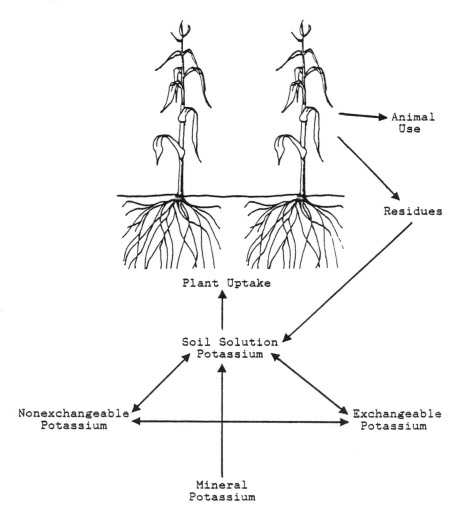

Figure 7-9. Potassium cycle in the soil - plant system.

Table 7-6. Common Sources of Potassium Fertilizers.

Fertilizers	Analysis M-P_2O_5-K_2O	Physical Form	Method of Application
Potassium chloride (KCl)	0-0-60	dry granules	broadcast, banded
Potassium Sulfate K_2SO_4	0-0-50	dry granules	broadcast, banded
Potassium Magnesium Sulfate ($K_2SO_4 2MgSO_4$)	0-0-22	dry granules	broadcast, banded

cess dissolves some of the mineral and releases potassium cations.

Soils deficient in potassium are usually older soils that have lost most of the mineral potassium to weathering, or soils that contain very few potassium-bearing minerals. In other words, sandy soils with very low mica clay content will normally be low in available potassium.

Most commercially available potassium fertilizers are dry granules. Since they are all very soluble salts, potassium fertilizers can cause seed and seedling damage if applied with the seed. Most potassium fertilizers are broadcast applied, although surface or subsurface banding can be more efficient in supplying potassium to standing crops.

CALCIUM, MAGNESIUM, AND SULFUR

Known as the secondary macronutrients, Ca, Mg, and S are normally sufficient in most soils. Indeed, in many soils they can be excessive and limit plant growth.

But in some soils, particularly acid soils, the addition of these nutrients can be required for some crops, particularly legumes. Legumes have a high requirement for calcium/magnesium. Indeed, in many soils without lime applications legumes cannot be grown at all.

Calcium is not mobile in the plant, so deficiency symptoms first appear in younger leaves and tissues. New leaves won't grow or unfold and leaf tips become necrotic. In addition, leaf margins don't grow, resulting in curled leaves.

Most soils contain large quantities of calcium

Figure 7-10. Response of wheat to liming in highly acid soils. Although calcium is often excessive in many soils, in highly acid soils the addition of lime can greatly improve performance (especially with legumes). The lime adds calcium (and/or magnesium) which may be deficient, but also changes the pH of the soil from acid to more neutral or basic.

Table 7-7. Common Sources of Calcium and Magnesium.

Material	Ca or Mg Content (%)	Physical Form	Method of Application
Slaked lime (Ca(OH)$_2$)	54% Ca	dry granules	broadcast
Calcite (CaCO$_3$)	40% Ca	crushed rock	broadcast
Dolomite (CaCO$_3$+MgCO$_3$)	22% Ca 12-14% Mg	crushed rock	broadcast
Gypsum (CaSO$_4$)	22% Ca	dry granules	broadcast
Normal Superphosphate (Ca(H$_2$PO$_4$)$_2$)	20% Ca	dry granules	broadcast
Triple Superphosphate Ca(H$_2$PO$_4$)$_2$	14% Ca	dry granules	broadcast
Potassium magnesium sulfate	11% Mg	dry granules	broadcast

(175-8800 lbs. of exchangeable Ca per acre). Many soil minerals also contain calcium. Soils deficient in calcium are generally found in very high rainfall regions where most of the calcium minerals have dissolved and leached out. These soils are generally very acid and require lime for optimum productivity.

Magnesium is the main constituent of chlorophyll, which is essential for photosynthesis. It also functions with calcium to regulate ion uptake. Magnesium is very mobile in the plant, so deficiency symptoms first appear on older leaves. The main symptom is chlorosis (yellowing of leaves) caused by the lack of chlorophyll in the leaves. The loss of green color begins on the leaf tips and margins and moves inward toward the midrib. Veins usually stay green. Like calcium, magnesium deficiency also normally occurs in highly leached, low pH (acid) soils.

The most common calcium and magnesium fertilizer materials are listed in Table 7-7. Since most calcium deficient soils are acidic, liming these soils will help correct the deficiency. The most common materials are limestone (calcite) and slake lime. Dolomite should be used instead of calcite if magnesium deficiencies also exist. (See the following section on Acid Soils and Liming for more details.) Gypsum or calcium sulfate can also be used to add calcium, but gypsum will not neutralize soil acidity. In the case of minor calcium deficiencies the application of super-phosphate fertilizers will also add calcium (super-phosphates contain Ca as well as P).

SULFUR

The sulfur content in plants ranges from .10 to .20%. Sulfur is a major constituent of plant proteins and is also essential for chlorophyll formation. Sulfur is very important to nitrogen fixation in legumes. Therefore, legume crops usually contain large quantities of sulfur (0.20-0.25%). Sulfur deficiency symptoms are similar to nitrogen in that leaves show pale green to yellow discoloration (chlorosis). Unlike nitrogen, sulfur is not as mobile in the plant and therefore, symptoms first appear on younger leaves.

Plants absorb sulfur from the soil as the sulfate anion. Sulfur dioxide can also be absorbed from the air directly through leaves. Most of the sulfur in soils is found in the organic matter fraction, although some soils contain sulfide inorganic minerals. The sulfur cycle which is shown in Figure 7-3 is similar to the nitrogen cycle. Sulfur containing crop residues are incorporated into the soil and broken down by microbial activity. The organic sulfur formed from these processes is mineralized to inorganic sulfate again by microbial activity. Many soils would accumulate sulfur if sulfate

Table 7-8. Common Sulfur Fertilizer Materials.

Fertilizer	Analysis $N-P_2O_3-K_2O$	Sulfur Content (%)	Physical Form
Elemental Sulfur	0-0-0	90-100	dry granules
Gypsum ($CaSO_4$)	0-0-0	17	dry granules
Ammonium Sulfate ($(NH_4)_2SO_4$)	21-0-0	24	dry granules
Ammonium Thiosulfate	12-0-0	24	liquid

were not removed by leaching.

Sandy soils that are subject to high leaching may be sulfur deficient. Likewise, soils deficient in sulfur are generally those with low organic matter content. Common sulfur fertilizer materials are shown in Table 8. The most common method of application is surface broadcast. Unlike phosphorus, sulfate is very mobile in soils and broadcast sulfur will move into the root zone with rainfall. Microorganisms must first oxidize sulfur to sulfate before it can be utilized by plants. Therefore, if elemental sulfur is used, it should be finely ground. The rate of oxidation is very slow with coarse ground sulfur. Some of these coarse fragments can still be visible five or more years after application.

Ammonium sulfate and ammonium thiosulfate can be banded, but the bands should be at least 1-1/4 inches away from the seed. Seedling damage and stand loss results if sulfur is placed with or near the seed. In some areas, industrial air pollutants put large quantities of sulfur dioxide into the atmosphere. Rainfall downwind from the source can add as much as 25 lbs. of sulfur per acre, which is many times that required for a crop. Irrigation water can also contain sufficient sulfur to meet crop demand. Therefore, growers should check sulfur contents from these sources before applying commercial fertilizers.

MICRONUTRIENTS

Micronutrients have been sometimes referred to as "minor" elements, however, this term does not mean that they are less important than the macronutrients or "major" elements. Since micronutrients are "essential" elements to plant growth, they are just as important.

Figure 7-11. Zinc deficiency in corn. Zinc is involved in growth regulation, and when deficient, growth is stunted, as indicated in this example.

Micronutrients are simply required in much smaller amounts. Most of the micronutrients are important components of numerous enzyme systems within the plant. Iron, copper, and manganese function in the formation of chlorophyll, and therefore plants deficient in iron, manganese, or copper exhibit chlorosis. These micronutrients are immobile in the plant and, therefore, deficiency symptoms usually appear on younger leaves first.

Zinc is important in amino acid and protein formation, and is also involved in plant growth regulation. For this reason, zinc deficient plants have shortened internodes in the stem, which causes the plant to be stunted. Yellow or white stripes on each side of the midrib are also a common deficiency symptom. Molybdenum deficiency symptoms are similar to those for nitrogen. Molybdenum is also very important for nitrogen fixation in nodules of legumes. Boron is essential for development of new cells in meristem tissues and also functions in flower and fruit development. Deficiency symptoms are cessation of terminal bud growth and death of young leaves. Chlorine is involved mainly in osmotic functions of plant cells and cell turgor. Partial wilting is a common deficiency symptom.

Soils low in organic matter, high pH, and/or calcareous subsoils are often deficient in some micronutrients. Micronutrient deficiencies often occur

Table 7-9. Common Micronutrient Fertilizer Sources.

Source	Micronutrient Content (%)	Physical Form	Method of Application
Ferrous Sulfate ($FeSO_4$)	19% Fe	dry granules	broadcast, band, foliar
Iron Chelate (FeEDDHA)	6% Fe	dry or liquid	foliar
Copper Sulfate ($CuSO_4$)	25% Co	dry granules	broadcast, band, foliar
Copper Chelate (CuEDTA)	13% Co	dry or liquid	foliar
Zinc Sulfate ($ZnSO_4$)	35% Zn	dry granules	broadcast, band, foliar
Zinc Chelate (ZnEDTA)	14% Zn	dry or liquid	foliar
Manganese Sulfate ($MnSO_4$)	27% Mn	dry granules	broadcast, band, foliar
Manganese Chelate (MnEDTA)	12% Mn	dry or liquid	foliar
Sodium Molybdate (Na_2MoO_4)	39% Mo	dry granules	broadcast, band, foliar
Ammonium Molybdate	54% Mo	dry granules	broadcast, band, foliar
Borax ($Na_2B_4O_7$)	11% B	dry granules	broadcast
Potassium Chloride (KCl)	46% Cl	dry granules	broadcast, band

on severely eroded soils, or where subsoils have been exposed due to removal of topsoils for land leveling, shaping, or terracing. Soils extremely high in organic matter can also exhibit micronutrient deficiencies. The organic matter can be either inherently low in micronutrients or strong organic-micronutrient complexes form which reduce availability.

Micronutrient deficiencies can also occur from the application of other nutrients. For example, large phosphorus applications can induce zinc, copper, or manganese deficiencies. (These micronutrients would initially have to be marginally deficient for the "interaction" to occur.) The application of one micronutrient can interact with and cause another to be deficient. Zinc-iron and iron-manganese interactions have been reported.

Once a micronutrient deficiency appears in a crop, it is usually too late to apply fertilizers to prevent yield loss. Thus, micronutrient fertilizers should be applied before or at planting. Table 7-9 gives several of the most common commercially available micronutrient fertilizer sources. Iron deficiencies generally occur on high pH, calcareous soils. There are many inorganic iron fertilizers although the most common is iron sulfate. Because of the very low solubility (availability) of iron in high pH soils, an extremely high rate of iron sulfate would be required. Fertilizer iron rapidly precipitates to insoluble or unavailable forms; therefore, banded or foliar applications are recommended. Even with banded iron sulfate, the iron stays in available form for only one year. Iron chelate is an organic iron source and can remain available for a longer time than inorganic iron. The disadvantage is that it is very expensive and thus, is not practical for soil application. Foliar applications (sprayed directly on the plant) at low rates (<1 lb Fe per acre) to young plants can be very effective. Iron sulfate can also be foliar applied.

Manganese and zinc deficiencies are common in calcareous soils, but can occur on organic soils. Band applications of inorganic sulfate forms of these minerals at rates of 4 to 9 lbs. per acre is an effective means of supplying zinc and manganese and can last for several years. These minerals can also be foliar applied. Broadcast application at high rates (20 lb. Zn per acre) can also be effective but is more costly.

Copper deficiencies have been observed in soils high in organic matter and also on sandy, highly-weathered soils. Copper sulfate is the most common fertilizer source. Broadcast applications have been shown to last for several cropping years, depending on the rate. The chelate form is an excellent fertilizer source when foliar applied.

Molybdenum deficiencies can be very easily corrected with small soil, foliar, or seed applications. Sodium and ammonium molybdate are the most common fertilizer sources. Seed treatments with molybdenum at 6 to 12 grams Mo per acre have been the most cost-effective means to correct deficiencies. **If soil or foliar applications of Molybdenum are made, caution should be given to apply the correct rate. Excessive molybdenum can be toxic to animals at low concentrations in feeds and forage.**

Boron deficiencies are observed mainly on light-textured, acid soils in humid regions. Alfalfa is very sensitive to boron deficiency. Sodium borate (borax) is the most common and economical boron source. Soil applied boron is the most popular method of application, although foliar applications are also effective. Boron readily leaches from sandy soils, thus, soil applied treatments generally do not last beyond the first or second year.

Very few soils are presently known to exhibit chlorine deficiencies. Recent data suggests that the application of chlorine to barley and winter wheat can reduce incidence of root and crown rot diseases. Potassium chloride is the best source for chloride. Broadcast or band applications are the most effective methods used to add chlorine to soils.

ACID SOILS AND LIMING

Liming is one of the oldest soil management practices. Application of 'ash' to soils was recorded as early as 300 B.C. Lime is applied to reduce soil acidity (raise soil pH). Soil acidity and pH were defined and discussed in the previous chapter on Soils. Acid soils are found in regions of high rainfall, where accelerated weathering

Table 7-10. Optimum Soil pH Levels for Various Crops.

Soil pH	Crop
> 6.5	Alfalfa Sweet Clover Sugar Beets
5.5-6.5	Red Clover Corn Wheat Barley Bluegrass
5.0-5.5	Blueberries Potatoes Tobacco
<5.0	Azalea Hydrangea Rhododendron

processes and leaching have removed a large portion of the calcium and magnesium from the root zone. Sandy soils found in humid regions can be very acidic with soil pH below 4.5. Most agricultural crops are not productive when soil pH is below 6.0 to 5.5. An example is shown in Figure 10 where wheat plants are severely stunted on a sandy soil of pH 5.3. Application of 2957 lbs. lime per acre, to raise soil pH to 6.2 results in normal wheat growth.

There are some horticultural crops like azaleas, rhododendrons, and hydrangeas that grow best when soil pH is below 5.0 (Table 7-10). Several soil-borne diseases which adversely affect potato growth are suppressed in acid soils. Potatoes produce best when soil pH is below 5.5. In contrast, most legume crops will grow only when soil pH is above 6.0 to 6.5. (The bacteria that fix nitrogen in the root nodules of legumes can not survive in acid soils.)

Chemical soil tests to determine soil acidity and the quantity of lime required to raise soil pH should be performed before applying lime. These "lime requirement" estimates are usually reliable. Once the lime rate required to raise soil pH has been established, the grower needs to select a lime material and apply it correctly. The most common material used for liming soils is ground limestone. Although the other liming sources can neutralize more acid per kg of material (neutralizing value), limestone is usually less expensive and more readily available. The neutralizing value depends on limestone purity and fineness. Most sources for limestone contain some impurities reducing the neutralizing value to 85-90%.

The common method of applying limestone is by surface broadcasting. Lime is very insoluble so incorporating the limestone with tillage will increase the rate of dissolution and acid neutralization. Plowing and discing are the common implements used to incorporate lime (4 to 6 inches deep). Recent studies have shown some benefit to banding lime below the surface. Generally, applications of 1 ton of lime per acre, if properly applied and incorporated, will last for 3 to 4 years, depending on the soil.

Liming acid soils is probably the single most important management tool to improve the productivity of acid soils. **In addition to raising soil pH and adding calcium, liming often increases phosphorus and potassium availability. Therefore, before expensive nutrients are added to acid soils, a sound liming program should be initiated.**

SALINE AND SODIC SOILS

Productivity of many soils around the world is severely hampered by excess salts. Generally, soil salinity problems exist in arid and semi-arid climates where evaporation exceeds precipitation. Basically there are two types of salt affected soils: 1. saline soils, and 2. sodic soils. Some soils may have a combination of saline and sodic problems.

Saline soils contain quantities of soluble salts that adversely affect seed germination and plant growth. The most common salts are the chlorides and sulfates of sodium, potassium, magnesium, and calcium. In severe cases the surface soil is white due to the precipitation of salts. There is essentially nothing that can be added to soils that will correct saline soils. The only means of reducing salt content is by leaching the salts below the

Table 7-11. COMMON LIME SOURCES AND RELATIVE NEUTRALIZING VALUE.

Source	Content(%)	Relative Neutralizing Value
Calcium Carbonate* ($CaCO_3$)	40% Ca	100
Calcium oxide (CaO)	71% Ca	178
Magnesium Carbonate ($MgCO_3$)	28% Mg	119
Magnesium oxide (MgO)	60% Mg	250
Dolomite ($CaCO_3$-$MgCO_3$)	21% Ca 13% Mg	109

*Limestone

Figure 7-12. An example of an extremely sodic (salty) soil. The white surface soil is due to salts of sodium.

root zone. That is, heavy applications of irrigation water with a low salt content, to leach the salts.

In sodic soils, 15 percent or more of the cation exchange capacity is occupied with Sodium (Na). Sodic soils have a high pH (8.5) and very poor soil structure. Generally, water will not infiltrate sodic soils because of the poor physical condition of the surface soil.

A shallow ground water table is a primary cause of saline and sodic soils. As water is evaporated from the soil surface, ground water moves up toward the surface by capillary action. Salts and free sodium are therefore concentrated near the soil surface and throughout the root zone. Irrigation water containing high levels of salts and free sodium can also cause saline and sodic soil conditions.

Although irrigation with low-salt water can help correct saline soils, a different approach must be taken to amend sodic or saline-sodic soils. The excess sodium must be removed before these soils can be amended. That is, the sodium is held in the soil with an electrostatic charge and cannot be removed by simple application of water. Therefore, soluble calcium is normally added to these soils which replaces the sodium on exchange. The replaced sodium can then be leached below the root zone. Gypsum (Calcium sulfate) is the most common calcium source used to amend sodic or saline-sodic soils. Limestone ($CaCO_3$) is not used because it is relatively insoluble at high pH compared to gypsum.

If the soil is calcareous (already high in calcium), acid or acid-forming materials can be added, which dissolve the calcium compounds and release free calcium to replace sodium on exchange. Elemental sulfur (Table 8) is acid forming and can be used to amend sodic or saline-sodic soils. The sulfur should be finely ground and broadcast on the surface. Sulfuric acid can also be added to high sodium soils. However, sulfuric acid is usually much more expensive than elemental sulfur. Since this material is a very strong acid, it is also difficult to apply and dangerous to work with.

SOIL TESTING AND FERTILIZER RECOMMENDATIONS

Soil testing involves (1) determining the relative availability of plant nutrients by means of simple chemical tests on the soil and (2) estimating the appropriate fertilizer rate based on the results of the chemical tests. It is very important to realize that making fertilizer recommendations from soil tests is not an exact science. However, most fertility evaluations are generally correct.

The first step in soil testing is to obtain a soil sample that represents the average fertility of the field. The soil sample should be taken from only the most uniform areas of the field. Low or depression areas, hilltops, etc., should be avoided or sampled separately if they make up a large area. Sampling depth is usually the tillage depth or 6 to 8 inches. A general guideline is to take 15 to 20 individual cores per 20 acres, with the cores combined into one sample. The final fertilizer recommendation is only as good as the soil sample. Extreme care should be exercised when soil sampling.

The analytical methods used in the laboratory are beyond the scope of this book. The important point is that the soil testing procedure results in an "estimate" or "index" of plant available nutrient. The fertilizer recommendation is then determined from the magnitude of this index. A high index or extractable level of nutrient would not result in a fertilizer recommendation. As the index decreases, the recommended fertilizer rate will increase. The exact recommendation depends on (1) the crop to be grown, (2) the climate and region the crop will be grown in, (3) whether or not the crop is to be irrigated, (4) yield goal, and (5) previous crop.

Fertilizer recommendations are based on numerous field experiments, which are designed to establish fertilizer response data for many crops over a wide range of soil test levels. An example of nitrogen response on winter wheat at three relative soil test levels is shown in Figure 7-13. As the soil test level increases, the crop becomes less responsive to applied nitrogen. Therefore,

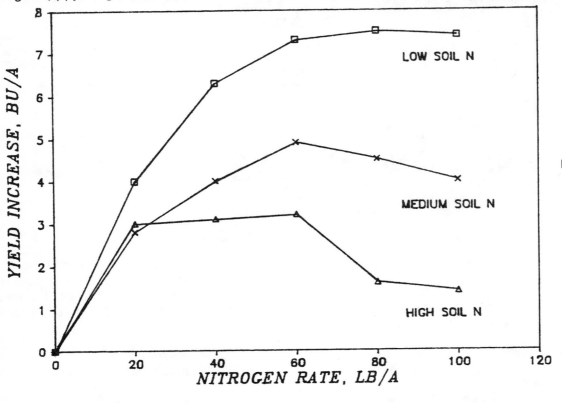

Figure 7-13. Nitrogen response on wheat at three soil test levels.

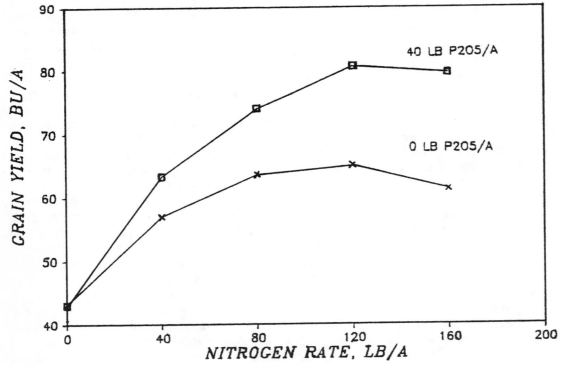

Figure 7-14. Nitrogen response on wheat at two levels of phosphorous fertility.

the recommended fertilizer rate for a given soil test is established at the level where fertilization above this rate will not increase yields.

Fertilizer recommendations should only be used as a guide. The exact or "best" rate which optimizes yields and profits should be established by the grower on his soil. It is important to apply all nutrients shown to be limiting by the soil test. Figure 7-14 illustrates that the wheat response to applied nitrogen was much greater when phosphorus was adequate than when no phosphorus was applied.

Many factors, including weather, insects, diseases, etc. can reduce the crop response to applied nutrients. Crop residues, soil structure, method of application, planting date, and many more can also affect yield response to fertilization. A very good "rule of thumb" to remember is that the cost of over-fertilizing is usually much less than the cost of reduced yield due to under-fertilizing.

CONCLUSION

Proper fertilization management involves identifying those nutrients that may be limiting crop yields. Ideally this should be done prior to observing nutrient deficiency symptoms. That is, fertilization after the appearance of nutrient deficiency symptoms may not be totally effective in preventing yield reductions. Therefore, soil testing should normally be used to identify nutrient needs.

The best fertilizer source to correct the deficiency is usually the least expensive source. The exception is when application equipment is not available for a given source. For example, if equipment is not readily available to apply liquid sources then dry sources must be used. When soil test levels are low, banding the fertilizers can improve efficiency. A regular soil testing program is an integral part of successful fertilizer management and sustaining soil and crop productivity.

CHAPTER 8   Tillage Principles and Practices

Dr. Dan Undersander
University of Wisconsin
Madison, Wisconsin

Man has tilled the soil and raised crops for over 9,000 years. While the tillage tools have changed greatly, the reasons for tillage and the problems associated with it have changed very little. Tillage is practiced to control weeds, prepare a seed bed, and reduce soil compaction. Tillage is expensive in terms of both labor and energy. Therefore, crop production should be managed with the least amount of tillage necessary for good crop yield. Also, if not practiced correctly, tillage may greatly increase soil erosion.

One of the primary functions of tillage is to prepare a seedbed. However, preparing a seedbed also coincides with another very important aspect of tillage . . . weed control.

Each field has a limited amount of water and nutrients which can be used either for crop, or for weed growth. Normally, each pound of weed growth reduces crop growth by at least one pound. Further, the crop establishment phase is a critical time for weed control. **If existing weeds are not killed prior to planting the crops,** they will recover faster than the crop can germinate and establish itself. The weeds can then outcompete the planted crop for water, nutrients and light. In some cases heavy weed growth may even cause crop plants to die.

The same thing can happen if tillage is performed too far prior to planting. Weed seeds in the surface may begin to germinate and grow. Thus, when the crop is planted it will be several days behind the weed growth and at a competitive disadvantage. Therefore, the functions of preparation of the seedbed and control of weeds are very nearly synonymous functions of tillage.

PLOWING

The first tillage operation for weed control is normally plowing, either moldboard or sweep. However, if large amounts of residue are present from the previous crop (or from weed growth), it may be necessary to disc or shred the residue prior to plowing.

The moldboard plow is normally used in higher rainfall areas. It cuts a furrow slice 6 inches deep and completely turns the soil over. This buries the plant growth that was on top of the soil, and exposes the root systems of the weeds. Moldboard plowing is very effective in killing many weeds except those with rhizomes, such as quack grass, bermudagrass, and Johnson grass, which will regenerate from the roots.

The sweep plow is the type of system typically used in arid regions such as much of the Western U.S. A sweep plow is a series of horizontal blades that are pulled 3-4 inches below ground level to cut off the plant (weed) root systems. The growing plants will generally die (in arid regions), where water demand is high because the plants no longer have a large enough root system to take up enough water. This operation may be done whenever weed growth is sufficient to justify it. There are 3 main advantages to sweep plowing in most of the Western U.S.: 1. residue is left on the surface to prevent erosion, 2. water from torrential rains is held on the field allowing better absorption (due to the residue), and 3. evaporative losses of water from the soil are reduced. (For a photo of a sweep plow, see Chap. XI, figure 11-6.)

Plowing is frequently followed by discing to break up soil clods, smooth the soil surface, and control weeds.

Fig. 8-1. A field of corn completely taken over by weeds (Johnsongrass).

Fig. 8-2. A four tier moldboard plow. The moldboard plow is very effective at tilling the soil, but leaves the soil surface vulnerable to erosion. For that reason, it is best only applied on flat and level ground where water movement (and subsequent erosion) will be minimal.

Fig. 8-3. A two-way set of discs. Each set turns the soil a different direction, which results in complete disruption. Very effective against weeds, but again the soil surface is exposed to erosion.

Discing kills weeds by burying them and exposing the root systems. Therefore, discing is more effective on hot, dry days when the plants will desiccate faster. (On cooler, more humid or rainy days, the root systems have a chance to reestablish themselves before the plant dies.) Discing is less effective against very large weeds and against rhizomatous weeds. However, continued discing will thin the weed stand and provide some degree of control.

Timing of plowing. To prevent erosion most plowing is done just prior to planting; that is, to leave the disturbed soil exposed for a minimum amount of time. However, some soils that are high in clay and organic matter, such as the corn belt soils, may be cloddy after plowing. In that case, fall plowing is normally recommended on these soil types so that the weathering forces of winter can begin to break down the clods.

CULTIVATION

Cultivation is the act of disturbing the soil alongside a row crop to disrupt weed growth. Cultivation of a row crop can be an effective means of weed control if done when both the weed and the crop are small enough. The cultivator uproots and buries weeds in-between the rows and can throw dirt into the row of crop plants covering small weeds. As with discing, cultivation is most effective in killing small weeds on hot, dry days. If cultivation is done when the crop plants are too big, or if the cultivator shovels are set too close to the crop plants, the cultivator may prune roots off of the crop plants and retard crop growth.

OTHER PURPOSES OF TILLAGE

Tillage is also used to provide a firm, smooth seedbed for planting. In some cases tillage may be required to make the soil either softer or more compact.

With small seeded crops it is very important that the soil not be too soft. This is because small seeded crops are generally planted shallow (less than 1 inch deep). If the ground is too soft, the planter may place the seeds too deep. This may mean that many seeds will germinate, but then die before they reach the soil surface.

As mentioned, tillage is also used to reduce the other extreme - soil compaction. It can occur either from the soil surface down, or in a well-defined layer below the soil surface (called a plow pan). Either may greatly restrict root growth. The restricted root growth will cause the plant to take up water and nutrients from a smaller volume of soil. This, of course, can result in less plant growth and reduced yield.

Soil compaction can also reduce the movement of water and air through the soil, further restricting plant growth. In some soils, compaction is a natural phenomenon but in most soils compaction can result from or be worsened by, driving field equipment over the field, excessive tillage, and, in some cases, heavy rains. Discing is the tillage operation most likely to cause compaction. It breaks down the soil structure and causes the soil particles to pack together. This action of discing is the reason why discs are used in road building to prepare a solid, well-packed road-bed. Discing should therefore be held to a minimum in soils that are prone to compaction.

Moldboard plowing is an excellent tillage device for reducing compaction in the surface soil (top 6 inches) because plowing loosens the soil as the furrow slice is turned over. However, as mentioned, plow pans can occur below the surface soil. Plow pans can be very restrictive of root growth. Where possible, plow pans should be loosened to allow the crop root systems to penetrate deeper. This can be done with a chisel or, if the plow pan is deeper, a subsoiler that has long curved, heavily reinforced knives.

Some soils may exhibit surface sealing. This occurs in soils of high clay content where, after being wetted, a very thin layer of clay particles form over the surface and inhibit further water uptake by the soil. A very shallow tillage such as harrowing, cultivating, or rotary hoeing will break up this layer and enhance soil water uptake.

Too much tillage, the wrong type of tillage, or tillage at the wrong time can be harmful. Excessive or improper tillage can greatly increase erosion from either wind or rain. In areas where either wind or rain erosion is a problem, fields should not be left without crop cover for ex-

Fig. 8-4 & 5. Fields in which water was allowed to flow along the slope. The resultant erosion is obvious. Terracing would have reduced or eliminated the erosion.

tended periods of time. A growing crop, crop residue, or even weed growth will help hold the soil in place until the field is to be used for crop production again. If a field must be left open (free of plant growth), it is best to have the soil be as cloddy as possible. Larger clods are less susceptible to either wind or rain erosion than finely worked soils. Highly erodible areas where water is the primary agent of erosion, should be managed so as not to let water aggregate and begin flowing in streams across a tilled field. This can be managed by several techniques. One is to put terraces across the slope of a field. This causes the water to be stopped at each terrace on the field and absorbed by the soil, rather than running off of the field and causing erosion. Another method of reducing erosion is to perform tillage on the contour or across hills rather than up and down hills. By this technique each planted row serves as a small terrace and helps hold water, preventing it from running off. Some areas of excessive slope, or where water naturally drains, may need to be left in a permanent sod because any tillage would result in severe erosion.

In recent years, minimum or no-till systems have become popular. These systems have generally involved maintaining crop residue on the soil surface for a larger portion of the year. They reduce or eliminate moldboard plowing and often involve planting into crop residue from the previous year. But ultimately, the key is exacting use of herbicides to control weeds. No-till systems have proven advantageous because they tend to result in fewer tillage operations (therefore less cost during crop establishment), equal or greater yields than conventional tillage systems, and reduced erosion. As mentioned, the key to the development of these systems has been the development of herbicides that provide good weed control. The major problems associated with no-till systems have been new weed problems and, in some cases, increased insect problems.

Fertilization During Tillage

Fertilizer is frequently incorporated during tillage, prior to planting. The bulk of the fertilizer is usually applied during the pre-planting tillage operations so that it may be incorporated into the soil. However, it is important to be aware that certain elements, particularly nitrogen and potassium, may be lost from the soil by leaching. Therefore, fertilizer should normally be applied as close to the time when it is needed by the crop as possible, in order to reduce losses in soils where leaching is a problem.

As a general rule, leaching is less of a problem with soils that have a high clay or organic matter content. Nitrogen and, in some cases, potassium may be applied to the crop in a split application. That is, some is applied before planting, and some after the crop has started growing.

Row Crop Spacing

The distance between planted rows (row spacing) is a major consideration in crop yield. The old standard was to plant crops in rows that were spaced 40 inches apart, because this was the space necessary for a horse to walk between rows and pull a cultivator. However, newer, higher yielding varieties and hybrids of crops have generally been shown to yield more when planted in narrower row spacings. The narrower row spacing means that if the number of plants per acre remains constant, plants are farther apart within the planted row. This results in less competition among plants within the row, because plants are better able to utilize water, nutrients and sunlight occurring between the rows.

Herbicides

One of the primary and major decisions in row crop planting is selection of the herbicide program. Most herbicides are fairly specific in that they can be used only on certain crops and to control certain types of weeds. Thus, the crop to be planted and the major weed problem will to a large extent, determine the herbicide used. Some herbicides work better on certain soil types than others. Also, some herbicides last longer (have longer residual times) in the soil than others. This residual time is influenced by the soil type and the environmental conditions, and may vary from field to field and year to year. It is desirable to have the residual be nearly as long as the growing season to provide season-long weed control. When using long-residual herbicides, it is important to consider the possibility of a carry-over. That is, damage might be done to the following crop. The large number of variables and choices in herbicide selection make it important to check with a reputable source for assistance in determining the best herbicide program. Herbicides can be applied prior to planting (pre-plant), between planting and seedling emergence (pre-emergence), and after emergence of the crop (post-emergence). The sensitivity of the crop to herbicide, whether or not the herbicide needs to be incorporated into the soil, and the time when herbicide control is most needed, will determine when the herbicide is applied.

Growing the same crop on the same field year after year will tend to reduce the soil condition and result in increased problems with insects, diseases, and nematodes. Therefore, crop rotation is advantageous in most situations. Many farmers will have a definite sequence of crops that is planted in each field over the years. The order of crops within the rotation will be determined by the planting and harvesting time of each crop; the potential for herbicide carry-over from one crop to the next; the opportunity to make use of nutrients left by the previous crop; and the potential for controlling troublesome weeds, insects, and diseases. A rotation will generally include both broadleaf and grassy crops because broadleaf weeds are easier to control in grassy crops, and grassy weeds are easier to control in broadleaf crops. Including both types of plants in the rotation allows for more complete weed control. The length of the rotation, usually 3-6 years, is determined by the amount of time necessary to reduce disease, insect, or nematode levels between plantings of the same crop on the same field.

SUMMARY

There are two main purposes for tillage: 1. preparation for a seedbed, and 2. weed control. Ironically, although some types of tillage such as moldboard plowing and chiseling are used to reduce soil compaction, some types of tillage (such as cultivation) tend to compact soil. Most types of tillage also expose the soil, and therefore make it more vulnerable to erosion. These factors, plus the labor and energy expenses, dictate that tillage be kept to a minimum.

In arid areas of high erosion potential (such as much of the western U.S.), sweep plowing will cause less damage than moldboard plowing. The sweep plow cuts off the weed roots without turning the soil over (as in moldboard plowing). In arid climates, this kills most weeds effectively without exposing the soil to wind and water erosion.

Herbicides can often be used to reduce tillage. However, the usefulness of herbicides depends upon a number of factors. Susceptibility of the crop to herbicide damage is, of course, highly important as is potential carryover and damage to succeeding crops in a rotation.

In short, tillage is a broad term that encompasses much detail. Detail that is specific to the crops grown, the rotation of crops, the climate and the soil type.

CHAPTER 9 AN INTRODUCTION TO IRRIGATION

by D.J. Undersander Ph.D.
University of Wisconsin
Madison, Wisconsin

In many parts of the world rainfall is insufficient, or occurs at the wrong time to provide the total water needs of crops being grown. Additional water is supplied to the crops of these regions through irrigation, either to supplement rainfall, or to meet the majority of the needs of the crop. The type of irrigation system used depends on the soil type, amount of water to be supplied to the crop, crop being irrigated, source of water for irrigation, and available physical and financial resources.

IRRIGATION SYSTEM

The type of irrigation system may be basin, furrow, sub-irrigation, drip, or sprinkler. Each type of system has applications where it works best.

Basin irrigation is the flooding of a large area or field. This type of irrigation system works well only on very level ground or on land that has been leveled. It is frequently used in rice production. This is an inexpensive method of irrigation if a large amount of land leveling is not required.

Furrow irrigation is practiced extensively in the Great Plains. Furrows are made the length of the field with some slope (usually 0.1 to 0.4%) and the crop, is generally planted on top of the beds between the furrows. Water is applied to each furrow at the upper end of the field and allowed to flow the entire length of the field. Furrow irrigation works well only in areas that have heavy textured soils and very small, uniform slopes.

Fig. 9-2. Drip irrigation in a grape vineyard.

Fig. 9-1. Furrow irrigation of wheat.

Drip irrigation is where pipes or tubing with emitters are used to irrigate the field. The emitters are spaced periodically to allow water to escape into the soil only at the base of the plants. For this reason it is vastly more efficient than either flood or sprinkler irrigation. Drip irrigation is becoming very popular for vineyards and orchards.

Sub-irrigation is similar to drip irrigation, only the pipes and emitters are placed underground. Sub-irrigation is very efficient because losses from evaporation and runoff are reduced to near zero. Also, sub-irrigation

Fig. 9-3. A sprinkler irrigation test area (Univ. of California at Fresno). By checking each of the receptacle gauges, the eveness of application of the sprinkler may be analyzed.

Fig. 9-4 & 5. The two most common types of sprinkler irrigation; the center pivot (above), and the roll line (below).

systems are usually operated at a low pressure, thereby reducing the energy requirement for pumping water. However, installation of the underground pipe and tubing are expensive and frequently must be removed from the field before tillage operations can be performed. Therefore, this irrigation system is used mainly on high value crops and in permanent installations such as orchards.

Sprinkler irrigation is different from the previous types of irrigation, because water is applied from overhead onto the crop through sprinklers. The systems can range in complexity from stationary systems with one or many sprinklers, to moving center pivot sprinkler systems that may cover as much as 500 acres. Sprinkler irrigation can be used on fields where furrow or basin irrigation will not work because the field is not level enough or the soil is too light (sandy). Sprinkler irrigation can be used to apply smaller amounts of water (as little as 0.1 inches/hr) than furrow or basin irrigation. This can be advantageous where the soil infiltration rate is low, the soil has a low water holding capacity, or there is danger of erosion.

Sprinkler irrigation systems can also be used for purposes other than irrigation. That is, the water can be used as a delivery medium for applying fertilizers, herbicides, or insecticides, frost protection, and temperature control. (Water can be sprinkled on to the plant to insulate against cold temperatures, or to reduce the effect of extremely hot temperatures.)

The major disadvantage of sprinklers is that they are expensive to purchase and operate. If a stationary sprinkler system is used, the labor involved in moving it is high. A second disadvantage of sprinkler irrigation is that wetting the crop itself (rather than just the soil) may increase the growth of fungal and bacterial diseases. Also, if poor quality water is used for irrigation, the water may leave undesirable deposits on leaves or fruit. Application of salty water to leaves when the temperature is high may cause leaf burn and death.

SOIL TYPE

The texture of the soil (proportion of sand, silt, and clay) and soil organic matter content greatly affect the soil water properties. The major properties of the soil affecting irrigation that will be considered here are: infiltration rate, hydraulic conductivity, and water holding capacity.

Fig. 9-7. An aerial view of a desert area in northern Mexico turned into cropland through irrigation. Note the large circles made by the center pivot irrigation systems.

Fig. 9-6. A single, high pressure pivot gun sprinkler. This type of sprinkler is commonly used when the water source contains debris that might stop up other types of sprinklers with small orifices. In this case the single gun is drawing water from a nearby stream. These types of systems are also commonly used to pump out manure retention ponds.

Infiltration rate is the rate at which a soil takes up water applied to the surface, either from rain or irrigation. This is an important factor to consider because if water is applied to the field faster than the infiltration rate of the soil, then either ponding of water will occur on the surface, or water will run off the field. Generally, soils composed of larger particles (sands rather than clays) have higher water infiltration rates (Table 9-1). Other factors can affect initial infiltration. Crusting decreases the initial infiltration rate. Wet soils have lower initial infiltration rates than dry soils. Layering of soils will affect infiltration rate. For instance, if a layer of clay soil occurs underneath a layer of sand, the final infiltration rate will be determined by the clay soil.

Table 9-1. Generalized infiltration rates of various soil textures.

Soil type	Infiltration rate (in/hr)	
	Average	Range
Sand	2.0	1.0 - 5.0
Sandy loam	1.0	0.6 - 3.0
Loam	0.5	0.4 - 0.8
Clay loam	0.3	0.1 - 0.6
Silty clay	0.1	0.001 - 0.2
Clay	0.2	0.04 - 0.06

Hydraulic conductivity is the rate of water movement in the soil profile. This factor is important because it determines the extent of water penetration into the profile. Coarse-textured soils (sands) transmit water faster than fine-textured soils (clays). However, clay soils will transmit water farther than coarse-textured soils. This also means that water applied to coarse textured soils moves straight down while water applied to clay soils will move horizontally in addition to moving down a profile. Hydraulic conductivity is greater through drier soils.

Soil water content is the total amount of water in the soil. It may be expressed either as **oven dry basis** (weight of water/weight of dry soil) or as **volume of water per volume of soil** (inch of water/inch of soil). Conversion of the oven dry information to the volumetric soil water may be done using the following equation:

$$\frac{12 \times \text{bulk density of soil} \times \text{percent water}}{100} = \text{inch of water/foot of soil}$$

Bulk density is the weight of the dry soil (in grams) per unit volume (cm) of soil. It is different for each soil type and varies with depth of soil.

The water holding capacity of the soil is the maximum amount of water the soil can hold. If more water is applied than the water holding capacity of the soil, the additional water will run through the soil profile and be lost for crop growth. The water holding capacity of the soil tends to increase as the proportion of finer textured particles (silts and clays) increase in the soil. Soil organic matter also tends to increase the water holding capacity of soil.

Available water is the total amount of the soil water content that is actually available for crop growth. It is important to realize that a proportion of the water in every soil is held so tightly by the soil particles that the plant is unable to use this water for growth. This is known as the unavailable water. A plant will use all the available water and then wilt. If not watered it will eventually die, even though some water (unavailable) remains in the soil.

The actual available water in the soil at any time depends on the total soil water content. When the soil is saturated (the soil water content equals the water holding capacity of the soil), the soil is said to be at **field capacity**. Generally, the available water in the soil at field capacity increases with finer textured soils (Table 9-2).

As Table 2 indicates, sands have the ability to hold less available water than finer textured soils. This means that sands must be irrigated more frequently to maintain sufficient water for good plant growth. Additionally, it is easier to over-irrigate on sandy soils. This is inefficient, because the extra water is not used for crop production. That is, it moves too deep into the soil profile for crops to reach. When this happens, the water may leach soil nutrients with it, and cause additional losses to the farmer.

Table 9-2. Available water of various soil textures at field capacity.

Texture	Available water/foot (inches)
Heavy clay	2.22
Silty clay (silty clay loam, clay loam)	2.34
Loam	2.58
Silty loam	3.12
Silty clay loam	2.10
Sandy clay loam	1.80
Sandy loam	1.58
Loamy fine sand	1.80
Medium fine sand	0.90

CROP WATER REQUIREMENTS

The total amount of water necessary to maintain soil moisture to produce a crop at the desired yield level is called the crop water requirement. Some crops have higher crop water requirements than others (Table 9-3). Additionally, the crop water requirement for a particular crop varies from year to year and location to location depending upon the climate.

The main component of crop water requirement is evapotranspiration which is the sum of water lost to the air from the soil surface (evaporation) and from the growing crop (transpiration). Evaporation is the major water loss from unplanted soils or from very young crops. As the crop gets bigger and most of the ground is covered by the crop, the major loss becomes transpiration. Both evaporation and transpiration vary from day to day and location and location, depending on climatic factors. Higher temperature, more sunshine, more wind, and lower humidity tend to increase both evaporation and transpiration. This means that the amount of water required for crop growth tends to increase. Under these conditions, more water must be supplied to the plant either from rain or irrigation to produce good growth.

The amount of water use (evapotranspiration) per day changes with the stage of plant growth (Fig. 9-8). The evapotranspiration from fields of very young plants is generally less than 0.1 inch per day while mature plants

Table 9-3. Pounds of water required by crop to produce a pound of dry matter.

Crop	Water required* (lb/lb dry matter)
Alfalfa	695-1047
Oats	448-876
Cowpeas	413-767
Cotton	443-612
Barley	404-664
Wheat	405-636
Corn	253-495
Blue gramagrass	290-389
Millet	202-367
Sorghum	203-298

*Range of water required over five years.

From Shanz, H.L. and L.N. Piemeisel. 1972. the water requirement of plants at Akron, Colorado. J. Agr. Res. 34:1093-1190.

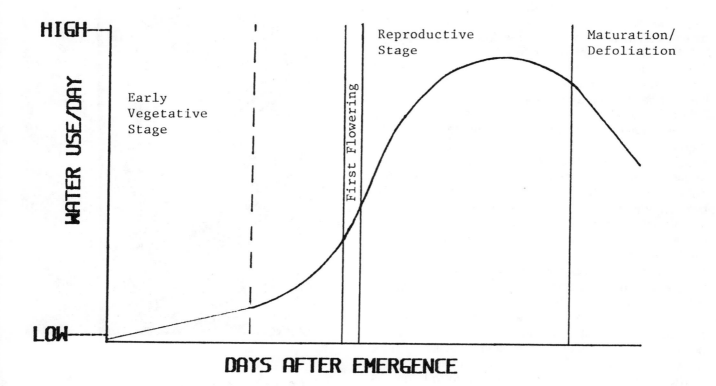

Fig. 9-8. Daily water use of a plant as it matures.

may use 0.3 to 0.4 inches of water per day until they begin to die.

The total dry matter crop production on a field increases with evapotranspiration until the maximum crop production per acre is reached. Beyond this point, some other factor, such as a nutrient or sunlight, is limiting and further increases in water will not produce increases in dry matter production. Thus, for a crop where all of the dry matter is harvested, (such as a forage), water applied at any stage of growth will produce approximately the same amount of crop.

In a grain or vegetable crop, where only the grain or fruit is harvested rather than the whole plant, timing of water application becomes very important. Most plants are especially sensitive to water stress during floral initiation. This is the stage of growth when the plant begins to develop kernels, grains, or fruits. During floral initiation, the number of kernels, grains or fruits that will develop on each plant is determined. Water stress at this time will cause fewer to develop, and therefore reduce yield. Stress at this time results in fewer ears of corn, heads of wheat, boles of cotton, or fruit or vegetables per plant. Water applied later in the plant life cycle, will tend to affect the size of the kernel, grain, or fruit, but will have less effect on the number produced per acre.

WHEN TO IRRIGATE

A very important consideration is when and how much to irrigate. If a crop is not irrigated when it needs water, yield loss will occur. On the other hand, if a crop is irrigated when there is no need, water and energy for pumping the water are wasted, and fertilizer nutrients may be leached from the soil. The most common techniques for determining when to irrigate are: visual estimates, use of weather data, and measurement of soil moisture.

Some producers irrigate by visual estimate of the plant water status. This is the least accurate method and requires the most skill to be successful. Many plant species have a slight change in color when they become stressed. This color change can be used as a criteria for determining when to irrigate. However, if too great a color change has occurred before irrigation is begun (especially later in the season), a significant yield loss will occur. Wilting of the plant is not a good criteria for determining when to irrigate. In many climates, plants wilt on the hottest days regardless of water status. Plants in good water status that have wilted during hot days, will recover overnight, and will not appear wilted the next morning. Those that are stressed severely, however, will not recover from wilting during the night. If plants are not recovering overnight, large yield losses may have already occurred, especially later in the season.

The second method to determine when to irrigate is through the use of weather data. Weather factors such as temperature, sunlight, wind speed, day length, and relative humidity have been combined into equations to determine how much water will evaporate from the surface. This equation has been combined with a crop coefficient specific for each crop climatic region to estimate the water use. The equation is:

$$\text{water use} = K \times U$$

where K = the crop coefficient for each specific crop
U = the evaporation determined from the weather factors.

This method is not very accurate for determining water use for short time periods (for several days or a week).

The third and **most accurate** way of determining when to irrigate is through measurement of soil water. The most common method for measuring soil water content is through the oven dry weight method. Soil samples are weighed, dried in an oven at 140 degrees F (60 degrees C) or higher, and then weighed again. The difference between the first and second weights, is the moisture. This technique, while inexpensive, is time and labor consuming.

Neutron probes (Fig. 9-9) are an instrument currently being used by researchers and consultants to determine soil water. They are very accurate, but expensive. An aluminum pipe is placed into the ground to the depth that the soil moisture content is to be determined. The

Fig. 9-9. A neutron probe. A high tech method of measuring soil moisture by using low energy radiation.

neutron probe is lowered down into the pipe and soil moisture readings are taken at various depths using low energy radiation.

Another instrument used to measure soil water is a tensiometer. The tensiometer consists of a porous ceramic tip attached to a hollow tube with a vacuum gauge at the top. The end with the ceramic tip is inserted into the soil to the depth where the reading is to be taken (usually 1 to 2 ft.). Water moves from the tube through the ceramic tip until equilibration is reached. The vacuum in the tube resulting from the water loss is read on the gauge. Tensiometers are accurate and inexpensive, but tensiometers do not work in very dry soils or in soils that crack as they dry.

Resistance blocks have also been used to schedule irrigation. A small gypsum block with two electric wires is buried in the soil. The other ends of the wires extend above the ground and can be attached to a potentiometer. The resistance between the two wires in the gypsum block is proportional to the soil water content. These blocks are most accurate when the soils are drier, therefore it is necessary to irrigate when the blocks first indicate that crop stress is occurring. The readings of the gypsum blocks are affected by salt, so it is necessary to be cautious in saline soils or when irrigating with salty water.

WATER SOURCE

The water used to irrigate a crop can come from either surface water or ground water. In either case it must be high quality water. High quality means it must have a low salt content.* If, however, salty water is applied continuously to soil without special precautions, the salt content of the soil will increase until plants can no longer be grown in it. The irrigation water must also be free of contaminants that would hurt plants.

Surface water for irrigation can come from lakes,

Fig. 9-10 & 11. An ancient gravity flow aquaduct built by the Roman Empire for irrigation (above), and a modern day aquaduct built and used in the same country (Portugal).

reservoirs, or rivers. Rivers and streams may be dammed to make reservoirs that hold the water until it is needed for irrigation. Frequently, streams contain a lot of water in the early spring when snow melts, but little water is needed at that time. When much water is needed for irrigation during the summer, the streams may have very little water in them. Building a reservoir allows water to be stored until the time it is needed for crop production.

---

* As will be seen in Chapter 10, in some cases irrigation water must be brought in from long distances away, because local ground water is too salty for use. Indeed in the Imperial Valley of Southern California, the ground water is so shallow and so salty, that proper drainage of irrigation water is almost as important as the application of the water. That is, to keep the water (salty water) table from rising into the crop root zones.

Throughout much of the western United States fields can be irrigated by gravity flow. The initial expense for the installation of canals and pipes that may extend many miles is high. However, since the water flows to the field by gravity the operating cost of such a system is lowered.

Another source of water for irrigation is ground water. Most of the eastern United States has a river of water flowing underneath it. The depth of this underground water varies from near the soil surface to several hundred feet below ground level. The thickness of the water layer may vary from a few feet to several hundred feet. Water is lost from this underground river by springs and wells. It is returned to this layer through rainfall and rivers. The Great Plains of the United States have an underground layer of water called an aquifer that can be used for irrigation. However, water is not returned to this aquifer once it is withdrawn, so it is gradually being used up.

PUMPING PLANT

If water is to be pumped onto a field the first consideration is the size of pump and motor required. The size of the pump required can be determined by the water flow rate needed. To determine the required flow rate, the producer needs to know the size of the field to be irrigated, and how much water is to be applied in a given time. Flow rate can be calculated by the equation:

$$gmp = \frac{\text{Water} \times \text{acres} \times 425.5 \text{ hours}}{\text{hours}}$$

where gmp = gallons per minute pumped;
water = inches of water to be applied;
acres = size of field to be irrigated;
hours = length of time irrigation is applied;

Example: If a producer wants to apply 1.5 inches of water to 80 acres in 4 days (96 hours), the required flow rate is:

$$gpm = \frac{1.5 \times 80 \times 452.5}{96} = 565.6^* \text{ gpm}$$

The size of the motor needed can be calculated by determining total dynamic head (TDH) and converting that to horsepower.

TDH = total static head + pressure head + friction head + velocity head

Total static head is the vertical distance the water must be raised. For instance, if the water is 70 feet below ground level and is being raised to the soil surface, then the total static head is 70 feet. If the same water is being raised to the top of a sprinkler 15 feet above ground level, then total static head = 85 feet (70 + 15).

Pressure head is the pressure at which the irrigation system is operating. This may range from near 0 to over 100 pounds per square inch (psi). For water, psi x 2.31 = pressure head in feet.

Friction head is the energy loss when water flows past the rough wall of a length of pipe. The pressure decreases due to friction that occurs. The amount of friction depends on the type and length of pipe. Friction losses also occur when the water flow passes through pipe fittings or the pipe diameter suddenly changes. Values for friction losses through pipe and fittings may be obtained from pipe friction tables.

Velocity head represents the energy required to maintain the velocity of the water in the pipe. The velocity head is small and is usually negligible unless a great deal of water is being forced through a small pipe.

The horsepower required from a pump can be computed from the following formula:

$$\text{horsepower required} = \frac{\text{gpm} \times \text{tdh}}{3960 \times \text{pump efficiency}}$$

where gpm = gallons/minute discharge from the pump,
tdh = total dynamic head in feet of water

Example: a pump is delivering 600 gpm from a water table 46.5 feet below ground level. There is a free discharge (0 pressure head). Friction loss in the pipe is 5 feet. Velocity head is assumed to be 0 and pump efficiency is 65%.

tdh = 46.5 + 0 + 5 + 0 = 51.5

$$\text{tdh} = \frac{600 \text{ gpm} \times 51.5 \text{ feet}}{3960 \times 0.65} = 12 \text{ horsepower}$$

DESIGN CRITERIA FOR IRRIGATION SYSTEMS

The two factors that must be taken into account when designing and installing irrigation systems are: 1. uniformity of application and, 2. efficiency of application.

Uniformity of application means that the same amount of water is applied to the entire field. When an irrigation is not uniform, parts of the field are over-watered and parts of the field are under-watered. Both situations result in economic loss to the producer. Disuniform irrigation

---

* As the high flow rate (565.6 gpm) indicates it is important to determine if sufficient water is available for irrigation before going to the expense of installing a system.

can result from unlevel fields for basin irrigation, disuniform slope for furrow irrigation, or, in sprinkler and underground irrigation, improper type or placement of nozzles, as well as worn and plugged nozzles.

Application efficiency is the second consideration in system design. Application efficiency is defined as the percent of the irrigation water applied, that is added to the soil profile.

$$\text{Application efficiency} = \frac{\text{change in soil water content}}{\text{irrigation water applied}}$$

Application efficiency of sub-irrigation systems is typically 100% (all of the water applied is held in the soil profile for plant growth). The application efficiency of basin, furrow, and sprinkler systems usually ranges from 40 to 90%. Well-designed and operated systems should range from 75 to 90% application efficiency. Low application efficiencies result from water running off the field, passing through the soil profile below the root zone, and from evaporation of water before it enters the soil either during application or from puddled water.

SUMMARY

Irrigation is essential to agriculture in many parts of the world. Water for irrigation may come from either surface water (rivers and lakes) or underground water. If water is to be pumped from underground, it is important to determine that the pump capacity can meet the water requirements of the crop.

There are five basic methods of irrigation: 1. Basin, 2. furrow, 3. underground pipe, 4. drip, and 5. sprinkler. Basin and furrow are known as flood irrigation, and require the greatest amount of water flow. Underground pipe and drip systems are more recently developed systems designed to minimize water usage. Sprinkler systems are very popular systems which need less total water flow than flood systems, but result in more evaporation (water loss) than drip or underground pipe. Sprinkler systems also require higher quality water (less salt) than the other systems, since water is deposited on the plant's leaves.

Chapter 10 Two Examples of Highly Successful Irrigation Projects
by Robert E. Moore P.E.
Agricultural Engineering Consultant, Phoenix, AZ

EXAMPLE DISTRICTS

This subchapter discusses two of the most successful irrigation projects in the world, the Imperial Valley Irrigation District (All American Canal) in southern California, and the Salt River Project in central Arizona. Average annual precipitation for the Imperial Irrigation District is 7.6 cm/year (3 inches/yr.), and average precipitation for the Salt River Project is 17.8 cm/year (7 inches/yr). Without irrigation, the Salt River area could support only seasonal livestock grazing, and the Imperial Valley area would be totally incapable of supporting any form of agriculture whatsoever. But through the use of these irrigation programs, both areas have evolved from wastelands into two of the most productive agricultural centers in the world.

Each district depends upon seasonal snow packs deposited far from the agricultural areas using the water, and upon a system of large storage reservoirs which regulate the irregular upstream flows. In the case of the Imperial system, the main storage reservoir (Lake Mead) is located nearly 200 miles (340 kilometers) from the center of the agricultural use of the water it stores.

Imperial Irrigation District

The Imperial Irrigation District was created out of desert where no evidence of prehistoric or historic irrigation existed.

"Juan Bautista de Anza, a captain in the Spanish Army, found only desert when he led an expedition across the area in the 18th Century. De Anza encountered a wasteland so forbidding that upon reaching San Gabriel Mission he declared that he had made 'La Jornada de los Muertos' (Journey of the Dead)."*

As early as 1849 other travelers to the area considered how water might be brought from the Colorado River flowing some 60 miles (97 kilometers) to the east.

"William P. Blake was one of the first to recognize the potential of the Imperial Valley. A geologist with a railroad survey party, Blake had passed through the desert in 1853 and had noticed the ancient shoreline at the foot of the mountains which proved that at one time there had been an inland sea where now there was desert. Barometric readings taken by Blake proved that the desert lay below sea level, and could be irrigated by a gravity flow canal diverted from the Colorado River."*

In 1896, the California Development Company was formed to reclaim the Imperial Valley with Colorado River water. Excavation of the canal began in August, 1900. The Imperial Valley received its first water through the Imperial Canal in June 1901, and some 1,500 acres (600 hectares) of land were put into crops in the fall of that year (currently nearly 500,000 acres are irrigated). In 1905, unanticipated winter floods poured out of the Gila River in Arizona and into the Colorado. This caused the

---
* Quoted material from "Welcome to the Imperial Irrigation District," I.I.D. Fact Book - Seventh Printing - June, 1982.

Fig. 10-1. Aerial view of the Imperial Valley of southern California. Without irrigation this area would be totally incapable of supporting any agriculture whatsoever (annual rainfall of only 3 inches per year). Note the Salton Sea in the background (where the highly saline drain water is deposited).

river to change its course and pour its full flow down the Imperial Canal into what is now the Salton Sea. It took two years to bring the river under control.

After more flooding in 1910, the original development company was forced into bankruptcy, and the Imperial Valley settlers formed their own organization. The main canal and levees were actually located just to the south in Mexico. The settlers therefore had little security in their water supply and flood defense. They therefore decided an "all-American canal" needed to be built north of the border.

The Bureau of Reclamation reported that such a canal would be impractical without a dam to control flooding. The result was passage of the Boulder Canyon Project Act in 1928, authorizing construction of Boulder Dam (now Hoover Dam), Imperial Dam, and the All-American Canal.

Excavation for the All-American Canal got underway in 1934. Boulder Dam and its reservoir, Lake Mead, were dedicated in 1935.

The first scheduled delivery of water by the new canal was in October of 1940, and by 1942 the Valley was drawing its entire supply through the All-American Canal.

System Orientation

The Imperial Project is relatively compact, with only small areas either not physically adjacent to the main area or somewhat set off by major topographic features. This physical arrangement allows a minimum of main canals to serve a relatively large area.

Extensive land leveling and drainage construction was necessary to bring the lands within the District into continuous production. Those efforts were worthwhile because the water supply had been secured and its reliability became assured, based on commitments by the government (Boulder Canyon Project Act).*

Ordering Water

The system is operated on a modified demand basis. This means that the irrigator (farmer) places his water order with the water office serving his portion of the District by a certain hour each day. That order is for a quantity of water for a number of hours, at a specific canal-lateral-gate location.

The system is a "demand" system rather than rotation, in that the irrigator must order the water, i.e., he makes a demand on the District staff for specified service. The Imperial System's operation is a "modified" demand system in that the duration of the order (total amount of water) may be varied by the District personnel, based upon physical system limitations, or the needs of other farmers in the area. This method of distributing the water is a key element in carrying out the equality of annual water allocation.

It should be noted that by physically devising the system into main canal, separate transmission canals, and spaced distribution laterals, the reliability and equality of service is enhanced. That is, no user is so physically removed from a lateral that his needs, however sensitive, cannot be reasonably met, both as to timing and quantity of water.

Continuing Service to the Irrigator

The Imperial District has a continuing program of installing concrete lined laterals which both conserve water and add to the reliability in timing and quality of individual water deliveries. The District has also constructed regulating reservoirs at key points on each of the transmission canals.

Fig. 10-2. Trenching machine used to dig the master drain trench. Imperial Irrigation District.

---

* This is extremely important, and is a major reason why many countries in the world do not have similar projects. That is, before a landowner can invest the large sums of money required to make ready otherwise useless land, he must have assurance of stability in the government, or other authority developing the irrigation system.

These reservoirs serve to store water not needed in the canals, or to release water back into the canals to meet specific needs during any fluctuation in the canal levels. Both actions, storage or release, are short-term and intended to stabilize lateral deliveries by "fine-tuning" the system. (More will be made of this point as maintenance is discussed in a subsequent section.)

On the point of responsibility, the canal-lateral-gate system provides a fixed, visible, clear point from which the District's and the farmer's responsibility is determined. The District has full responsibility from the river diversion works through the All-American Canal, the transmission canals and individual laterals and District headgates. It is the District's responsibility to design, operate and maintain all those elements of the system so that the irrigator receives the amount of water, at the time requested, and at the correct location. Beyond the District's gate, all responsibility belongs to the farmer.

In the Imperial System there is one additional service provided by the District to the landowners. Due to the character of the soils and the relatively shallow water table under these lands (water that is highly saline), a pattern of District-maintained drains are constructed across the District. The major drains serve as collection and routing channels from the landowner's farm drains.

For all practical purposes there is no reuse of drain waters, due to the high saline content of the leached water. All drains move water in a northerly direction in the District, with ultimate discharge into the Salton Sea.

Maintenance Concerns

No irrigation district can attain or sustain reliability, adequacy or accuracy, if there is not an active, maintenance program. The irrigation district must deal with

Fig. 10-3. Elmer Fudge Regulating Reservoir, Imperial Irrigation District. This is one of a series of small dams and reservoirs, designed to regulate water from the canals. During times when the farmers served by the particular canal do not need water (such as after a rain), it can be temporarily stored. Conversely should there be a sudden, unexpected need for water (during unusually warm weather), it can be released. These canal/reservoirs are particularly important in the Imperial Valley System since the main source of water, the Colorado River, is nearly 60 miles away (97 Km.), and the main storage reservoir, Hoover Dam, is over 200 miles (340 Km.) away.

Fig. 10-4. The regulating reservoir system applied on a much smaller scale, at a different irrigation project, in southern New Mexico (Elephant Butte Irrigation District). In this case, a series of locks have been constructed on the main delivery canal. These locks can be used to control the rate of flow (head pressure) as well as to store water.

Fig. 10-5. Drain inlet structure. An item unique to the Imperial Irrigation System. Due to the high salinity content of the soils, irrigation water cannot be reused. Also, it cannot be allowed to flow through to the highly saline ground water for fear of raising the already shallow groundwater into the root zone. It is therefore carried away by a series of drains constructed by the Irrigation District. Ultimately it is deposited in the Salton Sea (see fig. 10-1).

Fig. 10-6. Master drain line being installed in the Imperial System. Collects the flow from field drains and routes it to the Salton Sea.

sedimentation, moss and algae, burrowing animals, as well as normal wear on channels, gates, controls, heavy equipment, and vehicles.

The sediment load of the Colorado River is substantial and the Imperial District has formal procedures and facilities to deal with the problem. At the Imperial Dam trash screens and sedimentation basins are in place and used to combat silting as well as to prevent trash (trees, limbs, etc.) from entering the system (the diversion dam for water from the Colorado River to the All-American Canal).

The climate which promotes year-round crop production also promotes moss and algae growth. Moss screens are mandatory if precision control of water levels, gate openings, and service to the farms is to be

Fig. 10-7. Trash screen, placed across the Imperial Dam, at the intake to the All-American Canal. Large obstructions such as trees, are removed by the boom, in photo center.

reliable. Therefore several moss screens have been placed at four drops in the All-American Canal, In addition, the Imperial District is experimenting with the use of grass carp to consume and control hydrilla (water plants).

Burrowing animals are also a maintenance problem. Muskrat and gopher control is a necessity to maintain canal bank integrity. Constant patrol by the maintenance staff, supplemented by information from farmers, is mandatory if leaks and failure of the canal are to be prevented.

Salt River Project

The Salt River Irrigation Project of Arizona, is another very old and highly successful riparian (river) based system of irrigation. Like the Imperial Valley of California, this area of Arizona* was a very hot, arid desert . . . but now is a veritable horn of plenty, especially well-known for citrus, other subtropical fruits, vegetables, and cotton.

Early settlers (1860's) in the Salt River Valley of Arizona noted the presence of ancient canals which had conveyed water from the Salt River on to the surrounding lands. Evidence of crude diversion dams in the river, as well as lateral ditches serving extensive acreage, were left behind by prehistoric Indian tribes that lived in the Salt River Valley. So extensive was this network of diversion and conveyance of water, that over 65% of the present Salt River Project had been irrigated by people who had disappeared from the area hundreds of years earlier.

Physical Description

Land which could be irrigated lay on both sides of the Salt River, and farming moved away from the river lands as canals and laterals were shaped and expanded, many of them using the old Indian alignments and grades. The district is laid out parallel to the river, utilizing the natural grade (gravity flow) of the river valley for the flow of the canal system. Laterals are spaced at about 1 mile (1.6 Km) intervals out of each canal.

Early Salt River (SRP) irrigated acreage (in 1951) was about 242,000 acres (96,800 hectares). The ratio of this service area to the watershed of the Salt and Verde Rivers is 1 unit of farmland to 30 units of watershed. However, this ratio is not sufficient, and the systems must also be supplemented with ground water.

Currently, SRP's service to agricultural users has dwindled to only about 100,000 acres, (40,000 hectares). This reduced area is due to the conversion of farmland to residential and business. Arizona's capitol city, Phoenix, and other smaller cities have grown on the basis of the assured water supply, originally designed for agriculture.

Service to Irrigator

Water delivery to the farmer by the Salt River District is made on a demand basis. The farmer places an order with the field office serving his area and all orders are accumulated by a fixed hour each day. The district's watermasters then calculate changes required in the entire system by canal and lateral.

The individual farmer receives water in the quantity and for the duration he has requested. As with the Imperial System, the SRP operators have guidelines requiring that the district begin the ordered water delivery within a specified time after receiving the order. The Salt River Project, however, allows the farmer to continue to receive the water for as long as needed (the Imperial System allows the water authorities to stop the flow, in accordance to other needs in the system).

The district makes periodic checks of the amount of flow at each delivery gate, and records this information

---

* Arizona, named by the early Spanish conquistadors, means "arid zone".

Fig. 10-8. Sedimentation Basins at Imperial Dam. Constant velocity channel is left-center. Clarifier motor and drive mechanisms are in the basins. Early attempts at harnessing the Colorado River for irrigation were continually disrupted by sedimentation at the cuts made for the canals. Only with equipment such as this (and continual supervision) can these types of irrigation projects continue to function.

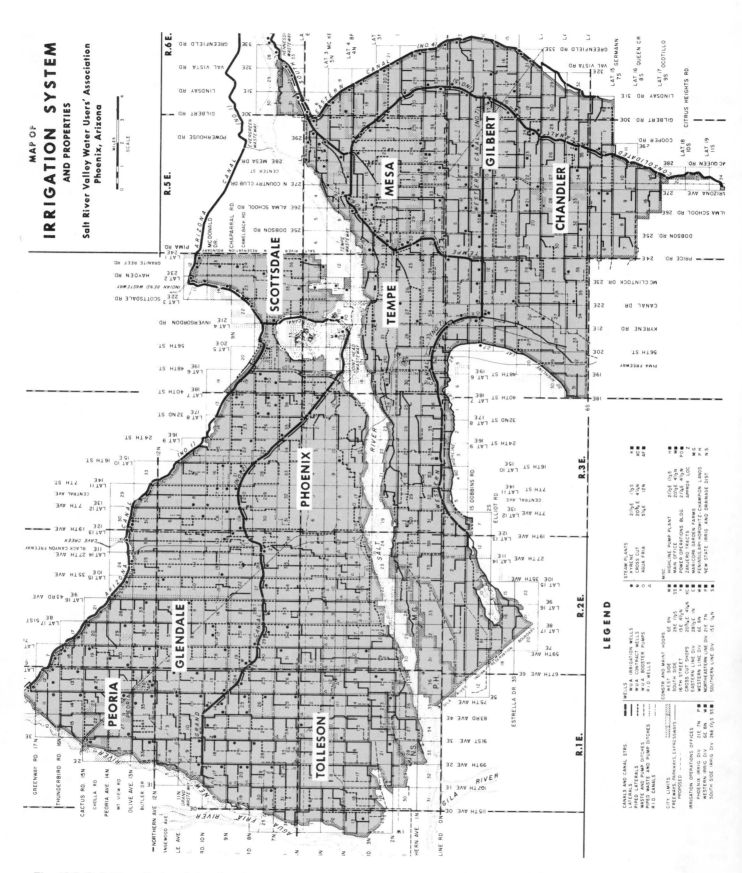

Fig. 10-9. Salt River Project Irrigation System. Courtesy Salt River Water User's Association. Phoenix, Arizona.

Fig. 10-10. Deepwell turbine lifting groundwater into a short pipe and thence to a Salt River Project lateral for delivery. The dotted alignment approximates that of the pipe from pump discharge to lateral.

as a part of its internal record-keeping. At the end of a delivery period, the volume of water delivered is recorded on the water accounts of that specific farmer, and deducted from his total allocation.

The farmer is then notified as to his water balance and his money credit on the delivery account. (The Salt River Project operates on the basis that the water must be paid for in advance.)

Crops Grown

The climate, soils and water availability within the SRP allows the production of a wide range of crops on a yearlong basis.

Nearly 40% of the area is double cropped. This is obviously a function of climate and water supply being favorable to such intensive agriculture.

Maintenance Concern

Much of the Salt River's system is lined with concrete, or in underground pipe. This work continues with the intent of adding more concrete and pipe each year, to conserve water and minimize channel and bank maintenance.

Water temperatures and relatively clear water (for the transmission of sunlight) promotes moss and algae growth, which interfere with water flow. Trash racks and automatic screens are utilized to control the aquatic weed problem.

District Direction and Politics

Policy and procedure decisions in both the Imperial and the Salt River Project Districts are determined by elected directors. These directors must meet local criteria; be landowners within the district, represent specific zones within the district, and be willing and capable of serving to discuss and determine district business.

Issues which normally must be acted upon by directors are:

1. Annual water service rates to be collected from each water user. These are normally set on a water volume or land area basis each year. The total amount to be collected is normally related to the expected expenditure budget for the district.
2. Annual allocation of water to each acre or owner for the year. This decision is made early enough before planting to allow the farmers to make crop selection decisions.
3. Selection and compensation of the technical and professional members of the staff.

Summary

Irrigation districts are a form of organizing landowners to work cooperatively, to maximize water supply reliability. In order to be successful, heavy capital expenditures are required both on the part of the landowners and the governing body (usually the country's government). *In order for landowners to be able to justify such expenditures, stability of the governing body is essential.*

The two irrigation districts outlined exist in extremely arid/desolate areas of the southwestern United States. Both districts are dependent upon storage of snow melt and precipitation from distant watersheds to provide the season-long crop requirements.

The districts differ in that the Salt River Project must pump from ground water to augment the surface water supply. The Imperial, California System must provide drainage to maintain proper soil aeration, and to minimize saline concentrations (due to poor internal soil drainage and the resulting concentration of salts in the root zone). In other respects, the basic operations are very similar.

One of the most similar aspects of the districts is the

strength and dedication of the people. That dedication is the basic key to development of assured water availability and the equitable distribution of that water supply. From that dedication came the progress exhibited by both systems and it is the catalyst for the future.

ACKNOWLEDGEMENTS

Imperial Irrigation District

The author is deeply indebted to Ron Hull, Director, Public Information & Community Services, and many other IID personnel for their interest and courtesies.

Salt River Project

Reid W. Teeples, Associate General Manager, E. C. Friar, Manager, Water Operations, and members of the "Water Users" staff.

CHAPTER 11 DRYLAND AGRICULTURE
John L. Havlin, Ph.D.
Department of Agronomy, Kansas State University

INTRODUCTION

In general terms, dryland agriculture refers to cropping practices that allow production without the aid of irrigation. Dryland agricultural production can be divided into two distinct climatic precipitation zones: 1. Semiarid zones in which water conservation practices are required and/or water availability may limit crop production to bi-annual or even tri-annual cropping, rather than annual, and 2. humid areas in which water does not limit annual production and may actually even create problems due to excess.

Although it is often used as a general guide, total annual rainfall is a poor indicator of available moisture required for crop growth. More precisely, rainfall during the growing season is a better indicator. Figure 11-1A, B & C illustrates 3 climates which differ in rainfall distribution, but have similar potential evaporation (PE) rates.

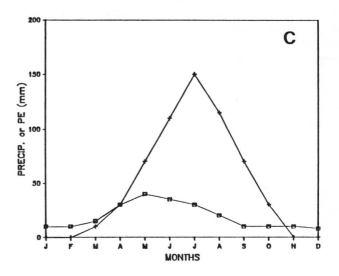

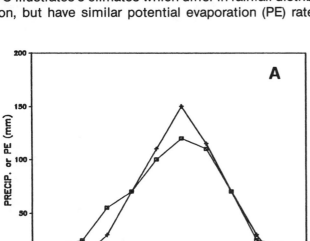

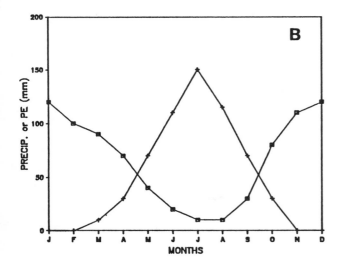

Fig. 11-1A, B & C: Precipitation (o) and potential evaporation (+) for three climates typical of dryland regions. (A) Represents a climate with similar precipitation and evaporation quantity and distribution. Dryland crop production is well adapted to this climate. Water should not be limiting. (B) In this climate total precipitation and evaporation are similar but most of the rainfall is received in the winter. Thus, water will be limiting during the crop months when potential evaporation is the greatest. (C) In this climate potential evaporation greatly exceeds precipitation, therefore moisture will limit dryland crop growth unless "fallowing" practices are adopted.

Generally, regions that receive most of the rainfall during the growing season, which is usually the period of maximum PE, are more suitable for crop production. For example, some environments, such as the semi-arid areas of the northwestern U.S., receive most of the precipitation in the winter months and very little in the summer months, during the crop growing period. Therefore inadequate available moisture precludes the growth of some crops even though total annual rainfall exceeds that required by the crop.

Generally, when rainfall is less than 600 mm (24 inches) per year, water is a limiting factor to crop yield. In these regions, one or more years of moisture must be stored in the soil to enable production of one crop. This practice is called "fallowing" and will be discussed later.

Quite obviously, the crop management practices and techniques in the two distinct precipitation zones are very different. As might be expected, the crop yield potentials are also quite different. Table 11-1 illustrates the effect of total annual rainfall on wheat and sorghum yields across the state of Kansas (USA). Rainfall nearly doubles from Tribune, KS to Manhattan, KS, with yields increasing about 40%. Actually, however, the yield differentials are even greater in Tribune and Garden City; the dryland wheat and sorghum yields are based on a 3 year wheat-sorghum-fallow rotation, and thus grain is produced in only 2 of 3 years.

In humid regions where water is not the limiting factor, soil fertility, disease, weed control, and other factors become limitations to crop yield. Table 11-2 illustrates

Table 11-1. INFLUENCE OF ANNUAL PRECIPITATION ON WHEAT YIELDS ACROSS KANSAS (USA).

Location	Annual Rainfall	Average Yield Wheat	Sorghum
	-- mm --	---------- kg/ha ----------	
Tribune	350	2200	2600
Garden City	450	2700	3100
Hays	550	3000	3500
Manhattan	750	3300	3900

Table 11-2. CROP MANAGEMENT FACTORS THAT AFFECT YIELDS IN DRYLAND CROPPING SYSTEMS.

Controllable Factors	Uncontrollable Factors
Soil Moisture	Growing Season
Fertility	Air & Soil Temperature
Variety/Hybrid	Light Intensity
Diseases	Humidity
Insects	Day Length
Weeds	Precipitation
Planting Rate	Wind
Planting Date	Rooting Depth
Row Spacing	Clay Content
Tillage	Organic Matter

those crop management factors which are most important to crop productivity in both types of dryland agriculture regions.

Crop yields in both humid and semi-arid regions have generally increased over time as crop management technologies have developed. The long-term average annual wheat yield increase has been nearly 35 kg per hectare per year (Figure 11-2). The primary soil and crop management factors that have contributed to those yield increases are listed in Table 11-3. In humid cropping regions, variety/hybrid development and fertilization have been the major contributors to yield advancements. Planting dates, weed control, and harvesting technologies have been somewhat less important.

In regions where moisture is a limiting factor to crop yield, as in semi-arid areas, those management factors which increase conservation of the limited moisture have contributed the most to yield increases. Hybrids and varieties are also important in semi-arid regions, along with optimum planting date, soil fertility, and weed control.

Fig. 11-2. Actual and average wheat yields in the Central Great Plains (USA).

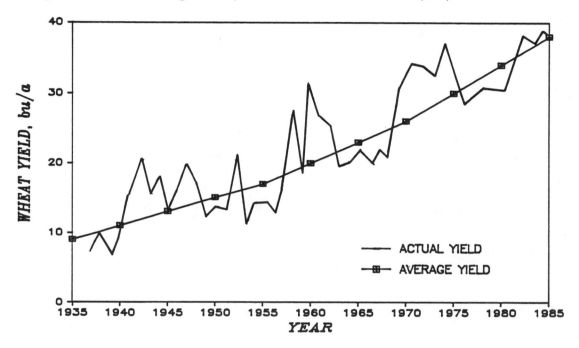

Table 11-3. FACTORS CONTRIBUTING TO GRAIN YIELD INCREASES OVER THE LAST FOUR DECADES.

Factor	Percent Contributions	
	Humid Regions	Semiarid Regions
Variety/Hybrid	35	25
Soil Fertility	25	5
Water Conservation	0	35
Planting Dates	20	20
Pest Control	10	10
Harvesting	10	5

DRYLAND AGRICULTURE: SEMI-ARID REGIONS

This is the type of agriculture that first comes to mind when the word "dryland" is mentioned. The semi-arid environment or climate in these dryland regions generally consists of 250 to 600 mm (9.8 to 23.6 inches) of annual precipitation. Regions that receive less than 250 mm of moisture are considered arid or desert climates, and crop production in these areas generally occurs only with irrigation.

Semi-arid regions typically have climates where water evaporation rates exceed rainfall during the growing season. For example, in the Central Great Plains (USA) average annual rainfall is 400 mm (15.7 inches), whereas potential evaporation exceeds 800 mm (31 inches) of water. Thus, a negative (-400 mm) water balance exists which restricts annual cropping options. Consequently, successful dryland crop production in semi-arid climates must utilize a "fallow" period to collect or store additional soil moisture.

Because annual precipitation is low, semi-arid region soils are slightly to moderately weathered. Most surface soils are calcareous (contain free lime) and have neutral or greater soil pH. Organic matter contents are generally lower than comparable soils developed under humid climates.

DRY LAND FARMING PRACTICES

When semi-arid regions were first put into production, soil and crop management practices were similar to those used in more humid regions. After drought periods resulted in crop failure, growers soon realized that new cropping practices were needed which would maximize soil moisture and enable successful crop production. Unfortunately, the problems in dryland farming were much easier to recognize than to solve. Many farmers were forced from the land before successful dryland farming technologies were developed.

Summer Fallow

Fallowing represents the most important cultural practice for semi-arid regions. Fallowing extends the period between crops to accumulate enough additional water in the soil to reduce the risks of drought and increase productivity. A threefold increase in grain yield can be realized by fallowing practices.

The length of the fallow period depends on annual rainfall and the water demand by the crop. As annual rainfall decreases from one region to another, the fallow period will generally increase. Coarse grain crops (corn and sorghum) have higher water requirements than small grain crops (wheat, barley, oats, etc.) and thus corn and sorghum crops require a longer fallow period. In the Great Plains region, where annual rainfall is less than 600 mm (23.6 inches), the typical fallow period is about 14 months in a wheat-fallow wheat cycle. Typically, wheat is planted in September and harvested the following July. The fallow period would run from wheat harvest through the fall, winter, and summer months until wheat is again planted in September or about 14 months after wheat harvest.

Fig. 11-3. Grassland prairie in the northern Texas Panhandle being plowed up for dry-land crop production. Such areas are subject to severe wind erosion if provisions for windbreaks, strip cropping, etc. are not practiced. (Dry-land farming is particularly susceptible to wind erosion due to the necessity of leaving land fallow for long periods.)

Fig. 11-4 and 11-5. Wheat fields blown over with sand in the Oklahoma Panhandle due to a lack of windbreaks or other protective measures.

Fallow Systems

As in all or most annual cropping systems, tillage is usually used to control weeds. Controlling weeds during the fallow period is extremely important because weeds use valuable soil water needed for the next crop. Indeed, the quantity of water stored during the fallow period is primarily determined by the success in controlling weeds during the fallow period.

Early tillage systems utilized the moldboard plow for complete burial of weeds and crop residues. Unfortunately, tillage also results in soil water loss through evaporation as the top soil is disturbed. However, tillage losses are negligible compared to weed losses. Tillage water loss is only about 0.5 to 1.0 cm (.2 - .4 inches) of water per **tillage operation**, compared to 3 to 4 cm (1.2 - 1.6 inches) water loss **per day** with a full stand of weeds.

The primary disadvantage with tillage or residue burial is that the dry, bare soil surface is subject to severe wind erosion losses. After the "Dust Bowl" era of the 1930's, strip cropping practices were adopted in the early 1940's to help reduce wind erosion losses. Strip cropping is a system where alternate narrow crop and fallow strips are placed in the field. An example of this is shown in Figure 11-10.

Tillage systems were eventually developed that maintained some surface residue cover while still controlling weeds. One of the first implements developed to maintain some surface residue was a one-way disk. A one-way disk would reduce residue levels approximately 20% for each pass over the field. Weeds would be controlled while leaving 75-80% of the residue still on the soil surface. Although this was an effective tool, if a grower used this implement three or four times (as was commonly done), most of the residue would eventually be incorporated. Thus, an implement was needed which would control weeds, yet after each pass, leave a substantial amount of residue on the soil surface.

In the mid 1940's a new implement was developed which would only reduce residues approximately 10% for each pass over the field. This implement is called a sweep plow or a V-blade (Figure 11-6). This implement is operated approximately three to four inches deep in the soil and as it passes through, it gently lifts the topsoil up and cuts off any live weed roots. Although the weed is still anchored into the surface soil, the plant will desiccate and die unless rain is received immediately after the tillage operation. This "undercutting" process is very effective in controlling weeds and also leaves approximately 60-70% of the residue still on the surface even after three to four operations in the field. Known as stubble mulching, the stubble left by sweep plowing helps to catch snow in the winter and therefore increase total water collected in a given field.

Another implement that was developed in the 1950's is called a rod-weeder (Figure 11-7). A rod weeder consists of a steel bar that is operated at approximately the same depth as the sweep plow. However, the steel bar rotates opposite to the direction of travel. In common stubble mulch fallowing systems, a grower might have a rod weeder attached to the rear of the sweep plow and thus undercut and rod weed simultaneously. Generally, several sweep and rod weed operations are performed throughout the fallow period.

Fig. 11-6. Sweep plow. Passed 3 to 4 inches deep, the purpose is to sever weed roots, while minimizing the disturbance (and subsequent erosion) of the surface soil.

Fig. 11-7. A rodweeder. Another device such as the sweep plow, designed to disrupt weed roots, while minimizing disturbance of the surface soil (see text for application).

Since the early to mid-1970's, new dryland fallowing practices have been developed. Known as no-till or chemical fallow dryland farming, all tillage is replaced with the use of herbicides for weed control. These types of chemical fallow systems will generally maximize the amount of water stored during the fallow period because no tillage is performed, and therefore evaporative water losses are reduced.

However, several problems exist with the chemical fallow system. First of all, many of the herbicides required for chemical fallow can be very expensive. Secondly, if the herbicides are not precisely applied in the field, poor weed control will occur and subsequent water loss will be experienced. Therefore, chemical fallow systems require precise and detailed knowledge of herbicides and herbicide application.

Table 11-4 shows the effect of the various fallow systems on total water stored, fallow efficiency, and winter wheat yields in the central Great Plains of the United States. By adopting chemical fallow and no-tillage systems, nearly 63mm (2.5 inches) of water could be stored during the fallow period, which translates into approximately 20 bushels of additional winter wheat yield.

The fallow system used during the fallow period will depend primarily on the management skills of the grower. No-till or chemical fallow systems require much more intensive management than either stubble mulch or conventional tillage systems. Currently, very few acres are managed under a chemical fallow system.

Table 11-4. EFFECT OF FALLOW SYSTEM ON SOIL WATER AND WINTER WHEAT YIELDS IN A WHEAT-FALLOW-WHEAT SYSTEM IN THE CENTRAL GREAT PLAINS (USA).

Fallow System	Water Stored in Fallow (mm)	Fallow[1] Efficiency (%)	Wheat Yield (kg/ha)	Water-use[2] Efficiency Kg/ha-mm
Convention Tillage (plow, disc, cultivator)	117	23	1350	11.5
Stubble Mulch Tillage (sweep plow, rod weeder)	155	33	2150	13.9
No-Till - Chemical Fallow	180	40	2700	15.0

[1] Fallow Efficiency - percent of precipitation received that is stored in the soil.

[2] Water-Use Efficiency - grain yield produced per unit (mm) of water used or transpired.

However, as the technology is further developed and refined, and as more growers become familiar with the new technology, it is conceivable that many more acres will be managed with chemical fallow.

FACTORS LIMITING DRYLAND CROP PRODUCTIVITY

MOISTURE

Available soil moisture is the greatest limiting factor to dryland crop productivity. In most of the semi-arid regions of the United States, about 75% of total annual precipitation occurs between April and October. The balance is received as either snowfall or rainfall in the winter months. Of course, not all the precipitation received is stored in the soil profile or used by the crop. Depending on the region, about 50-75% of the water received is lost by evaporation. The quantity of water evaporated from the soil surface accelerates in relation to increasing air temperature and wind speed. Weeds and volunteer crop plants remove the second largest quantity of water received. These losses can be as high as 25 to 30% of total annual precipitation. Significant losses can also occur from rainfall not infiltrating into the soil profile and moving over the soil surface to non-crop areas, commonly referred to as "runoff". In addition, losses also occur from snow blowing off bare soil surfaces. Generally, under conventional dryland cropping systems (complete burial of previous crop residues), the maximum amount of annual precipitation utilized by a crop is only about 22% or less.

As discussed previously, there is a wide variation in the amount of water required by various crops. Therefore, one must select a crop with a requirement that can be satisfied by the available water. With winter wheat and similar crops common to semi-arid regions, about 20 cm (7.9 inches) of soil water is required before any grain is produced (Fig. 11-8). Grain yields can then increase about 150 to 200 kg per hectare with each additional cm (.25 inch) of moisture.

When managing dryland cropping systems for maximum soil moisture storage and water use efficiency, it is very important to understand the value of various sources of available soil water in producing grain. Table 11-3 lists the various sources of water for plants and an arbitrary index of efficiency. Stored soil water or water that is present in the soil profile at seeding time is 100 per cent plant available, whereas growing season moisture is only 25% or less available. It is apparent that maximizing soil moisture at seeding time (usually in the fall for winter wheat) and maximizing the amount of snow captured are very important management goals for improving dryland crop productivity. These will be discussed more thoroughly in succeeding sections.

EROSION

Climatic conditions in semi-arid regions are generally optimum for soil losses by erosion. Topsoil losses by wind erosion are usually much more serious than soil losses by water erosion. For example, in the central Great Plains of the U.S., wind and water erosion represents approximately 75 and 25% of total soil erosion losses, respectively.

Wind erosion occurs whenever a strong turbulent wind blows across an unprotected soil surface that is dry, bare of residues, smooth, and/or finely granulated. With these conditions, a minimum wind speed of 20 km (12 mph) at a height of 30 cm (about a foot) above the soil surface is required for soil particle movement. Water erosion occurs when rainfall intensity exceeds the intake capacity of the soil. Generally, as field slope steepness increases and residue level decreases, water

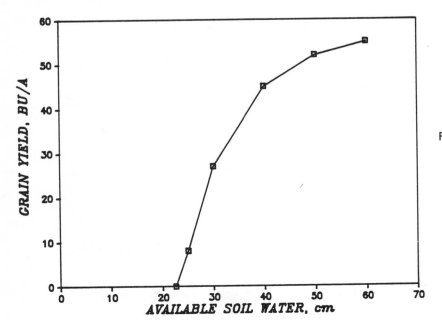

Fig. 11-8. Available soil water required to produce wheat yields in the Central Great Plains (USA).

infiltration decreases and subsequent erosion increases. Figure 11-9 shows an example of severe water erosion.

Soil erosion, especially wind erosion, can decrease soil and crop productivity by the gradual removal of silt, clay, and organic matter from the surface soil. The coarser sand particles that remain are infertile and have a very low water holding capacity. The United States Department of Agriculture Soil Conservation Service estimates soil erosion losses greater than 11 tons per hectare (4.5 tons per acre) per year will reduce soil productivity. Approximately 30% of cultivated semi-arid region soils in the U.S. exceed the critical soil erosion rate with about 60% exceeding twice the critical erosion rate, or 22 tons per hectare per year. The effects of soil erosion losses on crop yield losses are difficult to quantify. Recent calculations with soil loss data from the Great Plains region of the U.S. indicated annual erosion losses of 0.20 cm (about .08 inches) topsoil per year. At this rate of soil loss, annual wheat yield losses were estimated at 130 to 330 kg per hectare per year (120-300 lbs/acre).

Soil erosion can be controlled by (1) increasing soil surface roughness with larger soil clods or furrowing, (2) increased surface residue cover, (3) reduced field widths or adoption of strip cropping practices, and (4) establishing windbreaks (bushes, trees, etc.) on windward side of fields. Of the four methods, strip cropping and increasing surface residue cover may be the most practical. For example, 20 to 30% crop residue cover can reduce water erosion by 50% and wind erosion by 80%. Forty percent standing residue cover can effectively eliminate wind erosion losses.

ORGANIC MATTER

The third factor which may limit or reduce dryland crop productivity is loss in organic matter. In addition to losses caused by erosion, cultivation or tillage will cause organic matter to oxidize or mineralize. Recent studies in western Nebraska (U.S.) showed that organic matter declined 20 to 25% in only 10 years. Studies from western Kansas indicate that organic matter contents have decreased about 50% over 40 to 50 years of cultivation.

Organic matter losses will result in (1) reduced available water holding capacity, (2) reduced native fertility and plant nutrient availability, (3) increased erodibility, (4) increased degradation of good soil struc-

Fig. 11-9. Example of severe water erosion.

Fig. 11-10. A typical "strip" cropping system with wheat-fallow in the Central Great Plains (USA). The alternating rows of crop aftermath are used as a windbreak to reduce erosion.

ture, (5) reduced water infiltration, and (6) reduced micro- and macro-biological activity. All of the above factors reduce the productive capacity of the soil, while increasing production costs and reducing production efficiency. In some soils additions of inorganic fertilizers can compensate for the loss in productivity, while in other soils they can not.

DRYLAND AGRICULTURE: HUMID TEMPERATE REGIONS

INTRODUCTION

Dryland agricultural production in humid, temperate regions occupies nearly half of the total arable land throughout the world. Average annual precipitation ranges from 800 to 1600 mm (31.5 - 63 inches), and therefore moisture is not a limiting factor to crop growth and yield. (It is called "dryland" agriculture because irrigation is not used.) Precipitation generally exceeds evaporative losses in these regions, and therefore soils are generally not calcareous nor contain free lime. In regions where rainfall exceeds 40 to 50 inches per year, soil pH can be very acid, requiring routine applications of lime to maintain soil pH at optimum levels.

What humid areas do have in common with semi-arid areas is potential erosion. The only difference being that erosion in humid areas is more often caused by water than wind. Some of the same management practices, such as inter-seeding and no-till weed control, are often done to reduce erosion.

CROPS AND CROPPING SYSTEMS

Because moisture is not a limiting factor to crop growth in humid regions, the options for crops and cropping systems are much greater than in semi-arid regions where moisture is limiting. In general terms, cropping systems can be divided into two main categories: 1. continuous cropping or mono-culture, and, 2. crop rotations. The cropping system that would optimize production of food and fiber depends on soil and climate suitability, available farm equipment or machinery, local markets and means of transportation for the crops produced, and availability of inputs to properly raise the crops involved. In the Soil and Soil Properties chapter, the detrimental effect of continuous cropping or mono-culture was discussed at some length. Despite the potential long term loss in soil productivity, continuous cropping or monoculture systems are commonly practiced throughout most humid regions of the world. Several reasons for this include: (1) a particular soil may be adapted to one crop, for example, rice or forage legume, (2) the climate may be more favorable for a particular crop, (3) machinery costs are often lower than with crop rotations, (4) some crops may be more profitable than others, and (5) some farmers prefer growing single crops because the management skills required are less varied than for crop rotations.

Even though these advantages of continuous cropping are apparent, the value of crop rotations and long term maintenance of soil and crop productivity are equally apparent. The primary advantages of crop rotations are: (1) increased organic matter and improved tilth of the soil, (2) weed, insect, and diseases are easier to control, (3) some economic advantages exist with the diversification of products produced on the farm, (4) inclusion of legumes in the rotation can improve soil structure and add nitrogen to succeeding crops, and (5) soil erosion losses by wind and water can usually be reduced if forage crops are included in the rotations. Studies with long term cropping systems have demonstrated the distinct advantages of crop rotations over continuous cropping in relation to long term soil and crop productivity. Therefore, crop rotations should be utilized wherever and whenever economically feasible.

Table 11-5. COMMON GRAIN AND FORAGE CROPS PRODUCED IN HUMID REGIONS.

Crop	Ave. Growing Period --- days ---	Water Requirement --- mm ---	Rooting Depth -- m --
Corn	110	700	1.3
Sorghum	105	600	1.3
W. Wheat	200	450	1.3
S. Wheat	120	450	1.0
Cotton	140	900	1.3
Alfalfa	200	1200	2.0
Clover	200	1000	1.0
Beans, Dry	100	400	0.6
Soybeans	100	800	1.0
Sunflowers	130	1000	1.2
Barley	120	450	1.3
Oats	120	450	1.3
Rice	140	900	1.0

Several other cropping systems have gained popularity over the past several decades and may become extremely important systems in the future. These include multiple cropping, relay cropping, inter-cropping, and ratoon cropping. Multiple cropping is the production of two of more crops in the same year on the same area of land. This system generally occurs in regions that have long growing periods, such as subtropical and tropical climates. In the southern climates of the United States, soybeans can often be planted following winter wheat harvest, which are then harvested in the fall of the same season. With short season cultivars of some crops, three crops can be grown where growing season and moisture are not limiting. In most multiple cropping systems, some field operations must be shortened or eliminated to allow production of both crops in the same year. Usually, one or more tillage operations are eliminated and the second crop planted in the residue of the first crop. Generally, long season crops such as cotton, corn, sunflowers, and others are not well suited for multiple cropping systems unless the growing season exceeds 220 days.

Relay cropping is a system which involves planting the second crop into the first crop before the first crop has matured. As with multiple cropping systems, relay cropping is only successful in those regions where moisture is sufficient for both crops. Soybeans, peas, alfalfa, and other legume crops have been successfully seeded or inter-seeded into wheat and other crops prior to stem elongation.

Inter-cropping involves two or more crops grown simultaneously. These systems usually involve crops grown in alternate rows. As with the other cropping systems, sufficient soil moisture must be available for both crops and both crops must be able to exist with little loss in yield compared with either crop grown separately.

Ratoon cropping is simply planting those crops that have the ability to send up new shoots from the base of the crop after it is harvested. Sugar cane, pearl millet, sorghum, rice, and cotton are crops that can be adapted to a ratoon cropping system.

COMMON MANAGEMENT PRACTICES

Crop management in developed, industrial nations is highly mechanized and includes high inputs for maximum yields. In less developed countries, crops normally produce less than maximum yields, but fewer inputs are also utilized. In humid regions of North America, corn, soybeans, forage crops, and wheat are the major crops grown. Average yields and nutrient content of these crops are shown in Table 11-6. Average nitrogen, phosphorous, and potassium fertilizer rates in this region are 160, 60, and 50 kg/ha, respectively. In addition to these high fertilizer inputs, most producers in this region utilize chemical herbicides for control of annual and perennial broadleaf and grass weeds. Some cultivation of weeds is still employed to effect complete control.

Prior to the early 1970's the primary tillage operation used for preparing the seedbed was fall plowing; crop residues completely buried with the plow. Seedbed preparation would be completed in the following spring with a disk operation typically followed by another tillage to firm the seedbed prior to planting. Sometime during the spring, secondary tillage operations and/or herbicides would normally be applied.

Over the last 15 years, increased concern about soil erosion and reduced soil/crop productivity has led to development and adoption of minimum or reduced tillage systems. These systems primarily use a chisel plow or a disk to partially incorporate previous crop residues, while leaving between 30 and 50% surface residue cover prior to planting the next crop. Specialized planting equipment is normally required to plant the succeeding crop into these residues. These planters are equipped with a cutting disk or coulter which runs directly ahead of the seed opener on the planter. This cutting disk cuts the residues and allows the opener to pass through the residue, placing the seed in the soil at the proper depth. Post-emergence cultivation for weed control in these residue systems can be a problem and, therefore, herbicides are normally used for weed control.

In addition to reduced tillage cropping systems, contour farming can also reduce soil erosion by water. In this system, all tillage and planting operations are done along the contour of the hill or slope. In other words, crops are planted "around" the hill instead of "up and down" the hill. Planting up and down the hill in a plow tillage system can result in much more soil loss compared to planting across the slope or on the contour.

Complete no-till or zero tillage systems, where none of the previous crop's residues are incorporated into the soil, have also been developed over the last 10-15 years. Examples of some of the planting equipment commonly used in these cropping systems are shown in Figures 11-11 & 2. Reduced labor and tillage expenses are immediate advantages of no-till systems. However, the chief advantage to no-till systems in semi-arid climates does not apply to humid areas. That is, in semi-arid climates, the advantage to no-till systems is reduced evaporation loss of water. Conversely, in humid areas this decreased evaporation can result in delayed planting due to cool, wet soils. This can also lead to increased disease and fungal problems.

Table 11-6. AVERAGE YIELD AND NUTRIENT CONTENT OF THE MOST COMMON CROPS GROWN IN NORTH AMERICA.

Crop	Yield	Nutrient Content				
		N	P	K	Ca	Mg
	-- t/ha --	-------- kg/ha --------				
Alfalfa	18	500	40	450	220	45
Corn Grain	11	190	35	8	20	20
Stover	9	80	15	180	40	40
Wheat Grain	6	160	20	25	2	12
Straw	7	50	5	125	4	12
Soybean Grain	4	280	25	8	20	20
Straw	1	90	8	55	10	10

Fig. 11-11. Alfalfa planted and grown along the contour.

Fig. 11-12. Wheat planted up and down the slope of a hill, rather than along the contour. The resultant soil erosion is evident.

SUMMARY

Dryland agriculture simply means crop production without irrigation. As a result, the term "dryland" applies to humid cropping areas as well as semi-arid areas.

In semi-arid areas the basic strategy is to preserve moisture while combatting erosion. In most cases management practices aimed at the goal of preserving moisture can unfortunately aggravate erosion. The reason is that the primary goal in preserving moisture is to control weed growth. Weeds, of course, utilize water that is needed for crops.

Conventional methods of controlling weeds typically disturb the topsoil, thereby exposing it to erosion. To reduce this problem, means of cutting weed roots through sub-surface plows have been developed. These implements reduce, but do not eliminate, topsoil disturbance. Chemical means of weed control through herbicides eliminate soil disturbance, but require substantial expertise and usually greater expense (than mechanical control).

A common practice in semi-arid dryland agriculture is to fallow fields. The length of fallow can vary from a few months up to one or even two years. The actual length of the fallow depends upon the water requirements of the crop to be grown, and of course, the precipitation.

During the period of fallow, the soil is subject to erosion. A number of strategies may be undertaken to reduce that erosion. One of the simplest yet most effective is strip cropping. That is, planting alternate strips of crop with strips of fallow. Thus, the standing crop tends to slow down wind and/or water movement across the

Fig. 11-13 and 11-14. Corn planter (top) and wheat drill (bottom) for no-till and reduced tillage conditions.

land.

Another common method is to leave crop aftermath in place. Indeed, planting methods have even been developed to plant without having to remove or disrupt crop aftermath.

In humid areas this type of planting is also utilized. Normally it is used as part of a double cropping scheme, as well as to prevent erosion.

Another common practice in humid areas is what is known as ratoon cropping, which refers to perennial crops that do not need replanting. That is, crops such as sugar cane, millet, some sorghums, and rice which have the ability to send up new shoots from existing stalks.

Although quite different in terms of precipitation, humid areas often have the same problem as semi-arid areas in terms of erosion. In semi-arid areas the problem is just accentuated through the necessity to leave land fallow in order to build up moisture. In many humid areas, through special seeding techniques, secondary crops may be planted before the first crop matures. In that way, a cover crop is always on the land to deter erosion.

CHAPTER 12 COMMON FIELD CROPS
D.J. Undersander, Ph.D.
University of Wisconsin
Madison, Wisconsin

Agriculture, defined as the cultivation of crops, did not begin until about 20,000 years ago. At that time, in approximately the year 7500 B.C., it was discovered that the seeds from the grass which we now know as wheat, were not only good to eat, but could be planted to grow more seeds. These early "farmers" began cultivating this "grass", and gave birth to the concept of agriculture. This same scenario was repeated in other parts of the world with similar, but different crops. In the Americas it was corn, in Africa it was sorghum, and in Asia it was rice.

As the concept of agriculture spread, profound changes occurred in the human population. Prior to the advent of crop production, the human population primarily led a nomadic life, living off the land. The advent of crop production allowed settlement in permanent villages, and the development of specialized labor and services. More than that, agriculture allowed the human population to expand.

Prior to agriculture, periodic famines (such as the recent ones seen in East Africa) kept the human population in check. With the advent of agriculture, humans were no longer dependent solely on the limited bounty of nature. As a result, the world's population began its upward spiral.

What seems strange is that the enormous population that the world now supports, is still dependent upon the basic crops that were developed during pre-biblical times. Indeed, there are less than a half-dozen commodities (crops) that support the bulk of the earth's population. Without wheat, corn, sorghum, soybeans, rice, and possibly barley, the world could not begin to support as many people as it does.

CORN

Corn originally began as a crop in what is now Mexico, cultivated by various resident Indian tribes. Today, corn is generally grown in regions of the United States that have a mean summer temperature of 70-80 degrees F and a mean night temperature exceeding 58 degrees F. Corn is grown primarily for grain in regions that have at least 120 days of growing season, and mainly for silage or sweet corn in regions of shorter growing seasons. Corn requires a fertile, well-drained soil with a pH between 5.5 and 8.0

Several types of corn are produced in the United States. Yellow (dent) corn is produced on the largest acreage and is used primarily for feeding livestock. White corn (with white kernels instead of yellow) is used primarily for processed food such as corn chips, because yellow kernels turn black when roasted. Sweet corn lines have a higher sugar content in the kernels than other types of corn. Sweet corn is normally grown for human food purposes, sold for either fresh consumption or canning. Popcorn is characterized by small kernels with a very hard interior.

Corn hybrids are planted in most developed countries. A hybrid is a cross between two inbred lines. In corn and certain other crops, this cross results in higher yields than either parent plant due to hybrid vigor. The disadvantage of hybrids is that they do not breed true (produce seeds that will result in plants like the parent plants). Therefore, seed from one crop **cannot** be saved for planting the following year. Less developed regions of the world tend not to plant hybrids but rather use seed from the previous year, because new seed is not readily available each year.

The initiation of grain formation on corn is very sensitive to day length. Since day length is determined by distance from the equator as well as time of year, corn types are generally adapted to a 100 mile wide band, and will not grow properly if moved north or south of their areas of adaptation. A corn hybrid will flower about 1 day earlier for each 10 miles it is moved south of its area of adaptation. This sensitivity to day length is also one reason why corn that is planted later than recommended will tend to grow taller but yield less corn. Corn hybrids range in maturity from approximately 80 days for the northern types, to 190 days for the southern types. Longer maturity hybrids usually yield more, so it is advantageous to plant the longest possible maturity that will not be killed by frost.

Corn is normally planted in the spring after the soil temperature has reached at least 50 degrees F at a 4 inch depth. Planting earlier will cause the seedlings to germinate much slower and make them more susceptible to seedling diseases. Corn should be planted 1 to 2 inches deep in rows spaced 20-40 inches apart. Seeding rate varies from 12,000 to 30,000 plants per acre. The lower rates are used where water is limited for yield, and the higher rates are used with irrigation.

Corn is most sensitive to heat and water stress during the 3 to 5 day period of pollination. Stress at this time will cause the greatest yield reduction. The period during which stress has the next greatest yield-reducing effect is the grain-filling period, immediately following pollination.

At the end of the grain-filling process, a black layer forms at the base of each kernel. This black layer is visible if a kernel is removed from the ear and cut in half lengthwise. When the black layer has formed, the kernel is at its maximum dry weight, and all that remains is for the kernel to dry down so that it can be stored. During the dry-down process, a dent forms in the end of the kernel.

Corn is harvested for silage after it has begun to dent when the entire plant is still at 60-70% moisture. The corn is finely chopped and packed into a pile, pit, or upright silo to exclude air. In the absence of air, fermentation takes place that causes the silage to become acidic and prevents further decomposition by molds. In the presence of air, silage will begin to mold again, so it is important to keep silage in airtight storage until it is to be used.

Corn grain can be harvested from the field when the kernels are at 28-30% moisture. When harvested at higher moisture contents, excessive kernel damage will occur, and below 28-30% moisture, field losses begin to increase. Corn grain can be stored in a silo at 28-30% moisture. Most corn grain is harvested and stored in grain bins. Grain must be less than 16% moisture not to mold while in storage bins. Farmers like to harvest as early as possible in the fall to reduce harvesting losses, but they must either dry the corn after harvesting, or leave the corn on the stalk until it is sufficiently dry for storage in the bin.

SORGHUM

Sorghum, which was developed under the harsh environmental conditions of Africa, is one of the few crops that can survive in nature. Indeed, sorghum is adapted to warmer, drier regions and does relatively better than corn in regions where periodic moisture stress occurs. In the U.S., it is therefore grown extensively on the perimeter of the corn belt as well as in dry, arid areas of the western United States. Sorghum also has greater resistance than corn to grasshopper, rootworm and other pests. The chief drawbacks to sorghum are a potential for serious bird damage, and lower feeding and market value than corn.

Sorghums are grown for grain and forage around the world, though different types have evolved in different regions. Among these are kafir, hegari, milo, feterita, durra, shallu, kaoliang, sorgo or sweet sorghum, and broomcorn. Most of the sorghum grown in the United States are hybrids that have been derived from crossing several of these groups. The hybrids grown in the United States can be generally divided into the grain types which grow 3-4 feet tall, and the forage types which grow 6 feet or taller.

Sorghum is usually planted when the soil temperature exceeds 60 degrees F. Depending on the region, this normally occurs between February and late June. It is normally planted in rows spaced between 20 and 40 inches apart. The narrower rows resulting in higher yields under irrigation or in higher rainfall areas. Wider row spacings (40 inches) are used in lighter rainfall areas or where furrow irrigation is practiced. Sorghum is seeded to a stand of 15,000-20,000 plants per acre under semi-arid conditions and 100,000-150,000 plants per acre under irrigation. The maturity of most sorghum hybrids grown in the United States ranges between 120-180 days with the longer season hybrids generally being higher yielding. There is the potential for a ratoon (regrowth) crop of sorghum if sufficient time exists between harvest of the first crop and a killing frost. This is most

Fig. 12-1. Sorghum was developed under the harsh environmental conditions of Africa. Although the nutritive value of grain sorghum is slightly less than corn, resistance to drouth, insect damage and disease is superior. Therefore, in many regions sorghum is selected in preference to corn.

likely to occur when the sorghum is harvested for silage, and/or in the most southern parts of the United States.

If sorghum is grown for silage, tall forage types are planted to increase the tonnage per acre. The crop should be finely chopped when the grain is in the early dough stage and stored under similar conditions to corn silage. If the sorghum is to be harvested for grain, shorter types are generally planted to reduce the head height and the amount of plant material that must be run through the combine. The shorter types are less likely to lodge than the taller types. Grain is mature when the seeds are fully colored and have begun to harden. However, grain should not be combined for storage until the moisture content is 13% or lower. Otherwise the farmer will have to artificially dry the grain after harvesting.

SMALL GRAIN PRODUCTION

Production of all small grains will be considered together because of the similarity in practices. Small grains consist of wheat, barley, oats, and rye. All have similar origins, having been developed from wild-growing species in the Eurasian area of Mesopotamia. However, also included in the small grains group would be triticale, which is a modern hybrid of wheat and rye. Small grains are adapted to most soil types but prefer a pH range of 6-7.

Small grains are much different than most other crops in that they can grow under cooler temperatures, and are otherwise much more cold tolerant. Small grains have a temperature optimum of about 70 degrees F (39 degrees C). However, small grains can still continue to grow with temperatures in the 50's, and many varieties can survive temperatures well below freezing. This allows small grains to be grown in areas that otherwise could not grow a grain crop. Thus, some areas in the northern U.S. and Canada that would otherwise not be very productive agriculturally, are major grain (wheat) producing areas. In more southerly latitudes, the cold hardiness of small grains allows crops to be grown during the winter. In that case, cold tolerance is referred to as winter hardiness. The order of winter hardiness of small grains is: rye > wheat > barley > oats.

Another very important and useful aspect of small grains is that genetically they are grasses. As discussed in the introduction, the first small grains cultivated were derived from seeds collected from wild-growing grasses. Thus, small grains not only provide grain, but in many cases are also used to provide grazing for livestock.

Indeed, small grains are particularly valuable for grazing since they will actively grow during the cooler times of the year. That is, small grains can be used to provide high quality grazing when most other grass or forage crops are dormant. The most common practice in much of the southern U.S. is to graze wheat crops during the winter and early spring. As mid-spring approaches, the cattle (and in some cases sheep) are removed to allow the plant to mature, so that a grain crop may be cut in early summer. These winter types of small grain may be grazed until just prior to when the seed head approaches the soil surface, with little, if any loss of crop yield.

Small grains may also be used as hay or silage crops. There are however, some special considerations when using small grains for these purposes. Small grains mature over a much shorter period of time than other forage crops, and therefore time of harvest is much more critical.

Ideally, small grains harvested for forage should be cut at the mid-dough stage of grain formation. At this point in time a maximum tonnage of vegetative matter can be harvested, while the plant is still green and immature. If small grains are allowed to go past the mid-dough stage, however, the vegetative portion of the plants matures very quickly. Crude protein can decline from 12-15%, to as low as 5-8%, with similar reductions

Fig. 12-2. Wheat grain beginning to mature.

Fig. 12-3 & 4. Cereal grain plants such as wheat are grasses. As grasses they are very tolerant of grazing, and in the more southerly climates are often used for that purpose. When grazing animals are removed prior to emergence of the spikelet, a grain crop may be cut with very little reduction in yield. In the most common situation, cattle or sheep are grazed on cereal pasture during late fall and winter. They are removed in early spring, and a grain crop is cut 3 or 4 months later. In some cases (usually due to government programs) animals are left in the fields to graze cereals out completely. (Upper photo shows cattle in an immature wheat field during winter, and the lower photo depicts cattle "grazing out" maturing wheat during late spring.)

in overall digestibility. The thing to keep in mind is that this process can occur over only a few days.

When harvesting most varieties of wheat as a forage crop, timing becomes even more crucial, due to the formation of awns or beards (see Fig. 12-2). These awns or beards can invade the salivary glands of the consuming animals and cause infections. Therefore, with most varieties of wheat, cutting the plant before it matures takes on added significance.

Planting Small Grains. Winter varieties of small grains are normally planted in October and November, to allow establishment during the fall. If the intention is to graze the crop, planting usually begins earlier than that; September, and even August. In areas where the winter is too severe for survival of small grains, they may be planted in the spring when the soil temperature is about 40 degrees F or higher. Winter types must not be used for spring planting because most require some degree of vernalization (exposure to a certain amount of cold weather before flowering will occur), and therefore will not produce grain if seeded in the spring.

The crop is usually seeded with a grain drill in rows spaced from 6-12 inches apart. As with other crops, the spacing between rows is primarily determined by available moisture and/or irrigation. In heavy rainfall areas, or under sprinkler irrigation, the close 6 inch row spacing is used. In lower rainfall areas, or under furrow irrigation, the wider 12 inch spacings are usually used.

Seeding rates vary substantially with the specific location. Areas of high rainfall and humidity, such as the East Coast of the U.S. typically require higher rates. This is because humidity increases plant diseases and leads to more seedling mortality. In the eastern United States, the seeding rate for wheat, barley, triticale, and rye is usually 1 ½ bu./A for grain and 1 ½ to 2 bu./A for grazing. Oats are seeded anywhere from 1 to 3 bu./A.

In the western portion of the U.S., wheat and barley are seeded at ⅓ to ½ bu./A under dry land conditions and at 1 bu./A under irrigation when grown for grain only. Oats and triticale are seeded at 1 to 2 bu./A. The seeding rate for grazing is generally 1 ½ bu./A.

The combining of the grain takes place when the moisture content of the grain is approximately 13% - low enough to be stored without molding. As a general rule all of the small grains are sometimes used for both human and animal consumption. The following is a short synopsis of use:

Oats. The majority of oats grown are used as animal feed, horse feed in particular. Oats are more palatable than any other feed grain, and thus are very popular with horses and horse owners. Oats are also much lower in energy than the other feed grains, and therefore horses are much less likely to founder on oats than other feed grains (another reason for the popularity of oats).

Oats generally have a much lower yield per acre than other small grains, but their high market value tends to offset the lower yield. As a general rule, oats have a market value of 25 to 40% higher than other feed grains.

Barley. In some countries, barley is a very important feed grain. Barley is particularly adapted to cool, moist, maritime climates, and thus is a very important crop in the U.K. (Britain, Ireland, and Scotland), and the Pacific Coast of the U.S.

Barley has a thick outer husk or seed coat which makes it relatively high in fiber. Occasionally the husk is removed and sold as "pearl barley" for human consumption. There are also special varieties high in maltose (a sugar), known as malting varieties, which are used in the brewing of alcoholic beverages.

Rye. Rye grain has a very low yield compared to other small grains, and a somewhat bitter flavor. Therefore, only a relatively small acreage of rye is harvested for grain (either for human or animal use).

Technically a great deal of rye is grown worldwide for pasture. However, most of the rye used for pasture is a perennial variety of rye, whereas the varieties used for grain are annuals.

Wheat. For purposes of human consumption, wheat is probably the single most important grain or commodity. Worldwide there are six major types of wheat grown. These types are: Hard Red Spring, Durum, Red Durum, Hard Red Winter, Soft Red Winter, and White. Hard Red Spring wheat is grown in the North Central United States and Canada, Russia, and Poland; areas where the winter is too severe for winter wheat production. It is the standard wheat for bread flour.

Durum wheat is also a spring wheat (planted in the spring) grown primarily in the Red River Valley of Minnesota and North and South Dakota, as well as North Africa, Southern Europe, and the Soviet Union. It is used primarily for making macaroni, spaghetti, and similar products.

Hard Red Winter wheat is grown mainly in the central and southern Great Plains of the United States, the southern part of the Soviet Union, the Danube Valley of Europe, and Argentina. For bread making, only the best Hard Red Winter wheats are equal in quality to the Hard Red Spring wheats.

Soft Red Winter wheats are grown primarily in the eastern United States and western Europe. This wheat is softer in texture and lower in protein than the hard wheats, and is used to make cakes, biscuits, crackers, pastries, and other similar items.

White wheats are grown in the far western states and Northeastern United States, as well as in Northern, Eastern, and Southern Europe, and in Australia, South Africa, Western South America, and Asia. The flour of the white wheats is used mainly for pastries and breakfast cereals.

Within these major categories of grain varieties, there are also sub-varieties developed for forage or grazing purposes. These are varieties that withstand grazing quite well, and still produce a good grain crop.

SOYBEANS

Soybeans, which were probably developed from wild legumes growing in Asia, are adapted to similar climatic and soil conditions as corn. Soybeans will grow in most soil types but are best adapted to loamy soils. Soybeans are a legume, which means that the plant has rhizobium bacteria attached to the roots that take nitrogen from the air, and supply it to the soybean plant. This means that nitrogen fertilization is not required after establishment.

Soybeans are divided into several maturity groups for the United States. Varieties grown north of their area of adaptation tend to be killed by frost before maturity; and varieties grown south of their areas of adaptation will tend to mature too early for optimum yields. However, earlier maturing types may be used effectively in double cropping systems where two crops are produced in one year.

Soybeans are planted in the spring when the soil temperature has reached 50-55 degrees F. Seed must be inoculated with rhizobia bacteria before planting. This is done prior to planting by adding a sticking agent to the

Fig. 12-5. A field of soybeans in the San Joaquin Valley of California.

seed, and then mixing the seed with inoculant powder to coat the exterior of the seed. The rhizobia bacteria will attach themselves to the roots of soybeans after the plants begin to grow. Soybeans are planted at a depth of 1 to 2 inches in 20-40 inch rows. The seed is normally planted with a corn or cotton planter. However, a grain drill may be used, particularly if the soybeans are grown for hay.

A very small acreage of soybeans is planted for hay each year, but additional acreage is harvested for hay when a poor bean crop results from unfavorable weather conditions. Soybeans can be cut for hay anytime from pod formation up until the leaves begin to fall, but the best quality hay is obtained just after the seeds have begun to develop. Soybeans are harvested for grain (bean crop) with a combine when the moisture content of the seeds drop to 12%. Later harvesting increases pod shattering loss in the field, as well as splitting of the beans. Earlier combining can also result in field losses due to poor threshing.

COTTON

Originating in the Nile River Valley of North Africa, cotton was for many centuries the primary source of summer clothing fiber. Today, although many manmade fibers have been developed, cotton remains a very popular fiber/cloth and is a major commodity in world markets.

Cotton is grown in areas that have a growing season of 180-200 days and a mean temperature (for the summer months) of not less than 77 degrees F. Cotton is grown in semi-arid regions, both dry land and with irrigation, and in more humid areas where rainfall exceeds 60 inches annually. Cotton grows best in moderately fertile

Fig. 12-6. Long staple cotton grown in the Gila River Valley of Arizona.

soils, with a wide range of acidity (pH 5.2 to pH 8+). Cotton tolerates more salinity than most crop plants.

Cotton is planted in the spring when soil temperature exceeds 60 degrees F. Delinted seed is planted 1 to 2 inches deep at the rate of 8-15 pounds per acre. Most cotton is planted in rows spaced 36-40 inches apart.

Because cotton flowers for a longer time span, it is not as sensitive to water stress as many other crops. Water stress during early stages will have less effect on yield than water stress occurring later in the season. However, stress occurring early may cause plants to become very short and cause difficulty in harvesting.

Harvesting occurs in the fall after the boles have opened. Because green leaves will stain the lint and reduce the value of the crop, cotton is first sprayed with a desiccant or defoliant to kill the leaves, unless they have previously been killed by frost. Then mechanical pickers are driven over the field to remove the lint and seed from the plant. The cotton lint and seeds are then dumped into a wagon or compressed into a module and taken to the gin. At the gin, the seeds are removed and the lint is cleaned of trash and baled. The seed is used as a protein and oil source.

SUMMARY

The preceding has been a brief description of the general aspects and requirements of major world crops (with the exception of rice, which is covered in the chapter on Tropical Agriculture). Volumes could be written about each individual crop, depending upon the detail one wishes to consider. For the purposes of this text, two general concepts have been stressed: 1. the adaptation and application of the particular crop and, 2. the variation within each crop.

With respect to variation within crops, the student should realize that corn is not corn, and wheat is not wheat, etc. Using corn as an example, there are, of course, various types such as dent corn, sweet corn, popcorn, etc. But more than that, there are specific strains within types that have been developed for particular areas. As explained in the text, date of maturity is the big variable. In areas with shorter growing seasons, short maturing types will be required, and in warmer, longer growing season areas, late maturing types will be more appropriate (yield more grain). Other special adaptations, such as resistance to specific insect pests and diseases, must also be considered.

Unfortunately the decision as to which crop to plant cannot be derived from a textbook. This is something that must be discussed and considered with agronomists familiar within a specific farming area. The purpose of this chapter has therefore been to simply supply the student with a basic background of information, so as to enable him or her to intelligently gather specific local information concerning crop planting decisions.

CHAPTER 13   SEEDS AND SEEDING
by Charles R. Glover, Ph.D.
New Mexico State University

Webster defines a seed as: "the part of a flowering plant that contains the embryo and will develop into a new plant if sown". He continues to say that, in the broad sense, a seed is any propagative portion of the plant that will give rise to another plant. Examples of this are the true seeds, seed-like fruits, tubers, bulbs, rhizomes and stolons.

Plants are divided into two major groups, monocotyledons and dicotyledons. Seeds that contain two cotyledons are classified as dicotyledons. (In layman's terms this means that the seed consists of two equal halves that will split apart.) Monocotyledons are typified by plants that have seeds with a single cotyledon. Cotyledons, or seed leaves, are the first leaves to grow out of the seed.

Grass plants are typical examples of the monocotyledons. Crops such as corn, sorghum, wheat, oats and barley are examples of grass crops. Legumes are normally dicotyledons. Examples would be: beans, peas, and soybeans.

Monocotyledon seeds are made up of the germ (embryo), endosperm, and seed coat. The endosperm is primarily starch, stored as food material surrounding the embryo. Figure 13-1 illustrates the typical monocot seed. The plumule, the top part of the embryo, develops into the top growth and the permanent root. The lower part of the embryo, the radicle, forms the temporary roots which function until the permanent root is formed. The scutellum, or cotyledon, is the part of the embryo that is next to the endosperm. Through enzymatic activity, the scutellum dissolves or digests the food materials in the endosperm.

The cotyledons form the bulk of the seed. In dicotyledons there is no endosperm present. The embryo is located between the two cotyledons (Figure 13-2). Upon germination, the radicle forms the permanent roots and the plumule develops into a stem and, in most species, pushes the cotyledons upward above the surface of the soil. The cotyledons furnish their stored food to the developing plant and eventually drop off.

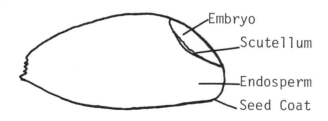

Figure 13-1. A typical monocotyledon seed.

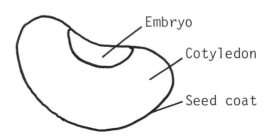

Figure 13-2. Representative Dicotyledon Seed.

OTHER FORMS OF SEEDS

As mentioned, variations of the true seed may also include vegetative parts such as tubers, bulbs, rhizomes and stolons. Some plants may be propagated by cuttings of stems which give rise to roots and shoots which develop into another plant.

Tubers are not roots; rather, tubers are short, thickened, underground stems. Tubers are capable of producing plants much as a seed would. The most common tuber crops we encounter are potatoes and the Jerusalem artichoke.

Potatoes are seeded by planting small tubers or by cutting larger tubers and planting the pieces. These pieces are usually cut in blocks of about two inches and weigh about two ounces. Seed pieces need to have from one to three eyes.

Bulbs are underground leaf buds. Ornamental flowers such as the dahlia and the crocus, are examples of plants that reproduce through bulbs.

Onions are a bulbous crop also. Although they usually are propagated by seeding, some farmers transplant small bulbs rather than direct seeding. Onion seeds take a relatively long time to germinate, and transplanting can be advantageous if time is short.

Rhizomes are underground stems that give rise to new plants. This usually occurs at what are known as the nodes. Some of the most difficult weeds are propagated by rhizomes; examples are Johnsongrass and Quackgrass. The growth of underground rhizomes into new plants makes removal by cultivation or even by hand pulling very difficult.

Stolons are another form of modified stem. These stems run along the surface or just below the surface of the soil. Like rhizomes, stolons have the capacity to produce new plants at the nodes where roots and shoots develop. Stolons are the most common means of propagating hybrid Bermudagrass.

Propagation from cuttings of stems and other plant parts is a common practice with some ornamentals.

Plants established from cuttings are genetically identical to the plant from which the cuttings came. For this reason, cuttings are often used in research, or in the development of special varieties. That is, plant breeders often use this method to establish a large population of genetically identical plants. This is helpful when only one plant has a desirable characteristic that they want incorporated into a line or variety.

VARIETY DEVELOPMENT

The ultimate potential of a variety is determined by the genetic material found in the seed. Many factors influence the performance of a variety, but the genetic potential determines the limits of that performance.

Genetic improvement within a crop or variety may be accomplished in several ways. Processes which have provided genetic improvement include natural selection, artificial selection and hybridization.

Natural selection has been in effect as long as plants have been in existence. Different strains grown near one another inevitably cross-pollinate and provide a source of variation. This variation allows the process of natural selection to occur. That is, the hardier, better adapted plants survive and the weaker plants are eventually replaced.

Man observed differences in plants and began to save seed from those plants that exhibited superior performance or had other desirable characteristics. As time went by, he began to cause desirable plants to intercross and expand the genetic variability. Through this process of artificial selection, he was able to select plants that were more productive, but not necessarily better adapted. (That is, in natural selection, only hardiness or adaptability is selected. Productivity is not usually a trait that enhances hardiness. Indeed, productivity is often a detriment, since more plant nutrients are often required to produce more grain, fruit, etc.)

In developing this system of artificial selection, the concept of hybridization came to light. Early man learned that while he could cross two different varieties, and sometimes produce a superior third plant, the seed from the third plant would many times not result in the same kind of plant. The reason is that the genes were not fixed. Therefore, the two parent varieties had to be grown to produce seed for the third, more desirable plant. The farmer could not simply hold back some of his grain (etc.) for use as seed (the next year).

Ultimately, directed hybridization has become one of the most effective means of increasing the productive capacity of many crops today. Corn and grain sorghum are excellent examples of the effect hybridization has had on crop production. Within a short period of time, varieties went from totally open pollination, to virtually all hybrids.

SEED QUALITY

Characteristics such as genetic purity and germination are basic requirements for good seed. External factors such as contamination with other seeds, foreign material and disease organisms can drastically reduce the quality of the seeds. Good seed is essential for successful crop production while poor seed can result in disaster.

SEED GERMINATION

Several factors, both internal and external, affect the seed's germination. The external factors include moisture, oxygen, temperature, and in some cases, light. Other factors such as broken seed, seed maturity, seed size, and dormancy also influence germination.

MOISTURE

Seeds require abundant moisture to germinate properly. Most field crop seeds begin to germinate with a moisture content range of about 25 to 75 percent, depending upon the type of seeds. For maximum germination it is important that once seed is exposed to moisture, subsequent moisture is not restricted. That is, under dryland agriculture there is always the danger that once the seed is planted, rainfall will be sufficient to allow the seed to swell and only partially germinate (sprout). Some seeds may re-sprout several times, but germination is lower after each sprouting.

OXYGEN

Oxygen is necessary for germination and yet some seeds have difficulty absorbing gases such as oxygen. However, when seeds absorb moisture, this apparently enables them to absorb oxygen as well. But too much moisture prevents adequate oxygen from entering the soil, and can thereby reduce germination. Therefore, oxygen absorption may be inhibited by planting seeds too deep or in too wet a soil. Rice is a notable exception, since it will germinate while completely covered with water.

TEMPERATURE

Germination of crop seeds occurs across a wide range of temperatures. Some cool season crops, such as winter wheat, may germinate at temperatures near the freezing point. Some warm season crops, such as sorghum, may germinate at temperatures near 120 degrees F. However, these are the extremes, and the percent which germinate will be greater at more moderate temperatures.

Obviously, warm season crops should be planted in the late spring or early summer; cool season crops in the fall, late summer, or, for some crops early spring. The peculiarity of each crop will dictate the exact time.

Figure 13-3 A common grain "drill" used for seeding crops. The seed is loaded into the hopper at the top and is automatically fed down into the individual "drills" for deposition into the ground. Adjustments are made internally for the various sizes of seeds, as well as the actual seeding rate. The lower drills are also adjustable for the depth of seeding.

As a practical matter, however, the availability of moisture is more critical than temperature. Therefore, under dryland farming planting to ensure adequate moisture is much more important than temperature. Under irrigation, simple management of water during planting is all that is required.

LIGHT

In some cases light is necessary for crop seeds to germinate. When light is necessary, the requirement is quite low, and as a practical matter not a serious consideration. Once the seeds have germinated, however, the need for light increases drastically.

BROKEN SEED

Some seeds will still germinate if the seeds have been broken. However, the germination is greatly reduced when the germ or embryo is injured. There will also be higher seedling mortality and the subsequent plants will be smaller than those produced by whole seeds. Breaks or cracks in the seed coat may also reduce or restrict germination. This is particularly true if mold develops in the cracks. Treatment with fungicides is helpful in giving protection, but cracked seeds are still much more susceptible to molds, diseases, and other infections.

SEED MATURITY

Immature seeds will often germinate, but performance and seedling development will be reduced. This is particularly true when conditions during germination are less than optimum. Immature seeds normally produce smaller sprouts, and are more susceptible to disease infection and frost injury.

SEED SIZE

Seed size is an important factor in germination of both large and small seeded crops. Seedling size is directly proportional to seed weight. Small seeds produce small seedlings. Large seeds produce larger, more vigorous seedlings that are better able to survive adverse conditions.

SEED DORMANCY

Some seeds will fail to germinate immediately after maturity, even when conditions are favorable for germination. These seeds are said to exhibit dormancy. Factors which may cause, or induce, dormancy in seeds are the seed coat, embryo maturity, germination inhibitors, and/or temperature.

Hard or exceptionally thick seed coats can prevent moisture absorption. Likewise, oxygen absorption may be curtailed. If the embryo is not fully developed when the seed is harvested, dormancy may result. High temperatures during the maturation stage of some crop seeds may tend to induce dormancy. In some cases germination inhibitors may be present. In those cases the seeds must undergo chemical changes before germination can occur.

Dormancy may be broken naturally or, in some cases, artificially. Natural means of breaking dormancy may be by aging and/or abrupt temperature changes. Dormancy in seeds with hard or thick seed coats may be broken by roughing or softening the seed coat. This is known as scarification. It may be accomplished through either mechanical or chemical means. (It should be pointed out, that scarification also leaves the seed more susceptible to diseases, fungus, and mold.)

STORAGE OF SEEDS

Proper storage of seeds is vital to ensure maximum germination when planted. The basic rule is to store seeds under dry, cool conditions. In addition, containers that prohibit contact with oxygen are highly desirable. This is particularly true in warm, high humidity areas. Indeed, in hot, humid climates it is imperative that seed be sealed in airtight containers.

A vitally important factor is proper drying of the seed. For maximum storage life, seeds should be dried to below 8% moisture. Since normal dry moisture contents of most seeds run 10-15%, this means that artificial drying is required. However, great care must be used in artificial drying, as applying heat too rapidly can damage the germ or embryo. If possible, it is best to air dry only (using fans without artificial heat). Unfortunately, in many climates the humidity in the air is too high and artificial heat is necessary. It should therefore be applied very slowly to keep from overheating and possibly damaging the seed.

There is no set rule as to how long seed may be stored. Germination after storage is highly dependent upon storage conditions, as well as the type of seed. Weed seeds have been known to germinate after 50 years of storage. However, most crop seeds will have a significantly reduced germination after 10 years of storage. Under less than ideal conditions, reduced germination may result after just 5 years of storage.

SEED CERTIFICATION

Each state has an agency that is responsible for the certification of seeds. This may be an agency of the state or an agency designated by the state. Many times this agency is a seed or crop improvement association. The seed certifying agency is responsible for inspecting and testing seed lots to insure they meet the qualifications of certification as set down by the Federal and State Seed Laws, and any additional regulations and requirements of the agency itself.

All classes of certified seed originate from *Breeder Seed*; the seed directly controlled by the originator. The originator may be an individual, institution or firm. This seed is provided to the certifying agency for increase and continuation of the variety. Breeder Seed is identified by a white tag with BREEDER SEED printed on it.

The first generation of seed away from Breeder Seed is *Foundation Seed*. Foundation Seed is that seed which most closely maintains the genetic identity and purity of the variety or strain. This is the source of seed for all classes of certified seed. The identifying tag on Foundation Seed is a white tag with FOUNDATION SEED printed on it.

The next generation of seed is classified as *Registered Seed*. Registered Seed is the second generation from Breeder Seed and is handled in such a manner to maintain the genetic purity and identity of the variety. Registered Seed has a light purple or lavender tag printed with REGISTERED SEED.

The last class of certified seed is *Certified Seed*. It may be produced from Foundation Seed, Registered Seed, and in some cases from Breeder Seed. To be classed as Certified Seed it must be planted, grown, harvested, and handled consistent with the specifications of the certifying agency. Certified Seed is tagged with a blue tag printed with CERTIFIED SEED.

All seed certified throughout the United States and Canada conforms to the tagging system. The tag color and class of certified seed is standard throughout. Also included on certification tags is the identity of the seed certifying agency, the kind of seed, variety and lot number. The back side of the tag may contain the germination, analysis, producer, etc. This information may be on another tag, but must be attached to the bag before it can be legally sold.

Vegetative materials such as stolons, tubers, etc. can be certified as well. Standards for certification must be met with these planting stocks just as with true seeds. Fields must be inspected for contamination and for genetic purity before the materials can be certified.

SEED SOURCES

Planting Stocks, both seeds and vegetative materials,

may come from a number of sources. They may be grown by the individual for his own use, or bought from a neighbor, seed association, seed dealer or commercial seed company.

Seeds bought from a neighbor can be good seed if care is taken in growing, harvesting and handling the seeds. Many times this is not the case and poorer quality seeds result. The price may be lower, but the farmer may find that this seed lacks the quality and purity needed for production of a good crop. In cases such as this, it is usually more costly to use the cheaper seed.

Seed associations and their certified growers usually offer an excellent source of quality seeds. Seed obtained from this source normally has gone through the certification process and the variety is identified. The seeds meet the standards set up by the seed certifying agency and seed law.

Seed dealers usually carry a line of good quality planting seed. However, they may also carry seeds of poor quality. It is essential that the farmer is familiar with the quality of the seeds that a dealer handles. (Even if the quality of the seed is good, it may be that varieties are not well-suited for the area.)

Seed obtained from commercial seed companies may or may not have been subjected to the certification process. However, reputable seed companies impose strict quality control of the seed they sell. Since the quality of the seeds reflect the reputation of the company, the quality is normally high.

SEED LAWS AND REGULATIONS

FEDERAL SEED ACT

The Federal Seed Act was passed in 1912 in an effort to protect the American farmer from low quality, contaminated seeds. It was amended in 1916 and again in 1926. Another Federal Seed Act was passed in 1939 which provided authority to penalize for mislabeling.

STATE SEED LAWS

All states have established seed laws to regulate the quality of seeds sold in the state and protect the interests of their farmers. This legislation has been an important means of reducing the spread of noxious weeds.

Enforcement of the states' seed laws is generally the responsibility of the State Department of Agriculture. Field inspectors regularly draw samples of seeds offered for sale to see if they are labeled properly. When a question arises, a "stop sale" is enforced until the question has been resolved. If a violation occurs, a penalty may be imposed.

SEEDING THE CROP

Seeding the crop requires specific tools or implements. Planting implements may be used to seed a hill, a single row or many rows. Some implements may be pushed by an individual, pulled by animals or by a tractor. Under systems where larger implements are used, the most common seeders used today are gain drills and row planters.

The method of planting varies with the type of operation, the type of crop and personal preference. Dryland conditions may dictate one form of operation, while irrigation another. Some crops are more adapted to broadcast plantings, some to row planting, and still others may adapt to either situation.

Planting equipment may require planting on a flat surface while others may require a lister bed. Specific conditions may dictate planting on the top of a bed, others on the side of the bed, or still others in the furrow.

VEGETATIVE PROPAGATION

Sprigging. Some hybrids fail to produce seeds or the seeds are unsatisfactory for establishing the crop. In such cases, the crop is propagated by vegetative methods. The most common method of vegetative propagation of field crops is by rhizomes and stolons.

More Bermuda grass is propagated by planting sprigs (rhizomes) than by seeding. Some varieties cannot be seeded, and therefore, there has been development of specialized machinery and methods to establish bermuda grass by sprigging.

One method that has had a good deal of success involves spreading the rhizomes and stolons and then disking. Methods of spreading the rhizomes and stolons include manure spreaders and machines adapted from old automobile parts. Once spread, the material is immediately covered by a tandem disk harrow followed by a cultipacker. Smaller areas may be sprigged by scattering by hand and roto-tilling.

Transplanting. Another form of vegetative propagation is transplanting. Crops that are especially susceptible to seedling damage from disease, insect damage, etc., are sometimes started in sterile beds and then transplanted.

Sodding. Sod production has become important in the turf grass industry. Golf courses, parks, athletic fields and home lawns are often established by sodding.

Sod is produced on sod farms either by seeding or sprigging. It is then removed by a machine that cuts a short distance below the surface and removes the sod as a roll. These rolls of sod are transported to the nursery or the area where it is to be laid. The rolls are laid out to form a solid turf without the time and effort it would otherwise take to establish a good turf.

Once the sod has been removed from the field, the area is re-seeded with the same variety to insure rapid regrowth of more sod. Vegetatively propagated turf grass such as Bermuda grass will re-establish from the roots remaining in the soil.

PLANTING TIME

Planting time is determined by the type of crop, the environmental conditions and the farm operation. Another factor that may come into play is planting to best offset potential insect and disease infestations. By planting at the proper time, the farmer may be able to establish the crop and allow it to develop while evading the peak occurrence of a pest.

Cool season crops such as small grains (wheat, oats, barley and rye) are planted early to allow maximum development before high temperatures restrict growth and development. Small grains are often used both for forage and grain production. When grown for forage, small grains are often planted somewhat earlier than when grown for grain only.

Warm season crops require a warm soil to properly germinate and develop. In many instances the planting time may be determined by calendar date. However, since the climatic conditions may vary from year to year, calendar date may not be completely satisfactory to determine the best planting date. A combination of calendar date, soil temperature, and long range weather forecasts may all be used to determine the best planting date for a particular operation.

SEEDING RATE

Seeding rates vary with seed size, type of crop, use of the crop, environmental conditions and even variety. Ultimately, the seeding rate will depend on the desired stand. Determination of the desired stand is influenced by moisture, row spacing and crop usage.

Tillage equipment may be designed for broadcast, narrow rows, conventional rows or wide rows. This factor plus the moisture availability can dictate a wide range of seeding rates.

The use of the crop may create a variation in seeding rates also. For example, small grains grown for forage are usually planted at higher rates than when grown for grain production. Likewise, corn grown for silage is planted at higher populations that grain corn.

PLANTING DEPTH

Several factors affect the optimum planting depth for crop seeds. Seed size, seed type, soil type and soil temperature all influence the planting depth. Planting into dry or moist soil also influences the depth of planting.

As a general rule, small seeds are planted at shallower depths than large seeds. Small seeded legumes and grasses are generally planted at depths of ¼ to ½ inches. The larger seeded crops may be planted at depths of 1 to 5 inches.

Seeds generally can be planted deeper in sandy soils than in clay soils. Seedlings can emerge through the sandy soils more readily than in clayey soils, and consequently, can emerge from greater depths.

Warm soil favors better emergence than cool soils as seedlings will generally emerge from greater depths in warm soil. If the soil is dry, oftentimes seeds are planted deeper in an effort to reach moist soil. Cool, heavy soils may result in poor stands when the seed is planted deeper than usual (as a means of reaching moisture). When irrigation is available or rainfall is expected soon, it may be best to seed at a shallower depth and wait for rain or irrigation.

SUMMARY

The planting seed is the key to good crop production. One should always select good quality seeds of a genetically superior variety that is adapted to the area. The variety should have resistance to as many of the prevalent insects and diseases as possible. Farmers should not try to cut their costs at the expense of using inferior varieties or otherwise poor quality seeds.

Once the specific variety and source of seed has been selected, decisions concerning seeding practices must be made. Actual time of planting, seeding rates, depth of seeding, equipment and tillage practices to be used will vary from area to area. Climate and soil type will be the two most important factors to consider. Since there is so much potential local variation, a textbook cannot supply enough information to enable farmers in different areas to make these decisions. Therefore, outside advice must be sought from sources familiar with the specific local area (extension services, consultants, other farmers, etc.).

CHAPTER 14 TREE FRUIT AND NUT PRODUCTION

By James H. La Rue
Univ. of California-Davis

Edible tree fruits and nuts are grown throughout the world north and south of the equator to about 60 degrees latitude. Severe winters restrict growth areas closer to the north and south poles where fruit production is nonexistent.

CLIMATIC ZONES

Within the wide belt of fruit and nut growing areas, both north and south of the equator, lie three separate regions.

TROPICAL REGION

The tropical region, the area nearest the equator, is characterized by high temperatures, rainfall, and humidity year-round with no frost. Many fruits are native to, and grown almost exclusively in, the tropics. The best known fruits are the papaya, mango, litchi, banana, pineapple, guava and cherimoya. Nuts include cashew, macadamia, Brazil nut, coffee and cocoa. In the continental United States, only the area of south Florida is climatically suited for growing tropical fruits. The Hawaiian Islands commercially produce papaya, pineapple and macadamia nuts.

Most tropical fruit and nut trees are evergreen. They do not have a chilling requirement, i.e., a "rest" period during the cooler winter months before growth can begin again in the spring. Tropical trees cannot tolerate frost, and temperatures of even 1 degree C below freezing can severely damage or kill leaves, limbs and fruit. In the tropics, the growing season lasts almost all year around.

SUBTROPICAL REGION

The next climatic belt, north and south of the tropical region is the subtropical area. Frost is seldom a problem in this area, but can occur frequently enough to prevent most tropical fruits and nuts from growing there. Many subtropical areas are characterized by low rainfall and humidity, while others may resemble the tropics in respect to high rainfall (and humidity). Citrus is the most commercially important fruit grown in most subtropical regions. This would be most true for the United States. Florida, parts of California, and the southern tier of states have subtropical areas.

Most subtropical fruit and nut trees are evergreen and require no winter chilling. In areas where occasional frosts occur, the use of orchard heaters or wind machines is essential to protect citrus from temperatures below -2 degrees C (28 F).

TEMPERATE REGION

The mild north and south temperate zones extend beyond the subtropical areas, and are closer to the areas of severe winters where fruit and nuts are not grown. Parts of the temperate zones blend into the subtropical areas or are moderated in their climate by their proximity to oceans or other large bodies of water.

There are four distinct seasons in the temperate zones. Summer temperatures are warm for tree growth and fruit production season. Winters are cold enough to provide a period of dormancy during which trees do not grow.

Most fruit and nut trees grown in temperate zones are deciduous; that is, they lose their leaves in the winter. Most all deciduous trees need a rest period or chilling requirement during the winter. Chilling requirements are usually expressed in terms of cold hours. Most deciduous fruit trees require between 500-1000 hours below 7 C (45F) during the dormant season.

Some fruit and nut species flourish in climatic zones other than those in which they are widely grown commercially. For example, citrus is usually associated with the subtropical zone but is also commonly grown in tropical regions of the world.

China is the country of origin for a large number of fruit and nut species grown in the United States, including citrus, peaches, nectarines and apricots. Avocados come from Mexico and Central America, plums from China and Europe, pears from southern Asia, olives from Europe, cherries from the region between the Black and Caspian Seas, English walnuts from ancient Persia, and almonds from western Asia. The most widely grown temperate zone fruit species in the United States is the apple, which is believed to be native to western Asia. The pecan, native to the south central United States is the most extensively planted nut species in the U.S. It is one of the very few important nut species grown for commercial purposes in the U.S. that originated in this country.

NURSERY PRODUCTION

The ability to reproduce a cultivated variety (commonly called "cultivar") is probably the single most important factor in the commercial fruit business. Most fruit and nut trees do not come "true to type" when grown from seed. Thus, most must be asexually propagated (grown by budding, grafting or from cuttings) in order to reproduce exactly the same as the parent tree. Superior cultivars have been developed over many years and are constantly being improved upon by plant breeders. Fruit cultivars are selected for their color, appearance, flavor, size, productivity, storage, and handling properties.

The cultivar is usually "grafted" or "budded" to a "rootstock". The rootstock is a seedling or cutting of a particular species that has certain desirable characteristics making it a suitable root upon which to grow cultivars of fruits and nuts. Most rootstocks are resistant to soil-borne diseases, impart dwarfing or vigorous growth to the top, and are close enough genetically to be compatible with a large number of cultivars. For example, peaches are grown on peach rootstocks, plums are grown on plum or peach rootstocks, and apples are grown on a wide selection of apple rootstocks with a wide range of vigor for size control. Citrus is also grown on a wide range of citrus seedling rootstocks, depending upon the species (orange, lemon, grapefruit, etc.), growing conditions, and soil types. Commercial nurserymen must be highly skilled propagators to provide good quality trees. They must depend upon the right rootstock and cultivar combination to yield true to type fruit for commercial growers.

Grafting is a process in which a short stick (called a scion) of the desired cultivar is placed on the rootstock, which has been cut off just above the soil level. The scion and rootstock must have their cambium layers crossing (see figure), and held tightly together and protected by wrapping or grafting wax to prevent drying. Grafting is usually done in the dormant season.

Budding is a process in which a single bud is removed from the desired cultivar and inserted in a slit cut in the bark or the rootstock. The bud is usually taken from the base of a leaf from the current season's growth. Budding is usually done in the summer.

Most evergreen fruit trees are dug from the nursery as a "balled" or container-grown tree. Since evergreen nursery trees have active leaves attached, they need moist soil around their roots to prevent desiccation while being transported from the nursery.

Most deciduous fruit trees are planted as "bare root" nursery stock. Trees are dug from the nursery during the dormant season, with roots exposed, and then transported in bundles to the orchard where they are planted while still in the dormant condition.

ORCHARD ESTABLISHMENT AND TRAINING

Commercial fruit growing is a highly skilled farming business. Orchard site selection is one of the most important decisions involved in the establishment of an orchard. A frost-free or low-frost risk area is essential for those fruits or nuts sensitive to low temperatures. Good soil drainage, water supply for irrigation, freedom from lime deposits or alkali salts, and other undesirable characteristics are essential for high production.

Both establishment and production costs are high for commercially produced fruit and nut crops, particularly those requiring high labor inputs such as pruning, thinning, and harvesting. Land preparation, irrigation systems, and trees amount to a sizeable investment for a crop that requires several years to come into full bearing.

Pruning, to train the tree into proper growth, is important when temperate zone fruit and nut trees are young. Proper training establishes a well distributed framework of limbs, which will have structural strength to support heavy crops, and will insure continued production over many years.

Each species of fruit or nut tree is trained to a system

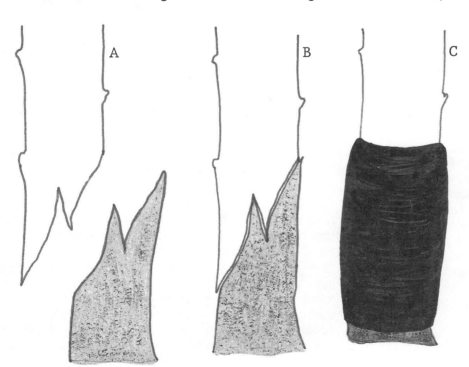

Figure 14-1. Whip graft. A.) The scion and rootstock cut for grafting. B.) Cambium layers fitted together. C.) Graft zone tightly wrapped and waxed.

which best suits it's growing and bearing habits. For example, upright growing trees, such as apples and pears, do best when trained to a "central leader"; that is, a system where one or more main upright limbs support smaller fruit bearing side branches. Other species, such as peaches, have a natural spreading growth habit. They bear best when trained to a "vase shape", where 3 to 4 main standard limbs are grown from the trunk giving the tree a vase-like appearance. In general, it is temperate zone deciduous trees that require the most training. Tropical and subtropical evergreen fruit and nut trees require little, if any, training when young.

PRUNING

Most commercially grown fruit trees require pruning. For evergreen subtropicals like citrus and avocados, pruning may consist only of removal of dead or broken limbs or occasional "topping" to limit the height of the tree for ease of picking. Most major temperate zone stone and pome fruit trees require annual dormant pruning. Annual dormant pruning of mature trees serves several purposes. Tree structure is maintained to support maximum production. Tree height is maintained at a level so fruit is not borne too far above the ground (usually below 12 feet for those to be picked by hand from ladders). Pruning opens the tree for better light filtration to the inside limbs, as well as removes old, weak branches that are unfruitful.

In addition, some "fruitwood" is thinned by pruning to support the maximum desired crop the following season. Fruitwood is the type of growth upon which fruit or nuts are produced. On peaches and nectarines, for example, fruit is produced on previous season's growth. On plums and some apples, fruit is produced on "spurs" (short shoots) that bear flower buds on the end or sides. Other trees, such as persimmon, bear fruit on current season's growth. The purpose of pruning for each type of tree should be to renew and invigorate the fruit-bearing area for continued production.

POLLINATION

Most fruit and nut trees grown for commercial production require pollination of flowers so fruit will set and grow to maturity. In some cases, trees are capable of pollinating themselves, and in other cases, a second cultivar is required.

A *perfect* flower is one which contains both a female pistil and pollen producing anthers.

Dioecious is a term used to describe a tree that bears only staminate flowers or pistillate flowers. In this case, two different trees (one male and one female) are required for pollination to occur.

Monoecious is a term applied where separate stamen-bearing (male-pollen bearing) and pistil bearing (female) flowers are on the same tree.

Self-fruitful species of fruits and nuts are those having pollen which sets fruit in the same flower or on the same tree. Those cultivars of species, such as apple and almond, that produce pollen which does not set fruit on the same tree are called *self-unfruitful*. In order to obtain fruit set, it is often necessary to use other cultivars of the same fruit species to supply pollen. Where dioecious species are grown, trees with staminate-bearing flowers must be placed in a regular pattern throughout the orchard in order to supply pollen for the pistillate, or fruit-bearing trees.

Figure 14-2. A commercial nursery yard where deciduous bare-root nursery trees are "heeled" in bundles of ten. Deciduous trees can only be transported with bare roots during late fall and winter -when they are dormant.

Honey bees are the major source of transferring pollen from flower to flower or tree to tree. Other insects, along with wind, also transfer pollen. An exception to pollination takes place where fruit growth and development occurs in some species without fertilization (i.e. navel orange) and is known as *parthenocarpy*.

FRUIT GROWTH AND DEVELOPMENT

Once fertilization (or parthenocarpic development) takes place, fruit growth and development follows. Some species develop their fruit rapidly, within a few weeks after bloom. Other species take a year or longer for fruit to mature. For example, a few early maturing peach cultivars produce mature fruit about two months after bloom, while other, later-maturing cultivars take as long as seven months to ripen. Valencia oranges and some avocado cultivars, although reaching market maturity in about one year, may not be picked until 18 months after bloom; thus, for at least part of the season, there are two crops on the tree.

The fleshy, edible portion of fruit is actually the wall of the ripened ovary (pericarp) or other vegetative tissue surrounding the seed cavity. The seed may be small and inconspicuous (i.e. apple) or large and prominent (i.e. avocado). In the case of nuts, the edible portion is the cotyledons, or "seed leaves", which store food for the germinating embryo.

Thinning (removal of a portion of the young developing fruit from the tree) is an essential practice on many

Figure 14-3 A & B. Some species of fruit trees require pollination, which in some cases requires beekeeping.

Figure 14-4. In semi-tropial areas such as southern California and Florida, orchard heaters are often required to protect orchards from severe frosts.

Figure 14-5 A & B. Not all tree crops are fruit or nuts. Cork (an outer covering of tree bark) is a major crop in Portugal.

fruit trees, particularly the pome (apple, pear) and stone (peach, nectarine, apricot, cherry, plum) fruits. Excess, immature fruit is removed to allow those remaining on the tree to grow to a larger size. Thinning is most often done by hand for stone fruits, but is sometimes accomplished by the use of chemical thinners for pome fruits.

NUTRITION AND IRRIGATION

Like other horticultural crops, fruit and nut trees need certain elements available to them to be used for growth and development. Three of these elements are carbon, hydrogen, and oxygen, which are the constituents of organic substances in the tree. Other necessary elements are nitrogen, phosphorous, potassium, sulfur, calcium, magnesium, iron, boron, zinc, copper, chlorine, manganese, molybdenum, and possibly cobalt, sodium and silicon. With the exception of carbon, hydrogen and oxygen, essential elements may occur naturally in the soil or may be supplied to the tree by adding inorganic (chemical) or organic (manure) fertilizer to the soil, or in the form of a foliage spray.

The main essential element is nitrogen which must be supplied to commercial fruit and nut trees in most growing areas to insure good tree growth and production. Phosphorous and potassium must also be applied to the soil in many areas. Zinc is a common element sometimes needed to be added to the tree at regular intervals (usually it is applied as a spray). Other mineral elements occasionally must be applied when the orchard soil does not supply sufficient quantities to the tree.

Most mature orchards require about 2.5 to 3 feet (75-90 cm) of water during each growing season. Most fruit-growing areas of the world, (except for the tropics) cannot produce commercial crops of fruit or nuts without supplementary irrigation. In other areas, some irrigation water must be added when rainfall is insufficient to supply needs. In the arid, southwestern United States and other desert areas of the world, nearly all of the water used by commercially grown fruit and nut trees is supplied by irrigation. Where gravity flow water is available from nearby rivers or streams, water cost is minimal. When water must be pumped from deep wells, the power charge represents a significant part of production costs.

INSECTS AND DISEASES

Insects and diseases are a constant threat to commercially produced fruits and nuts. Crops can be made totally unmarketable within a short time if preventive or control measures are not taken. When necessary, both the tree (including roots, limbs and leaves) and the crop must be protected by fungicides or insecticides or both. Integrated pest management (IPM) methods combine natural pest-predator systems with chemical and cultural controls to reduce the amount of chemicals applied and also lower production costs (see Chap. 16). However, because in many areas of the world the market demands totally insect and disease free fruit, some insecticides and fungicides must be applied.

HARVESTING AND MARKETING

FRESH SHIPPING FRUIT

For commercial purposes, fresh shipping fruit matures and is ready to harvest when it reaches "horticultural maturity". At this stage, fruit has its best color and flavor, yet is still firm enough to be picked, packed, and shipped to market. Since not all fruit on the tree is mature and ready to be picked at the same time, 2 to 4 picks, each a few days apart, are necessary in some cultivars. In the case of other fresh fruits (e.g. oranges), trees are stripped (all the fruit is picked at one time).

Fruit is carefully picked by hand and transported to a packinghouse in bins (or field boxes) where it is packed and cooled immediately. It is important to immediately cool fruit to slightly above freezing to slow down the respiration (ripening) process. This allows the fruit to remain in good condition while it is transported to domestic or foreign markets.

Some fruits (e.g. apples) are stored in controlled atmosphere (C.A.) storage. This type of storage includes a low oxygen, high carbon dioxide atmosphere, as well as refrigeration. This altered atmosphere further reduces respiration, and (along with refrigeration) allows storage for up to one year.

In the United States most fresh shipping fruit is grown in farming areas far removed from areas of dense population. Thus, marketing and transportation are vitally important. Most fruit is sold through independent or cooperative packer - shipper agencies. Growers are charged a packing fee and sales commission, based on the gross sales price received by the selling agency. Most fruit sales are made on an F.O.B. basis, but some is sold at auction. The current trend is to sell directly to purchasing agents representing individual marketing firms, chain stores, or groups of chain stores. A much smaller volume of fruit is grown and packed on the farm with direct sales from the farm, or in nearby towns or cities.

Fresh fruit prices are established by a direct supply - demand relationship. Prices tend to fluctuate as day-to-day supplies change. Producers and handlers can adjust supplies to some extent by placing fruit in storage, but alternatives are restricted by the relative short storage life of fresh fruit.

In marketing fresh fruit, the term "quality" has many meanings: rich flavor, attractive appearance, handling without bruising, long shelf life, large fruit size, or a desired stage of maturity. Superior quality can result in

Figure 14-6 A & B. In tropical areas rainfall is often sufficient to fully support orchards. In most other areas of the world, at least supplementary irrigation is required. Photo above is a developing nectarine orchard utilizing furrow irrigation, and photo below is an almond grove utilizing a drip irrigation system.

prices well above the going market price on a specific day. (This is proved by the confidence buyers have in certain labels or brands).

PROCESSING FRUIT

Many species of fruit are grown for processing (i.e. cling peaches, olives, juice oranges, sour cherries, prunes etc.). Each is harvested and handled in a way somewhat different fruit for fresh shipping. Some processed fruit is mechanically harvested into bulk bins and then transported to the cannery, dehydrator, or processing plant. Quality is still as important as in the case of fresh fruit; thus, the processing fruit must also be picked and processed at peak horticultural maturity.

Processed fruit and nut products also are packaged and sold by both private and cooperative organizations. Some cultivated varieties of most fruit species are also processed into canned fruit, dried products, juices, etc. Some are by-product outlets for fruit that is not sold fresh because, although of good eating quality, it is cosmetically inferior, or overripe. Other fruit is grown exclusively for processing. Processing fruit must be of high quality in order to result in a superior processed product. Quality standards of the raw product for processing must be as high as those for fresh shipping fruit. This is particularly important for fruit which is dried as a method of preservation, including prunes, figs, peaches, pears, apricots, apples and many others.

NUTS

Although almost all fresh and much of the processed fruit is harvested by hand; most nut crops in the United States are shaken from trees with mechanical shakers. These shakers are attached to tree trunks or limbs and vibrate until the nuts are shaken loose and fall to the ground or into a catching frame. On the ground they are swept into a windrow and picked up by machine (or hand) and transported to an area where they are cleaned (hulled) and "dehydrated". Excess moisture is removed by blowing warm air through the nuts in large bins. They may also be exposed to the sun until excess moisture is removed. Drying extends storage life by preventing mold and rancidity from developing on the edible portion of the nut.

Some nuts such as walnuts, pistachios, and pecans may have a portion of the production marketed unshelled. Others, including a large percentage of walnuts, almonds and pecans are shelled and placed in sealed containers. Some are further processed by roasting or seasoned with spices before being sealed in cans, jars, or other containers.

Figure 14-7 A & B. Because of the potential for bruising, or other damage most fruit must be hand picked. Nuts on the other hand, can usually be machine harvested. Photos are of equipment used in harvesting English walnuts. In above photo, a tree shaker dislodges walnuts from the tree. Picture below is a windrower which readies walnuts for mechanical retrieval.

OVERVIEW

Tree crops involve extensive capital investment. Investment that will not yield any income for several years (during development of the orchard). For that reason, extraordinary planning must go into each decision concerning orchard development.

As discussed in this chapter, cultivars for each particular fruit vary greatly from area to area. The selected cultivar is then grafted to a rootstock adapted to the particular soil type and area. The rootstock may or may not be from the same species of fruit or nuts. Since the selected cultivar and rootstock will result in an orchard that may be in place for many decades, very serious thought must be put into selecting that cultivar and rootstock.

Since fruit is a highly perishable product, prior marketing arrangements are a must. That is, common field crops such as grain and hay may be stored on-farm for a considerable time after harvest. Fruit cannot, and therefore marketing takes on added significance with most tree crops.

Figure 14-8. Fruit is an extremely perishable crop, which dictates that marketing be a major consideration before a crop is ever produced (or orchard established).

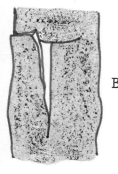
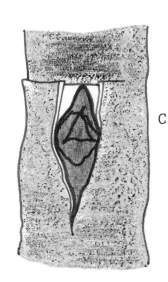

Figure 14-9. Budding. A.) Single bud is removed from desired cultivar. B.) Slit cut in bark of rootstock to receive bud. C.) Bud inserted into new rootstock.

Figure 14-11. Plum tree scions grafted to new rootstock.

Figure 14-10. Three types of grafts used to join scions and rootstocks. A.) Bark graft B.) Wedge graft C.) Cleft graft

CHAPTER 15   TROPICAL CROP PRODUCTION SYSTEMS
by Stephen C. Mason, Ph.D.
University of Nebraska

INTRODUCTION

Crop production practices in tropical areas are quite diverse, principally as a result of the large differences in climate, soils, and human cultures present in the tropics. This chapter provides an overview of crop production systems found in the tropics, recognizing that many local variations of these systems commonly occur. For purposes of this chapter, the area lying between the Tropic of Cancer and the Tropic of Capricorn (between 23-1/2 latitude north and south) will be considered the tropics. The tropics contain 36% of the total land area of the world and approximately 40% of the world's population.

CLIMATE

The tropics are characterized by only small variations in quantity of solar radiation received, temperature, and length of day throughout the calendar year. The most important climactic difference between regions in the tropics is the quantity and distribution of rainfall. This directly influences the type of vegetation that predominates, the types of soils found, and the crop species produced. It is a common misconception that rainy climates with tropical rain forest vegetation predominate in the tropics. This type of climate is present in only 24% of the tropical land area. A seasonal climate with 2.5 to 6.5 month long dry seasons occupies 49% of the land area. This land area is covered mainly with grasses and sedges. The rainy climate is more common in tropical America, while the dry and desert climates are more common to Africa and Asia. However, all four of the major tropical climates are present in Tropical America, Africa and Asia.

Figure 15-1. Climatic regions of the tropics.

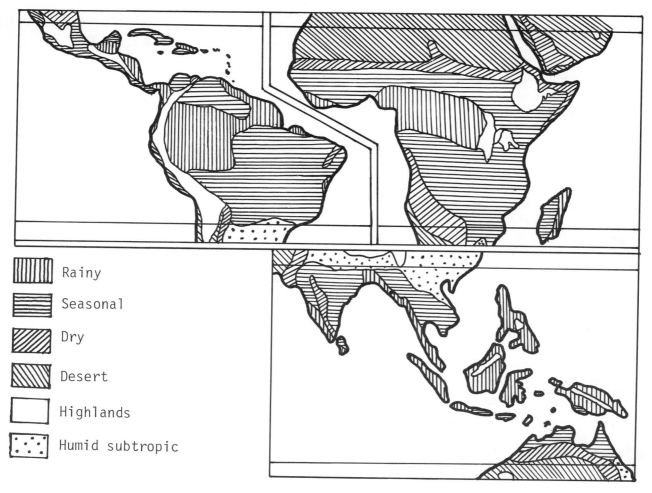

SOILS

Great diversity in soil types exist in the tropics as a result of differences in parent material, climate, and the vegetation that predominated during the period of soil formation. Acid, infertile soils are commonly found in high rainfall areas, while basic, low organic matter, high calcium and/or salt content soils are common in areas with dry and desert climates. The major exception in the desert regions is in areas of the Sub-Sahara desert, where acid soils are sometimes in evidence. In this case the acid soils are a result of high rainfall periods that existed during prehistoric times. During this period tropical rain forest was probably the predominate vegetation. Soils with clay accumulation in the subsoil and a slightly basic pH predominate in areas with seasonal climates without high annual rainfall.

Mountainous areas are also of considerable importance in tropical America and Asia. In the case of the mountainous regions, soils are similar to the soils found in temperate regions. The major exception is when the soils are of volcanic origin.

All major soil types are found in tropical areas and are locally important. In general, however, tropical soils have more limitations for crop production than do soils of temperate areas.

CROPPING SYSTEMS

Many different cropping systems are practiced as a result of extreme variations in climate and soil. Likewise, the availability of markets and local customs also influence cropping systems.

Shifting cultivation of agronomic crops is practiced on more of the land area in the tropics than any other system. However, other major cropping systems are of greater economic importance. Pastures are used for ranching and nomadic herding, while grain crop production is widely based on multiple cropping. Plantation systems are practiced on only 4% of the land area, but provide valuable economic returns for many countries. In this chapter each of the cropping systems, plus rice production are discussed.

SHIFTING CULTIVATION

In shifting cultivation systems, areas of land are cleared (usually burned) and cropped for one to five years before they are fallowed for up to 30 years. During this fallow period the natural vegetation is allowed to return and dominate the land area.

At the end of the fallow period large quantities of nutrients are present in the forest vegetation with smaller amounts stored in litter on the soil surface and in the soil. Near the end of the dry season, the vegetation is cut, and after allowing it to dry, it is burned. The burn-

Figure 15-2. Upland rice growing under shifting cultivation in the Amazon basin of Brazil.

ing process is rapid and does not consume all the plant material. It produces an ash that contains phosphorous, potassium, calcium, magnesium and other nutrients for use by the succeeding crops. (These nutrients also tend to neutralize some of the acidity of rain forest soils.) The contribution of nitrogen and sulfur depends upon the severity of the burn, since these elements are lost to the atmosphere when plant tissue is completely burned. Partially charred or burned material from a rapid burn can contribute sizeable quantities of these elements to succeeding crops.

Immediately after burning, the first crop(s) is (are) planted directly into the soil, covered with unburned portions of the trees and the ash. It is important to get the first crop established rapidly to provide cover over the soil in order to prevent loss of the nutrients in the ash before they can be leached or eroded away by heavy rains. Since the nutrient status of the soil declines with time during the cropping phase, the most highly valued crops are planted initially, such as rice and corn in tropical America and yams in Africa.

In subsequent years of cropping, the nutrient supply declines and weed problems become more serious. In response to this, farmers tend to gradually shift to tall-growing crops which will produce well under nutrient-limiting situations and compete vigorously with weeds. Common crops with these characteristics are plantain and cassava. Once the nutrient supply becomes so low or weed problems so severe that crop production is not possible, the land is fallowed and allowed to return to its natural vegetation. Usually the cropping phase is shorter when the natural fertility of the soil is low, and in grassland situations where the nutrient contribution from burning the plant matter is small.

Shifting cultivation is the predominate cropping system of the humid tropics and supports approximately 20% of the human population (in the tropics). Historically, shifting cultivation was probably the first cropping system used by man. In early years man discovered that fire was a useful tool to clear land and that nutrients were made available to succeeding crops in the ash produced by burning.

As the human population increased, and improved technologies for crop production evolved, shifting cultivation became a system primarily used on acid and/or infertile soils of the tropics. Since these soils tend to occur in areas with rainy climates and rain forest vegetation, shifting cultivation has become associated with these tropical regions around the world. (It should be noted that shifting cultivation is also widely practiced in the seasonal climatic areas of Africa where grasslands predominate.)

Shifting cultivation is a system that requires little capital and, as discussed, is ecologically sound as long as the fallow period is maintained. However, it requires a huge expenditure of labor to clear the land, produces low yields, and leaves a large proportion of the land in fallow at any one time. Efforts to improve productivity on land with relatively high natural soil fertility have been successfully accomplished by gradual shifts to continuous cropping, often in the form of intercropping systems. On soils of low natural fertility, efforts to improve productivity have had minimal impact, with the greatest success occurring with permanent tree crops often produced in plantation systems.

Shifting cultivation is an ecologically sound system as long as its three basic principles are followed:

(1) The fallow period is sufficiently long to allow accumulation of adequate nutrients for the cropping phase, and that it is long enough to break pest cycles (especially weed cycles).

(2) Wise selection of crops is made to efficiently use nutrients provided by clearing and burning, and to effectively compete with increasing weed problems during the cropping phase.

(3) The use of intercropping is practiced (pg.   ) to maintain soil cover to prevent soil erosion and destruction of the soil's physical properties. (Also, to minimize weed growth during the cropping phase.)

When these principles are followed the land resource base is maintained and a minimum sustenance for farmers is produced. However, as the population pressure increases, or efforts are made to more intensively farm these marginal lands, ecological disasters due to soil erosion and depletion of soil fertility have commonly occurred.

Intercropping, the production of several crops at once, is a common practice in shifting cultivation. Intercropping of up to 10 crops at once often occurs in an effort to provide soil cover and thereby reduce erosion and destruction of the soil's physical properties. In addition, intercropping tends to reduce risk of crop failure and pest problems associated with single crops, as well as to provide a diverse and nutritionally adequate diet on a small land area. It is also used to fully exploit the different soil and micro-environmental conditions present within a small plot of land. This includes differences in soil fertility, drainage, temperature, and moisture availability, (discussed in more detail later on).

MULTIPLE CROPPING

Multiple cropping is the production of two or more crops on the same land area during the calendar year. For centuries farmers in the tropics have used multiple cropping to take advantage of the desirable year-round temperatures and the constant quantity of sunlight. This practice appears to be a hold-over from shifting cultivation systems, and is especially useful for farmers working on small tracts of land. In multiple cropping two distinctly different types of cropping systems are used; the above mentioned system of intercropping and sequential cropping.

Sequential cropping. In sequential cropping one crop is produced during the same calendar year. Since only one crop is present in the field at any time, no competition occurs between the crops. As discussed, in intercropping two or more crops are grown in the field at the same time, so competition between the crops for nutrients and water occur during all or part of the cropping season.

Sequential cropping systems of the tropics are similar to the double-crop production systems of the temperate areas. Many sequential cropping systems in the tropics extend the cropping season into the dry season by planting, near the end of the rainy season a drought-tolerant crop, such as grain sorghum or millet, that grows well on stored soil moisture. In both of these systems the influence of the preceding crop and the short time interval between crops influences soil fertility status, soil physical properties, stored soil moisture, and pest infestation. In both the temperate zones and the tropics, however, depletion of soil fertility or stored soil moisture by one crop can greatly reduce the yield of the succeeding crop.

Intercropping is practiced when higher yields can be obtained by growing a mixture of crops rather than a single crop, and/or to minimize risk and provide income stability. The advantage often associated with intercropping is usually the result of more efficient use of solar radiation, nutrients, and water. As a general rule, crops with growth cycles of different lengths are used in intercropping. As an example, rapid growing, early maturing crops such as cowpeas, peanuts and dry beans use solar radiation, nutrients and water during early portions of the cropping season; while slower growing, later maturing crops such as corn, grain sorghum, cassava and plantain require these in the greatest amounts during the latter portions of the cropping season. The end result is greater efficiency in the use of solar radiation, nutrients and water by planting two or more of these crops together.

In addition to leading to more efficient use of the solar radiation, water, and nutrients, intercropping also often leads to reduced pest problems and soil erosion. Rapid formation of the crop canopy and maintenance of the canopy for all, or most, of the cropping season reduces weed problems by shading; this also protects the soil surface from beating rains. Population buildup and dissemination of insects and diseases are retarded since the different crops planted together are seldom attacked by the same insects or diseases.

Although intercropping offers many advantages in terms of yield and risk aversion, mechanization of intercropping systems is difficult. Therefore, it remains a system used on farms where land is scarce and labor is relatively plentiful. The traditional intercropping systems have produced low yields; but recent innovations in production technologies for intercropping offer opportunities for large increases in productivity.

RICE SYSTEMS

Rice is the most important food crop in the tropics, both in terms of land area planted as well as grain production. More than 90% of the rice produced in the tropics occurs in Asia, but it is also very important in Africa and tropical America. Rice is the only major food crop that can be grown under flooded soil conditions. This is because rice has the ability to transfer oxygen through the leaves and stems to the roots. Rice is normally produced in three principle systems: upland, deep-water or floating, and lowland or paddy.

Upland rice is a system in which rice is produced in a manner similar to other crops. This system is characterized by conventional tillage and direct seeding into dry soil, and then complete dependency on rainfall for moisture. Upland rice is grown in areas with seasonally large quantities of rainfall often on clayey soils. Worldwide, approximately 25% of the tropical lands used in rice production are upland rice. In tropical America and Africa, however, over 70% of the area planted to rice is in this system. Rice yields are lower in

Figure 15-3. Transplanting of lowland rice in Taiwan.

Figure 15-4. Upland rice production in Japan.

upland systems than in lowland systems, primarily as a result of periods of moisture stress during the growing season. Yield reduction from lodging and pest infestations can also be greater than in other rice systems. Upland rice is produced under a wide range of management intensities ranging from shifting cultivation to completely mechanized farming systems.

Deep-water, or floating rice, is produced along rivers that flood during the rainy season or in tidal marsh areas. In this system, rice is sown into dry soil at the start of the rainy season. After the start of heavy rains the area is naturally flooded to depths ranging from 2 to 12 feet. Varieties used in this system tolerate being submerged and keep growing rapidly to keep the leaves and the panicle above the water surface. In most areas, deep-water rice varieties are photoperiod sensitive to insure that flowering and grain-fill occur when the plant is least susceptible to submergence. Harvesting is done either from boats or by hand after the flood waters have receded. Yields of deep-water rice are low, but significant research gains are being made. Deep-water rice accounts for approximately 9% of tropical rice production.

Lowland, or paddy rice production systems are the most common systems used in Asia, comprising approximately 65% of the world's tropical rice production. This system is the one most commonly associated with rice, in that the water level during the growing season is maintained between zero and 20 inches deep. After the first rains fall or after the initial irrigation, dikes are constructed to capture rainfall or hold irrigation water. The soil is tilled when it is wet to "puddle" the soil. This eliminates most of the large pore spaces in the soil, reducing percolation losses during the growing season.

Soaked, pre-germinated seed is seeded directly onto the puddled soils, or 10-30 day-old seedlings are transplanted from beds into the puddled soil. Flooding is done either by irrigation or captured rainfall. Flooding is done not only to prevent moisture stress, but also to reduce weed problems, and sometimes to alter soil chemical properties. Flooding leaves soil pH levels near 6.5-7.0 in most soils, and increases phosphorous availability. When poor management of the water level occurs alternate wetting and drying of the soil allows large quantities of nitrogen to be lost from the soil by denitrification.

In recent years spectacular increases in rice yields in lowland systems have taken place as a result of the development of special dwarf varieties that produce many tillers. Traditional varieties respond to nitrogen application by increasing their height, resulting in increased vegetative growth and lodging without a large yield increase. The newer dwarf varieties respond to nitrogen application by producing more grain (rather than vegetation). To produce maximum yields with the dwarf varieties, precise water control is essential to minimize nitrogen losses from the soil.

PASTURE SYSTEMS

Approximately half of the world's permanent pastures and half of the world's cattle population are located in the tropics, but the productivity of livestock in tropical areas is quite low by U.S. standards. This low productivity has been attributed to heat stress and animal diseases, but most animal scientists agree that the feed supply, which is mainly pasture, is the greatest limiting

Figure 15-5 & 15-6. Rice straw drying in Japan. Rice straw is a valuable commodity in much of Asia as a roughage source for cattle and water buffalo.

factor. Pasture production in the tropics can be classified into three systems: (1) grazing of native pastures, (2) grazing of improved grass-legume mixtures, and (3) high production systems based on fertilization of grass species. Each of these systems requires different management schemes and has different potentials for producing forage.

Utilization of native pastures in arid areas is principally done by nomads. In this system the livestock herds are moved according to rainfall to areas where pasture is available. In the Sahara region of Africa, as the rainy season progresses, the nomads move their animals further into the desert areas to utilize the available forage. Later, during the dry season when forage is unavailable in the desert they retreat to higher rainfall areas or to the banks of permanent streams. Nomadic systems work well in these dry areas where pasture production is low and erratic and when grazing is controlled. However, social and economic pressures cause nomadic herders to increase animal populations as much as possible. This has led to serious overgrazing of the marginal grasslands. When this occurs the pasture resource base is rapidly degraded, leading to huge losses in human and animal life.

Ranching systems are used to exploit native pasture on approximately 11% of the tropical land area, primarily in the grassland areas with seasonal climates. Low animal productivity in these systems primarily results from inadequate soil moisture during part of the year and low levels of soil fertility. During the rainy season, large quantities of forage are available for livestock consumption. However, as the dry season starts, the quantity and quality of forage declines rapidly. This cycle of forage availability leads to rapid weight gain by animals during the rainy season, with loss of weight occurring during the dry season.

Fire is the most commonly used management tool to increase productivity in this system. Burning destroys old, coarse, and fibrous grass and promotes re-growth of young, more nutritious grass. In addition, it destroys many insect pests, and the ash contributes phosphorous, calcium, magnesium and other nutrients to the soil. In areas with higher rainfall, burning is common every year while in lower rainfall areas it is used less frequently. Productivity of native pastures can be improved by rotational burning, rotational grazing, fertilization, and adjustment of stocking rate to the quantity of pasture produced. These practices require careful management and capital investment so they have not been widely accepted.

The most successful way to improve productivity of native pastures has been to incorporate legumes into the native pastures. Incorporation of legumes increases the total production, lengthens the portion of the year during which production occurs, and elevates the nutritional quality of the pasture. In addition, tropical pasture legumes are capable of fixing between 100 and 300 pounds of nitrogen per acre per year.

In order to generate a productive pasture situation, often it is necessary to replace both the native grass and incorporate a legume. Establishment of a productive grass-legume mixture requires adaptation of both the grass and the legume to the same climatic and soil conditions, plus compatability of the growth habits. Therefore, the best mixture is usually site specific. Tall grass species that grow in bunches such as Jaragua, Guinea grass and Elephant grass, and low-growing or runner legumes such as Stylo, Centro and Kudzu have given the most successful results in the tropics. Grasses that form a thick mat can also be used in mixtures, but careful management is required to keep them from severely competing with the legume. In addition to fertilization and seeding, effective establishment in native grasslands also requires heavy burning during the latter part of the dry season and tillage to eliminate the native grass and to prepare a seedbed. Once established, careful grazing management is required to keep the legume in the system, and fertilization to replace nutrients removed by grazing is required to maintain productivity.

Intensive forage systems, using the production of tropical grass, require high fertilizer application rates. This is due to the fact that legumes seldom provide sufficient nitrogen for tropical grasses to achieve their maximum yield potential. For example, heavily fertilized elephant or guinea grass can achieve yields as high as 90 tons per acre a year (in rainy climatic areas or under irrigation). Therefore, yearly application rates of up to 800 pounds per acre of nitrogen are needed (applied in equal rates after each cutting or grazing). Since the costs of production, especially for fertilizer, are high, this system is only feasible for production of highly valued animal products such as dairy cattle.

All three pasture production systems in the tropics have limitations in their ability to adequately feed livestock during the dry season. Storage as hay is not a common practice in the tropics. This is primarily due to the difficulty of drying hay in the field during the rainy season. Silage is seldom made because tropical species of grass do not lend themselves well to fermentation (see pg.   ). However, as explained on pg.   , tropical grasses can be used to make satisfactory silage, but the addition of molasses or grain is required. Development of this knowledge could do much to aid the seasonality of tropical animal production. Therefore, mature grass pasture standing in the field, or small, irrigated pasture areas are usually used to provide the needed forage during the dry season.

PLANTATION SYSTEMS

Only 4 percent of the tropical land area is used for production of plantation crops, but the economic value of these crops is great for many tropical countries. The plantation production system is characterized by private or government ownership of large farms on which one crop is produced for export. Usually, plantation crops are crops that can only be produced in the tropics and require processing or rapid transportation to the market. These systems are sophisticated using modern, high technology practices and large capital investments. A small administrative body makes use of large quantities of "unskilled" cheap labor. Historically the Romans first used a plantation system for wine and oil production in North Africa. Growth of this system expanded in the 16th and 17th centuries in tropical America, where European settlers, plus large numbers of slaves brought from Africa, started producing sugar, coffee, cotton, and tobacco for the European markets. Later the plantation system spread to Asia and Africa. Today the most important plantation crops produced in the tropics are sugar cane, pineapple, banana, coconut, cocoa, coffee, tea, rubber, palm oil, and sisal. Most of these crops are also produced for local consumption by small farmers using traditional intercropping practices.

SUMMARY

Tropical climates have little variation in terms of length of day, temperature, and solar radiation. Seasonal changes as known in temperate climates (spring, summer, fall and winter) do not occur in the tropics. If seasonal variation is present, it occurs as alternate dry and wet (rainy) periods. (Some tropical climates, however, are perennially dry deserts, or perennially wet rain forests. The wet season/dry deason climate is simply the most common.)

During the rainy or wet portion of the year, total precipitation is often quite high. During the dry portion, complete drought is often the norm. The intense variation of these extremes limit both crop and animal production in most tropical areas. The extreme and heavy rainfall during the wet season leaches away soil nutrients and creates acid-type soils.

Shifting cultivation was the first crop production system to develop in the tropics. In shifting cultivation, areas of natural vegetation are cleared and then burned during the dry season. The ashes left from the burning provide nutrients for the soil. Crops are grown until soil nutrients are exhausted. These areas are then left fallow for extended periods of time (sometimes as long as 30 yrs.). During the fallow period native vegetation is allowed to reclaim the area, but is slashed and burned in the next crop cycle.

Shifting cultivation requires little capital investment and has therefore been popular in many third world nations. If practiced properly it is ecologically sound, but attempts to intensify production can produce severe erosion.

Growing more than one crop at a time is often necessitated by tropical environments. Known as intercropping, up to ten crops may be grown at one time. This reduces pressure from insect pests and diseases, as well as reduces the risk of crop failure. Where much of tropical crop production is for subsistence, reduced risk is highly important.

Intensifying crop production in many tropical areas is difficult in terms of both physical and ecological restraints. Physically, heavy fertilization and other capital investment is required. Ecologically, the intensity of the rainy/wet seasons can cause severe erosion.

The most practical and feasible types of intensive cropping in highly erodable rainforest climates usually involve tree crops. Often developed as plantation type systems, erosion problems are minimized due to the continued presence of the trees and minimal disturbance of the soil.

Much of the land area in tropical environments is used for livestock grazing. This is primarily due to lowly fertile soils that could not otherwise be used for crop production. The limiting factor is most often the dry season. Grazing animals gain weight through the wet season, but during the dry season often lose weight.

Rice is the major staple crop in the tropical areas of Asia, and is otherwise quite important in tropical Africa and America. Rice is well suited for the wet/dry climate of the tropics. This is because of the ability of rice to transfer oxygen from the leaves to the roots. Thus, the flooding of the wet season is not a detriment. During the dry season, however, fields may be worked and rice planted.

CHAPTER 16 ECONOMIC ENTOMOLOGY
by Mark Mayse, Ph.D.
Fresno State University
Fresno, California

INTRODUCTION

Economic entomology can be defined as the scientific study of insects and their relatives (known collectively as arthropods), with particular emphasis on interactions among insects and humans. Of prime consideration are those situations where any of the resources which humans consider important (e.g., crops, livestock, forests, human health and even peace-of-mind) are seriously affected by various arthropods. In many cases the effects of these organisms on human existence are detrimental. However, the beneficial impact of many insects must also be recognized. Today only about 1% of all named insect species are actually considered pests (i.e., approximately 10,000 out of nearly 1,000,000).

This chapter consists of three sections: 1) Morphology and Physiology, which deals with basic aspects of insect structure and function, 2) Taxonomy, involving identification and patterns of relationship among insects, and 3) Applied Insect Ecology, which includes general types of interaction among insects and the living and nonliving parts of their environment. Special emphasis is placed on the field of integrated pest management (IPM).

MORPHOLOGY AND PHYSIOLOGY

Morphology is the study of bodily structure, and physiology is the study of bodily function. Several important adaptations in both these areas have helped to give insects a stunning preeminence in the animal kingdom; about three out of every four named animal species in the world today are insects. (Many insects today are remarkably similar to their ancestors of 250 million years ago.)

Insects and their relatives share the same basic body plan. Their bodies are divided into segments and their appendages are jointed for flexibility. Much of the tremendous diversity of types of arthropods results from adaptations and modifications of the basic segmented body form and jointed appendages. Consider the similarities among such diverse groups of arthropods as centipedes, sowbugs, lobsters, spiders, scorpions, and grasshoppers. An insect's body is divided into three generally distinct regions: head, thorax, and abdomen. Each region consists of a group of modified segments along with their specialized appendages. In many cases, segments have been modified to the extent that appendages have been lost altogether.

The insect head contains a specialized ganglion which serves as a brain. Although capable of coordinating surprisingly complex activities associated with the nervous system, no clear evidence indicates any sort of rational thought processes. Sensory structures on the insect head include the compound and simple eyes for vision, a pair of antennae used primarily as organs for smelling but also for touching, and the often highly-modified mouthparts which mechanically manipulate the food, and also serve the important function of taste reception. The major insect mouthparts, one pair each of mandibles and maxillae along with the labium, have become modified to serve many feeding functions. These range from the generalized chewing mouthparts of a cockroach to the hypodermic needle-like structures of mosquitoes.

The insect thorax is the body region primarily concerned with locomotion. All three pairs of an insect's legs, as well as up to two pairs of wings, will be found on the thorax. Understandably, a major portion of the inner volume of the thorax is filled by elaborate muscle systems which are necessary to move the insect's locomotive structures.

It is interesting to consider the unique nature of insect flight when compared to all other groups of animals able to fly. All birds, bats, and the now-extinct pterosaurs fly/flew by means of forelimbs which became modified as wings. However the wings of insects are outgrowths of the integument or body wall, and are entirely independent from the three pairs of limbs (thoracic legs) which have been retained.

The insect abdomen is comprised of the remaining body segments behind the thorax which serve a number of important functions. Major portions of virtually all the insect's essential organ systems are found in or on the abdomen. Included are the digestive, respiratory, circulatory, excretory, and reproductive systems, along with important parts of the nervous, endocrine, and integumentary (outer body) systems. Although there are a number of basic similarities between the operational aspects of human and insect organ systems, there are also many fundamental differences. For example, oxygen reaches appropriate tissues in an insect's body by diffusion without any direct involvement with a bloodstream. The insect respiratory system consists of paired external openings (spiracles) which lead into an extensive system of finely-divided tubes, which deliver oxygen and carry away carbon dioxide.

Insects do not have a "bloodstream" as such. Their major body fluid, known as hemolymph, circulates openly throughout the inside of the body. Thus, insects generally lack the extensive system of transporting vessels (arteries and veins) as found in higher animals.

The sex organs of insects are paired testes in the male and ovaries in the female. Females can store sperm for long periods of time in an organ called the spermatheca. Many female insects can release these sperm at will, thereby controlling fertilization at the actual time of oviposition, or egg-laying.

Patterns of growth and development in insects reflect the fact that their "skeleton" is on the outside. That is, the relatively inflexible insect exoskeleton operates much like a medieval suit of armor; it is very effective for protection, but it imposes serious restrictions on bodily growth and expansion. These problems have been overcome through the molting process, whereby a growing insect splits its old "skin" and emerges with a new skin. The new skin is soft and folded, with room for expansion. It then hardens and darkens as a larger version of its former self. However, until the new skin hardens to become a true exoskeleton, the insect is quite vulnerable to attack from various natural enemies.

Growth and development among insects involve two major patterns which may be described as gradual and complete metamorphosis. Metamorphosis means a marked change in form (such as a caterpillar turning into a moth or butterfly). Insects with gradual metamorphosis pass through three life stages: egg, nymph, and adult. The nymphal stages of these insects for the most part resemble the adults, but are smaller, have undeveloped wings and generally are not sexually mature. Examples include grasshoppers, earwigs, termites, true lice, thrips, leafhoppers, and stink bugs. Approximately 15% of all insect species show gradual metamorphosis.

The other 85% which have complete metamorphosis exhibit four distinct life stages: egg, larva, pupa, and adult. Insect larvae are generally quite different from their corresponding adults. Examples would be caterpillars vs. moths, maggots vs. flies, and grubs vs. beetles. The pupal stage of insects has often been referred to as a resting stage. This description is generally true as the insects normally are somewhat immobile. From the standpoint of development, however, the pupal stage is spectacularly active. It is during this stage that the tissues of a caterpillar are broken down into a soupy liquid, from which eventually develops the adult moth or butterfly.

TAXONOMY

Taxonomy involves the principles which relate to the naming and classification of organisms. In agriculture, it is vitally important to know the general characteristics common to particular groups of insects. This is necessary in order to be able to make timely and accurate identification of pests and beneficials. Unfortunately, many serious mistakes in insect identification are made each year, with detrimental and sometimes even devastating consequences for farmers.

As mentioned at the beginning of the chapter, insects belong to a particularly diverse group of invertebrate organisms known as arthropods. The arthropods can further be grouped into a number of important taxonomic classes: the arachnids, which include spiders, mites, ticks, and scorpions; the crustaceans, which include crayfish, lobsters, crabs, sowbugs, and barnacles; the centipedes; the millipedes; and of course the insects, which comprise about 90% of all known arthropod species.

In a book with such broad agricultural scope as this one, it would seem appropriate to condense a substantial amount of insect taxonomy information into a comprehensive table which includes order names, common names, type of metamorphosis, type of mouthparts, type of wings, distinctive characteristics, and general agricultural importance. (See Table 1.) For more detailed information, the reader is encouraged to refer to any of the numerous books which deal with the taxonomy and biology of insects. (See Further Reading section at end of chapter.)

APPLIED INSECT ECOLOGY

Ecology is the study of how organisms relate to both the living and nonliving parts of their environment. Insect ecology examines how insects and related organisms interact with their environment; and applied insect ecology focuses on the practical aspects of insect ecology which are of primary concern to humans. Thus, most of the activities of economic entomologists are actually directed toward this field of applied insect ecology.

During the past 25 years or so, the term "integrated pest management" has become prevalent among economic entomologists. This term, conveniently shortened to IPM, can perhaps be most concisely defined as scientific pest control. A much more comprehensive definition was developed by a task force of the National Academy of Sciences. IPM was described as the "evaluation and consolidation of all available techniques into a unified program to manage pest populations so that economic damage is avoided and adverse side effects on the environment minimized". One of the most significant aspects of this view of IPM is that growers should try to **manage** pest populations, rather than try to annihilate all individuals in a given area. IPM provides a philosophical approach to dealing with pest pressures which requires consideration of both the economic and ecological consequences of various types of pest control actions. To better understand what IPM means today, we should investigate the historical development of what has traditionally been called pest control.

HISTORY OF PEST CONTROL

Somewhat surprisingly, major advances in the field of pest control had already been made several thousand

Table 16-1. GENERAL SUMMARY OF ECONOMICALLY IMPORTANT ORDERS OF INSECTS.

Order name	Common names	Metamorphosis	Mouthparts	Wings*	Pest Impact	Beneficial Impact
Thysanura	silverfish, firebrats	gradual	chewing	primitively wingless	starch feeders	x
Collembola	springtails	gradual	chewing	primitively wingless	a few plant feeders	nutrient recycling
Odonata	dragonflies, damselflies	gradual	chewing	2 pr. membranous	x	predators
Orthoptera	grasshoppers, crickets, cockroaches, mantids	gradual	chewing	2 pr. FW (tegmina) leathery HW membranous	plant feeders	predatory mantids
Dermaptera	earwigs	gradual	chewing	some wingless, others 2 pr. FW leathery, HW membranous	some plant feeders	some predators, detritus feeders
Isoptera	termites	gradual	chewing	2 pr. membranous reproductives only)	damage wood structures	nutrient recycling
Mallophaga	chewing lice	gradual	chewing	secondarly wingless	external parasites (esp. birds)	x
Anoplura	sucking lice	gradual	piercing-sucking	secondarily wingless	external parasites (esp. mammals)	x
Thysanoptera	thrips	gradual	rasping-sucking	2 pr. fringed	plant feeders	a few predators
Hemiptera	true bugs	gradual	piercing-sucking	2 pr. FW (hemelytra) partially sclerotized HW membranous	many plant feeders	many predators
Homoptera	aphids, scales leafhoppers, whiteflies	gradual	piercing-sucking	2 pr. variable sclerotized/ membranous	plant feeders	shellac from lac insect
Neuroptera	lacewings, antlions	complete	modified sucking (larvae) chewing (adults)	2 pr. membranous	x	predators
Coleoptera	beetles, weevils	complete	chewing	2 pr. FW (elytra) sclerotized HW membranous	many plant feeders	many predators
Lepidoptera	butterflies, moths	complete	chewing (larvae) siphoning (adult)	2 pr. covered with scales	caterpillars major plant feeders	adults prized by collectors
Hymenoptera	sawflies, ants, bees, wasps	complete	chewing	2 pr. membranous	several plant feeders	many predators, parasitoids; honey bee
Diptera	mosquitoes, true flies	complete	modified sucking	1 pr. membranous	vectors of major human disease pathogens	some predators, parasitoids
Siphonaptera	fleas	complete	chewing (larvae) piercing-sucking (adults)	secondarily wingless	external parasites (mammals)	

*FW = forewings
HW = hingwings
sclerotized = hardened and darkened
primitively wingless = evolved from wingless ancestors
secondarily wingless = evolved from wingled ancestors

years ago. The ancient Chinese were using insecticidal chemicals derived from plants (known as botanicals) to treat crop seeds. They had also achieved a highly sophisticated program of biological control for citrus insect pests by introducing predatory ants into their orchards. They facilitated tree-to-tree movement of these natural control agents by tying bamboo runways between twigs of different trees.

Unfortunately, in European countries, for hundreds of years during the Middle Ages, the most popular approach to dealing with pests centered upon mysticism, superstition, and even legalistic pronouncements. However, major breakthroughs during the Renaissance helped to vastly improve methods for dealing with pest problems. Developments, such as the compound microscope and Redi's experiments dispelling the notion that life arises from inanimate objects, helped promote a general return to more rational and scientific thought.

By the second half of the 19th century, several major pest outbreaks in various parts of the world were successfully fought by rational, human-directed activities. A major breakthrough in pest control occurred around the turn of the 20th century with discoveries that linked diseases such as yellow fever and malaria to mosquitoes. Furthermore, it was established that by destroying the habitats in which such insect pests bred, the disease cycles in humans could be broken. One important global consequence of this advance in dealing with human disease vectors (insects that transmit diseases to man), was that the healthy labor force needed to complete construction of the Panama Canal could finally be maintained.

By the early part of the 20th century, pest control had advanced to become a relatively sophisticated, multifaceted system. Growers relied heavily upon a diverse array of tactics. There were few insecticides and the timing of application had become a science. Also, knowledge of pest biology, accurate identification of pests, and monitoring of pest population levels were recognized for their importance. In addition to chemical control, cultural practices such as rotation and trap cropping were used to help avoid pest problems.

Near the end of World War II, however, the field of pest control was dramatically altered almost overnight. Development of the so-called miracle insecticides such as DDT and other chlorinated hydrocarbons provided chemical tools which were regarded by many as "magic bullets" against insect pests. The unbridled enthusiasm generated by these chemicals was somewhat understandable in view of their low cost and long-term effectiveness against a remarkably wide range of insect pests.

Insect pest control, which before had been considered primarily an ecological problem, suddenly came to be regarded as more of a chemical and engineering problem. Many growers traded careful monitoring of pest population levels and use of effective cultural practices for essentially a "washday" program of scheduled insecticide applications. Applications often were made as insurance, regardless of whether or not pest problems actually existed.

Serious shortcomings of this switch in attitude began to surface very rapidly. Pest populations, including house flies in dairy operations where DDT had been used intensively for less than a year, quickly began to show resistance. Higher and higher doses of such chemicals were applied, with the result that some pests could survive being literally covered with a chemical which would have been lethal to their ancestors in previous generations.

Most intelligent entomologists quickly recognized this resistance phenomenon as a predictable consequence of applying strong selection pressure (i.e., mortality) on the susceptible individuals of each pest generation. Those individual pests which were not susceptible to the chemical were the ones surviving to produce offspring, which generally inherited their parents' resistance.

Other problems which arose from the widespread over reliance on chemical insecticides after WWII included what are known as target pest resurgence and secondary pest outbreaks. In other words, after the initial kill, larger and more devastating pest populations recurred to pose an even greater threat. What was happening was that the natural enemies of the pests (predators, parasites, pathogens), which had previously been regulating pest population levels, were especially susceptible to mortality from chemical applications. Thus, beneficial insects which controlled pests naturally, were killed off and controlled to a greater extent than the pests themselves.

In addition, a series of problems which might be described broadly as environmental contamination have clearly developed with the unwise use of chemical insecticides. Included among these difficulties are: insecticide residues on fruits and vegetables; contamination of air, soil, and water resources; exposure of factory and field workers to toxic materials; and destruction of nontarget organisms, including wildlife.

Integrated pest management (IPM) represents a philosophy which can help alleviate a major portion of these problems. This scientific approach to pest control utilizes all available techniques. IPM takes into consideration important information which has accumulated during the past 40 years concerning pest biology and behavior, along with the patterns of interaction between insects, their host plants, and the environment.

SAMPLING CONSIDERATIONS

One of the most important fundamental aspects of any IPM program is the use of sampling techniques to provide population density estimates for both pest and beneficial insects. The critical IPM decisions which involve control tactics are integrally tied to such population estimates.

Most insect monitoring techniques used in IPM decision-making are relative methods, yielding data which should reflect a consistent, although unknown proportion of the actual population numbers which exist. The standard insect sweep net is one of the most widely-used relative sampling tools in IPM work. However, the efficiency of sweepnet sampling is strongly affected by such factors as personal style, moisture level on foliage, and physical attributes of the crop being sampled.

Another important IPM monitoring technique is the visual count, where insect numbers are evaluated by direct observation on plant material where they occur. This method has numerous obvious advantages, including simplicity and cost.

Recently, IPM researchers at the University of California have developed what are called "presence/absence" techniques for sampling various tiny arthropods such as mites on almonds, cotton, and citrus. It is very difficult to see these arthropods with the naked eye. But with these methods, actual mite numbers need not be recorded. Instead, leaves are scored as either P (mites present) or A (mites absent), and treatment decisions are made based upon the overall proportion of leaves infested. However, these techniques clearly can be used only in systems where a predictable relationship has been established between different pest population densities and the respective proportions of leaves infested.

ECONOMIC INJURY LEVEL AND ECONOMIC THRESHOLD

For IPM decision-making, sampling in order to estimate pest population density is an essential first step. However, baseline information about pest numbers and their impact on the crop must be available in order to interpret and utilize the data gathered by pest monitoring. These baseline values are known as the **economic injury level** and the **economic threshold** for a given pest in a particular crop.

Economic injury level can be defined as the lowest pest population density that will cause economic damage. Stated simply, economic damage occurs when the amount of pest injury to the crop is balanced by the cost of pest control. However, since crop pests make up dynamic biological systems, it would be unwise for a grower to wait to attempt control until economic damage is actually occurring. Instead, there must exist some density at which control measures should be applied to prevent an increasing pest population from reaching the economic injury level. This pest density is known as the economic threshold.

Development of economic threshold values is a major goal and challenge in the research area of integrated pest management. Clearly, an IPM practitioner must have some benchmark pest density against which to compare sample counts in order to make rational treatment decisions.

Economic threshold (ET) values exist for numerous important crop pests, but most of these values could be refined with further research. Today we rely heavily on "working economic thresholds" which are rather conservative and generally tend toward over-treatment with insecticides. Technically, whenever the cost of control or the value of the crop changes, the ET value should also change.

INSECT DAMAGE AND INJURY TO CROPS

The diverse types of damage and injury which various insect pests may inflict on crops can be generally classified as either direct or indirect. Direct damage includes detrimental effects which the pest exerts on the actual commodity to be marketed, as shown in Figures 1-3.

Many crop pests exert indirect damage by injuring plant parts other than the primary marketable commodity, which are usually the leaves, stalks, or roots. Examples of indirect pests are shown in Figures 4 & 5. As indicated in the photographs, damage may involve partial or complete leaf destruction due to the activity of chewing or piercing-sucking mouthparts.

Several important groups of insects, including various aphids, leafhoppers, and mosquitoes, fall into the special pest category of pathogen vectors. These insects carry a wide range of bacteria, protozoa, mycoplasmas, and viruses which cause numerous plant and animal diseases. The actual feeding damage which these pests exert on crop plants is relatively insignificant compared to their impact by spreading important plant pathogens. As mentioned earlier, some of the most significant pests in the history of the world are vectors of important human disease agents, including mosquitoes, fleas, and a variety of non-insect arthropods (especially mites and ticks).

SURVEY OF INTEGRATED PEST MANAGEMENT TACTICS

By definition, IPM involves the evaluation and consolidation of all available techniques into a unified program for managing pest population levels. The range of these techniques is extremely broad, including simple approaches like crop rotation, as well as very complex methods like sterile-male release. Most methods used in IPM programs are included in the following categories: 1) cultural control, 2) physical/mechanical control, 3) biological control, 4) chemical control, 5) genetic control, and 6) regulatory control.

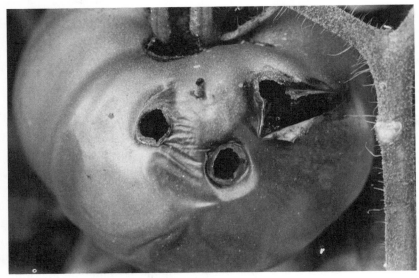

Figure 16-1 & 2. *Heliothis zea* has several common names and is one of the most devastating direct insect pests in North America. It preferentially feeds on the fruiting structures of a wide variety of important crops. For example, in sweet corn this species is known as the corn earworm (Fig. 16-1), in tomato as the tomato fruitworm (Fig. 16-2), in cotton as the cotton bollworm, in soybean as the soybean podworm, and in sorghum as the sorghum headworm. (Photos by M.A. Mayse and K.W. Thorpe)

Figure 16-3. Carrots damaged by feeding of western grapeleaf and larvae. (Photo by M.A. Mayse and K.W. Thorpe)

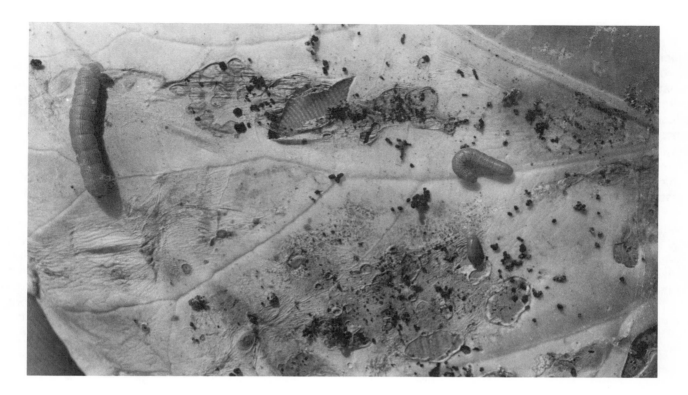

Figure 16-4. Grape leaf damaged by feeding of western grapeleaf skeltonizer larvae. (Photo by M.A. Mayse and K.W. Thorpe)

Figure 16-5. Watermelon leaf damaged by feeding of isopods (commonly called sowbugs or pillbugs), which are actually non-insect arthropods in the crustacean class. (Photo by M.A. Mayse and K.W. Thorpe)

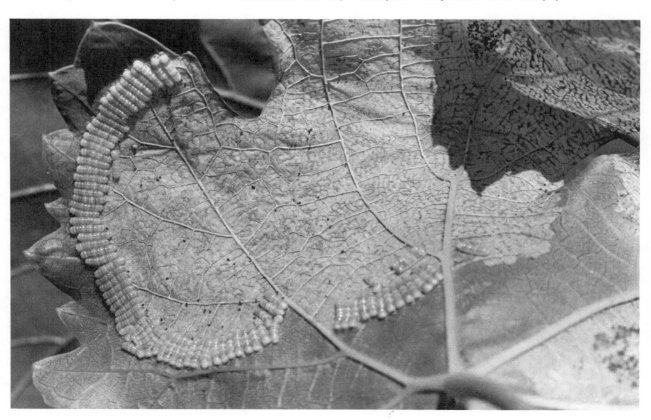

Cultural methods involve various agronomic practices which may be modified to the detriment of certain pest insects. An especially important technique is the use of resistant varieties of crop plants and livestock. Other cultural methods include rotating crops, destroying crop refuse, changing the time of planting or harvest, and managing water/fertilizer inputs in such a way as to thwart the pests.

Biological control involves the use of predators, parasitoids (insects which parasitize other insects), and/or pathogens in order to manage pest population levels. Several different strategies may be employed in this area. Such strategies include enhancement of natural enemies which are already present, introduction and establishment of natural enemies from foreign countries (often from the native land of an accidentally introduced pest species), and artificial culture of natural enemies with periodic release.

Examples of Biological Control. Biological control can be defined as the regulation of a pest through the use of a natural enemy. Natural enemies may be predators which feed directly on the insect pest, or pathogens which produce disease among the insect pests.

Predators - Biological control of cottony cushion scale by vedalia beetle on citrus in California shines as the first major success for importation and establishment of non-native natural enemies used in insect control (i.e., classical biological control). Since Australia was determined to be the home of the accidentally introduced pest insect, during the late 1800's researchers traveled there to find the predatory lady beetle which generally kept the scale in check on its native continent. In less than two years after 140 vedalia beetles were introduced into California, their descendants had brought the cottony cushion scale under control throughout the state's citrus growing areas. From 1890 until the present, vedalia beetle has prevented cottony cushion scale from rising to economic pest status except in a few areas where DDT temporarily eliminated the predator.

Parasitoid - A tiny almost microscopic wasp (*Anagrus epos*) serves as a major biological control agent against the grape leafhopper on grapes in California. The parasitoid adult (i.e., wasp) has the amazing ability to locate and attack grape leafhopper eggs which are laid within the epidermal tissues of grape leaves. With a relatively short life cycle allowing nine to ten generations of the wasp during the season, up to 95% of all leafhopper eggs deposited after July may be parasitized. This clearly represents significant mortality for the leafhopper pest.

Pathogen - Various types of microorganisms produce disease among insect pests, including bacteria, viruses, fungi, and protozoa. Two bacteria are currently registered and under commercial production for use in controlling insect pests. *Bacillus popilliae* has been used to suppress outbreaks of Japanese beetle over wide areas of the U.S., and *B. thuringiensis* is used much like a conventional insecticide against such insects as caterpillars and mosquito larvae.

Chemical control relies primarily on the use of insecticides, but also includes such materials as attractants, repellents, chemosterilants, and insect growth-inhibiting substances. Insecticides are generally considered the key tactic for use when pest population levels are near economic threshold. This is because of their ability to bring about rapid population decline. However, it should be reiterated that such chemicals should be used only when necessary, and even then in combination with bio-environmental techniques (e.g., cultural and biological tactics) designed to make background conditions less favorable for insect pests.

Genetic control includes the production and release of insect pests which are sterile or otherwise genetically incompatible. A major success of this particular IPM approach was the eradication of the screwworm fly, a devastating pest of livestock, from the southwestern United States. Other genetic control tactics include production of pests with altered genetic backgrounds such as lethal genes and chromosomal translocations. Insects with these genetic difficulties operate as "reproductive time bombs" and may bring about major population decline.

Physical/mechanical control aim to modify the environment or to directly subvert some life activity of pests in order to keep their population numbers at an acceptable level. Included are such techniques as varying temperature and humidity conditions, as well as using various traps, screens, barriers, or even hand destruction to regulate pest numbers.

Regulatory control involves the use of quarantines for plants and animals, along with eradication and suppression programs. These approaches clearly necessitate the intervention of agencies like the Department of Agriculture as well as state departments of food and agriculture. They generally involve far-reaching policies which are often controversial, with sides determined by what one may personally have to gain or lose. Perhaps no other type of insect control program is more susceptible to the foibles and limitations of human nature than the regulatory arena.

FUTURE CHALLENGES IN INTEGRATED PEST MANAGEMENT

There is tremendous room for improvement in the area of IPM decision-making. As mentioned previously, most Economic Threshold (ET) values currently available for crop pests are very conservative, often leading to unnecessary application of insecticides. In many other instances, no functional ET has been established whatsoever. Furthermore, the remarkably stringent standards of cosmetic perfection (i.e., eye-appeal to consumers) for most of our fruits and vegetables, force many growers to strive for virtually total elimination of certain insect pests from their fields.

In establishing ET values, the sole consideration for cost of control has often been simply the cost of insecticide plus the cost of application. This ignorance of external factors, such as the economic impact of resistance development among pests, and environmental contamination with pesticides, has frequently resulted in unrealistic estimates of cost/benefit ratios in insect pest control programs.

A further challenge with regard to ET values involves the fact that normally more than one pest species occurs in an agro-ecosystem at any given time. That is, since crops are generally being stressed by more than one pest species, ET values should ideally take into consideration the relative importance of each species. This would result in decision-making criteria which would better represent the actual overall insect pest threat to a crop, rather than just a portion (single-species ET) of the potential economic impact.

Regulatory aspects of IPM decision-making may play a major role in the long-term success or failure of programs. Currently, it is legal for individuals who derive direct financial benefit from the sale of insecticides to make treatment recommendations to growers. This has been compared to a medical system in which a pharmacist would be in the position of actually diagnosing illnesses, for which he/she could then prescribe drugs to be sold to the patient. Surely such conflicts of interest are no more appropriate in the field of integrated pest management than in medicine. In other words, an IPM practitioner should not be biased in making a treatment recommendation by the chance to make more money with one decision as compared to another.

Finally, perhaps the greatest future challenge in the entire field of economic entomology is to properly integrate the discipline of plant protection with that of plant and food production. Today, crop production managers far too often make decisions which may be advantageous from the standpoint of yield, but which may have a detrimental effect on the pest situations. Both types of considerations are of major importance, but there should be much more effort to make decisions in agriculture which are mutually beneficial to production and protection aspects. Students should receive broad training in what might be called "optimal plant and food enterprise management" in order for this important future challenge to be successfully met.

FURTHER READING

Borror, D.J. and R.E. White. 1970. A field guide to the insects of America north of Mexico. Houghton Mifflin, Boston. 404 pp.

Borror, D.J., D.M. DeLong, and C.A. Triplehorn. 1976. An introduction to the study of insects, 4th ed. Holt, Rinehart and Winston, NY. 852 pp.

Chapman, R.F. 1969. The insects: structure and function. Elsevier, NY. 819 pp.

Clark, L.R., P.W. Geier, R.D. Hughes, and R.F. Morris. 1967. The ecology of insect populations in theory and practice. Chapman and Hall, London. 232 pp.

Davidson, R. H. and W.F. Lyon. 1979. Insect pests of farm, garden, and orchard, 7th ed. Wiley, NY. 596 pp.

Dethier, V.G. 1976. Man's plague?: insects and agriculture. Darwin Press, Princeton. 237 pp.

Evans, H.E. 1984. Insect biology: a textbook of entomology. Addison-Wesley, Reading, MA. 436 pp.

*Flint, M.L. and R. van den Bosch. 1981. Introduction to integrated pest management. Plenum, NY. 240 pp.

Huffaker, C.B. (ed.). 1971. Biological control. Plenum, NY. 511 pp.

Huffaker, C.B. (ed.). 1980. New technology of pest control. Wiley, NY. 500 pp.

Knipling, E.F. 1979. The basic principles of insect population suppression and management. USDA, Washington, DC. 659 pp.

Little, V.A. 1972. General and applied entomology, 3rd ed. Harper, NY. 527 pp.

Matthews, R.W. and J.R. Matthews. 1978. Insect behavior. Wiley, NY. 507 pp.

Metcalf, C.L., W.P. Flint, and R.L. Metcalf. 1962. Destructive and useful insects, 4th ed. McGraw-Hill, NY. 1087 pp.

*Metcalf, R.L. and W.H. Luckmann (eds.). 1982. Introduction to insect pest management, 2nd ed. Wiley, NY. 577 pp.

National Academy of Sciences. 1969. Insect-pest management and control. NAS, Washington, DC. 508 pp.

Perkins, J.H. 1982. Insects, experts, and the insecticide crisis. Plenum, NY. 304 pp.

*Pfadt, R.E. (ed.). 1985. Fundamentals of applied entomology, 4th ed. Macmillan, NY. 284 pp.

Pimentel, D. (ed.). 1975. Insects, science, and society. Academic, NY 284 pp.

Price, P.W. 1984. Insect ecology, 2nd ed. Wiley, NY. 607 pp.

*Ross, H.H., C.A. Ross, and J.R. P. Ross. 1982. A textbook of entomology, 4th ed. Wiley, NY. 666 pp.

Snodgrass, R.E. 1935. Principles of insect morphology. McGraw-Hill, NY. 667 pp.

van den Bosch, R. 1978. The pesticide conspiracy. Doubleday, Garden City, NY. 226 pp.

van Emden, F. 1974. Pest control and its ecology. Edward Arnold Publ., London. 59 pp.

Varley, G.C., G.R. Gradwell, and M.P. Hassell. 1973. Insect population ecology: and analytical approach. Univ. of Calif. Press, Berkeley. 212 pp.

Watson, T.F., L. Moore, and G.W. Ware. 1976. Practical insect pest management: a self-instruction manual. Freeman, San Francisco. 196 pp.

Wigglesworth, V.B. 1972. The principles of insect physiology. Chapman and Hall, London. 827 pp.

Wilson, M.C., D.B. Broersma, and A.V. Provonsha. 1984. Practical insect pest management: 1. Fundamentals of applied entomology, 2nd ed. Waveland Press, Prospect Heights, IL. 216 pp.

BOOK III   FEED AND FEEDING
(excerpted from *Modern, Practical Feeds, Feeding and Animal Nutrition* by D. Porter Price)

CHAPTER 17   BASIC ANIMAL FEEDS AND NUTRIENTS

Classification. Although there are many different ways to classify animal feeds, the most basic classifications are roughages and concentrates. From there, many different subcategories emerge.

Roughages are feeds with high concentrations of fiber. Because of their fiber content, only ruminant animals (cattle, sheep, goats, etc.) or very specialized non-ruminants (horses, rabbits, etc.) can digest and use them to any extent. Hay, silages, and straws are the most common examples. Such by-products as vegetable waste, peanut hulls, cottonseed hulls, soybean hulls, etc. are also considered roughages.

Concentrates are feeds that are high in starch, protein, sugar or fat. Examples would be grains, oil seeds, molasses, edible fats, and oils. The name "concentrates" comes from the fact that they have a much higher concentration of energy. On the average, concentrates typically contain 30-50% more energy than roughages. With respect to extremely coarse roughages such as straw, concentrates may have 200 to 300% more digestible energy.

Factors affecting the quality of roughages. Feeds classified as roughages can have enormous variations in digestibility. The biggest factor affecting digestibility is **maturity.**

As a plant matures, the digestibility steadily declines. This is primarily because the amount of fiber increases as a plant matures. In addition, the fiber already present becomes more coarse and tough.

With maturity, the soluble sugars and highly digestible carbohydrates contained in the juice of the plant dry up. (The sugars and highly digestible carbohydrates are usually transformed into fiber.)

Much of the protein and mineral content also declines with maturity. (With grazing animals, this is one of the most significant changes, as animals pastured on dry, mature forage must usually be supplemented with protein and several minerals).

Therefore, because maturity has such an effect on quality, there can be as much difference in digestibility for a single plant species as there can be between species. For example, lush, green, ryegrass can be as high as 20% crude protein, with a Total Digestible Nutrient (TDN) value of nearly 70%. Dry, yellow, mature ryegrass will normally run 5-6% protein, with a TDN of only 40-45%.

MAJOR NUTRIENTS CONTAINED IN FEEDS

WATER

By weight, livestock consume more water than any other nutrient. Water may be consumed separately, or as

Figure 17-1. Water is one of the most overlooked of all nutrients. While the quantity of water available is usually provided for, the quality of the water is often overlooked or left to chance. Problems with quality often include excessive levels of sodium, calcium, magnesium, sulfates, nitrates, or bacterial contamination.

a constituent of feeds. All feeds contain water. The amount can be quite variable. The range can run from about 10% for dry grains, to over 90% for many vegetable by-products.

Because the amount of water contained in feeds is so variable, most analyses of feed nutrients are reported on a dry matter basis. Likewise, animal consumption is often reported on a dry matter basis.

But the amount of water contained in feeds is important for more than just technical reasons. When purchasing feeds, the moisture content relates to how much feed you actually get for your money.

For example, fully cured hay normally runs about 10 to 12% moisture. Freshly baled hay, however, will often run 20-24% moisture. Thus, if you were to buy freshly baled hay, you would pay for twice as much water.

Of course, while water is found in feeds, livestock also need access to water for drinking. While this may sound very simple, in actual practice it may not be. Problems with water quality, quantity, and/or availability are quite common.

Water quality problems often include dissolved salts of sodium, calcium, and/or magnesium, sulfates, nitrates, or bacterial contamination. Problems with

Table 17-1. DAILY WATER REQUIREMENTS FOR VARIOUS CLASSES OF LIVESTOCK.

Species	Am't Required at Normal Temps. (Gallons)	Am't. Required at Excessively Hot Temps. (Gallons)	Notes
Dry Beef Cow	10-12	15-20	Moisture content of the feed greatly influences water consumption of cattle. Moisture content of pasture grass can vary from about 12% moisture to 80% moisture. These figures are for dry, mature feeds. For lush succulent feed, water consumption will be 40 to 60% less.
Lactating Beef Cow	12-15	18-24	
Feedlot Steer	8-10	14-18	
Horse	8-12	12-16	Horses have sweat glands which increases their water requirement when working. Horses are also susceptible to water founder, and therefore must be allowed to "cool down" before being given unlimited access to water after heavy work.
Lactating Mare	10-14	16-18	
Sheep	.8-1.2	1.5-1.8	Sheep are able to absorb large amounts of water from the intestinal tract, and thereby have much less fecal loss than other species (feces are much drier than other species). Still, sheep do require unrestricted access to water.
Lactating Ewe	1.0-1.4	1.7-2.1	
Hog	1-3 (depending on size)	1.5-5 (depending on size)	During hot weather swine, like other animals require more water. It must be pointed out, however, that swine are much more subject to heat stress than other livestock.
Lactating Sow	4-9	6-12	
Dairy Cow	12-14 plus 3 to 4 gallons for each 10 lbs. of milk produced.	20-26 plus 3 to 4 gallons for each 10 lbs. of milk produced.	The total amount of water required by a dairy cow varies greatly with the amount of milk produced. As with beef cows, the amount of moisture in the feed has a great effect on subsequent water consumption. These figures are for dry feeds.

these types of contaminants can render water toxic, or more commonly, so unpalatable that livestock performance is reduced. Therefore, a water sample should always be analyzed before construction of a livestock facility is begun.

Water quantity problems are usually more solvable than quality (provided water is available). Water quantity problems are usually a matter of estimating the maximum amount needed, and then providing the size well, pump, and water delivery system that can supply the required amount. Table 17-1 shows the approximate amount of water required by different species of livestock.

CARBOHYDRATES

Carbohydrates make up the major portion of animal diets. However, it is important to realize that there are enormous differences between the various types of carbohydrates. That is, carbohydrates is a very broad term which refers to all compounds which contain carbon (C), hydrogen (H), and oxygen (O). The hydrogen and oxygen are found in the same ratios as water (H_2O). Starch which is found in grains; cellulose which is found in roughages; and sugars, are all carbohydrates. Together, carbohydrates supply the bulk of the energy supplied by animal feeds.

Carbohydrates all have similar chemical structures as they are all formed in plants by photosynthesis. However, their digestibilities vary greatly.

Starch is primarily used as a storage form of energy in seeds. Starch is, therefore, primarily found in grains and is very digestible.

Cellulose is a structural component of plants, which we commonly refer to as fiber. Cellulose is not digestible to any significant degree to simple stomached animals. Only ruminants (cattle, sheep, and goats) and animals with similar specialized digestive organs (horses, rabbits, etc.) can utilize cellulose as a source of energy.

Sugars are sometimes used as a storage form of energy in fruit, and in some specialized plants such as sugar cane and sugar beets. In other cases sugars are apparently intermediate compounds created during the formation of structural carbohydrates. They are therefore found in the juices of green, growing plants, but are absent when the plant is yellow and mature. Sugars are digestible to most animal species.

FATS

Fats are very concentrated forms of energy. As a general rule, fats are said to contain 2.25 times as much energy as carbohydrates. In some ruminant rations, however, the value of fat may be greater than 2.25.

Included in the fat category are edible oils which are very similar in composition. The major physical difference is that oils are liquid at room temperature, whereas fats are solid. Normally, oils are of vegetable origin and fats from animal origin. Both are sometimes used in farm animal diets.

As a practical matter, fats must be used sparingly in farm animal diets. This is particularly true for ruminants and poultry, and, to a lesser extent for swine. The reason is that fat does not normally comprise much of the normal diet for these animals, and thus their digestive and metabolism systems have not developed for high fat diets.

PROTEIN

Protein is a high cost item in most animal feeding, and therefore it is important to thoroughly understand the protein requirement of the species being fed.

Protein is made up of amino acids, which may be called the building blocks of life. The animal breaks feed protein down into its basic amino acids, and then resynthesizes the amino acids into body proteins (muscle, skin, etc.). Without adequate protein, (amino acids) the animal will not grow at the maximum rate. Likewise, in a pregnant animal, the fetus will be born undersized, weak, and more susceptible to disease.

Animals cannot be deprived of protein for any significant length of time. If they are, the result will be reduced resistance to disease (antibodies, which are a main line of defense against disease, are constructed almost entirely of protein). Growing animals may or may not be stunted, depending upon how severe the protein restriction is and at what age it occurs. The younger the animal, the greater the chance for permanent growth retardation.

Obviously then, it is important that the animal's requirement for protein be known and provided for. But it should be realized that exceeding the protein requirement is a waste. It is not only a waste of money, since protein feeds are typically more expensive than energy feeds, but it puts a physiological stress on the animal and can reduce performance.

TERMS USED IN RELATION TO LIVESTOCK FEEDS

DRY MATTER

Dry matter (DM) is expressed as a percentage and refers to the weight of the feed after all moisture or water is removed. With many feeds the amount of moisture can be surprisingly variable. As mentioned previously, this is extremely important because the DM that is left (after the water is removed) is what you actually pay for when you buy a feed. If there is excess moisture in a feed, then you will be paying for more water, and less actual feed (dry matter).

The term "air dry feed" refers to 90% dry matter, or 10% moisture. This is the amount of moisture most dry feeds will retain if exposed to environmental air. This is also the moisture that is generally too low to allow mold growth.

Most grains and hays will run about 90% dry matter. However there can be exceptions. Corn will keep at 85.5% dry matter, which is the standard dry matter (usually abreviated as DM) for corn. Hay can be quite variable running from about 92 to about 82% DM.

In actual use, the term "air dry" is sometimes confused with the term "as-is" or "as-fed". These two terms, "as-is and "as-fed", refer to whatever moisture is actually in the feed (for example corn at 85.5% DM). The term "air-dry" is a technical term and refers to 90% DM.

When a laboratory is reporting the nutrient analysis of a feed, there are three ways it can be reported: 1. "as-is" or "as-fed" which includes all the moisture; 2. "air-dry" which adjusts everything to 90% DM (10% moisture); and/or 3. "dry-matter basis" which adjusts everything back to 100% dry (no moisture). In order to understand the meaning of the laboratory's report, it is imperative to know on what "basis" it is being reported.

FIBER

As mentioned in the carbohydrate section, fiber is a form of carbohydrate. However, because fiber is relatively indigestible, it has special significance in relation to animal feeds.

Actually fiber is a catchall term that refers to the compounds that perform the structural functions of plants and seeds. That is, the seed coats in grains, the structural material that provides support for stalks, stems, etc., and cell walls are made of various forms of fiber. Most of what would be considered fiber is relatively indigestible to simple stomached animals. Ruminants, however, can digest at least a portion of what would be included in the fiber fraction of most feeds. How large a portion depends upon the type of fiber.

Fiber is usually broken down into three components; cellulose, hemicellulose, and lignin. Cellulose is the most common and is generally considered to be digestible to ruminants. Lignin is totally indigestible. Cellulose is usually complexed with hemicellulose and lignin, and the amount of lignin present (usually referred to as lignification) will determine the overall digestibility of the cellulose.

Lignin is a very complex compound, but can best be described as the woody portion of the plant. Lignin is used to give strength to form to the fiber action of the plant. Lignin is frequently complexed with cellulose to give it strength. As the amount of lignin complexed with cellulose increases, the digestibility of the cellulose decreases. The relative strength or coarseness of the fiber is an indication of the amount of lignin present. Likewise, the strength or coarseness of the fiber is an indication of the digestibility of the fiber. For example, straw is much stronger and more coarse than tender young grass and, therefore, the fiber would be much less digestible. Wood is almost completely lignified, and therefore is almost completely indigestible, whereas straw is at least partially digestible.

As a general rule, the taller the plant, the less digestible the stalk will be; that is because stronger (more lignified) fibers will be required to hold it up. Thus, the extremely tall forage sorghum type plants are less digestible than shorter grass type sorghum plants. Also, as a plant grows older, the fiber content increases in lignification and the digestibility declines accordingly. This is usually because the fiber is needed to support the developing grain head. Without a strong fibrous stalk, the plant would collapse.

CRUDE PROTEIN

Crude protein is a term commonly used in laboratory analysis of protein feeds. It's a relatively simple analysis, and has therefore, become a standard throughout the world.

It is extremely important to realize that **the "crude protein" rating of a feed has nothing to do with the digestibility of the protein in that feed.** Actually, crude protein is simply an analysis for nitrogen (N), which is a major constituent of protein. The percent of nitrogen is then multiplied by a factor (6.25) which represents the ratio by which N is present in most proteins. (Nitrogen is present in most protein at a level of 16%; 6.25 is the reciprocal of 16%).

Therefore, a laboratory analysis of rained-on hay, overheated silage, overcooked soybean meal, etc., will report the same crude protein level as if it weren't damaged. The same level of N will be present, and therefore the same level of crude protein will be reported.

DIGESTIBLE PROTEIN

The term digestible protein means the amount of protein that is actually digestible. Not all protein contained in a feed will be digestible, and therefore research is often conducted to determine how much of a particular type of protein is digested by animals. This type of research is often included in feed tables in which both crude protein and digestible protein values are reported for feeds. In some cases, however, the levels reported will be calculated, rather than derived from direct observation with animals.*

---

* What can be very misleading is that commercial laboratories often report "digestible protein" as if it were actually analyzed chemically. In reality, however, it is usually calculated mathematically from crude protein. That is, a factor which would be an average for that type of feed is multiplied times percent crude protein. For example, alfalfa protein is routinely considered to be 80% digestible (for ruminants). Many laboratories therefore routinely take the crude protein they obtain by analysis, multiply it by .80 and report the results as digestible protein. This leads whomever sent in the sample to believe that digestible protein has been analyzed, when in fact, it has not. If the alfalfa had been cured properly, then the figure will be reasonably accurate. If the alfalfa has been damaged (rained-on and overheated, etc.), the calculated digestibility will overvalue it.

There is, however, a laboratory test known as "apparent digestible protein". This is an actual analysis (with enzymes) which is reasonably accurate and does, therefore, give a true indication of protein digestibility (see also footnote on previous page).

NON-PROTEIN NITROGEN (NPN)

As discussed in the previous sections, nitrogen is a major constituent of protein. Indeed, nitrogen is the one indispensible element in protein.

Microorganisms, such as those found in the rumen, or paunch of ruminant animals (cattle, sheep, goats, etc.), can take elemental nitrogen and synthesize it into protein. Therefore, to a large degree, ruminants can utilize elemental nitrogen to replace natural or preformed protein in the diet. This is important, since products that can be used as protein substitutes in ruminants are usually much less expensive than true or natural proteins. Known as non-protein-nitrogen (NPN), there are a number of nitrogen containing compounds that can be used for this purpose. The compound known as urea, however, is by far the most widely used.

Urea contains much more N than is found in protein. Therefore, on a crude protein basis, urea has several times the value of true protein. For example, soybean meal usually has a crude protein value 44 or 50%; whereas urea has a crude protein value of either 265 or 281% (depending on the concentration).

In most cases urea (NPN) may be used to supply only a portion of the supplementary protein required by a ruminant. The rule of thumb is that urea should supply no more than 30% of a ruminant's total protein intake. As a practical matter, however, most consulting nutritionists usually restrict urea to no more than 20% of total protein.

Care must be exercised in using urea and most other NPN compounds as excessive levels can cause toxicity and death.

ASH

Ash is representative of the total mineral content of a feed. That is, in the analysis of feeds, a sample is totally burned (in an atmosphere of pure oxygen) and what is left over is the ash, or mineral portion of a feed.

TDN (TOTAL DIGESTIBLE NUTRIENTS)

Total Digestible Nutrients, commonly spoken of as TDN, is used as a measure of the energy contained in a feed. It is actually a measure of the digestibility of all the organic constituents of a feedstuff (which would include protein). However, since the digestibility and energy value are so closely related, the term is considered synonymous with energy. As will be explained in the next section, the TDN concept has a tendency to overestimate the value of roughages.

NET ENERGY (NE)

Net energy is a system of measuring the energy content of feeds which is available to the animal for productive purposes. To arrive at a value, the animals are given the feed in question under carefully monitored environmental conditions.

In practical application, net energy has its greatest advantage over the TDN system in evaluating roughages. TDN is only a measure of what an animal takes in, versus what is left in the feces. There is no allowance made for the gas and heat loss. In net energy evaluations the gas and heat losses are not included, whereas in TDN they are. That is, TDN simply means total digestible nutrients, and since the gas and heat losses occur as a part of digestion they are included, even though they provide no benefit to the animal. Since roughages contain much more fiber than concentrates, much more fermentation is required to digest them, and therefore there are much larger gas and heat losses with roughages than with concentrates. The TDN system therefore over-estimates the energy value for roughages.

For example, the net energy and TDN systems rate the following feedstuffs accordingly:

Feedstuff	TDN[1]	NE[1]	% Difference
Corn	94.1	94.23	0.1%
Milo	89.21	87.41	2.0%
Alfalfa hay	56.02	44.86	19.0%
Wheat Straw	43.84	10.79	75.4%

[1] Dry Matter Basis

As would be expected, there is very little difference between the TDN and net energy systems for the grains, but note the wide variation for the roughages. Note also the great difference within roughages.

The TDN system rates wheat straw as having 78% the energy of alfalfa hay. That is certainly not true. For example, if cattle or sheep are fed wheat straw they will usually lose weight; whereas if fed alfalfa they can gain weight. The Net Energy System rates wheat straw as having only 24% the energy of alfalfa hay, which is much more nearly correct.

MINERALS

A WORD ABOUT MINERAL NUTRITION

Minerals play an extremely important role in animal nutrition. Gross evidence of this is exemplified by the many deficiency symptoms that manifest themselves when certain minerals are not present in adequate amounts. Except for a few isolated areas, however, clinical symptoms of deficiencies do not ordinarily occur.*

The problem is, that in most instances, the deficiencies that do occur are usually subclinical in nature; i.e. the animals appear to be perfectly normal and healthy, but less than optimum amounts of minerals are being supplied, and therefore the animals are not as productive as they should or could be. Adding to the problem is the fact that requirements for most minerals are not precisely known in all situations.

Probably the biggest reason that precise requirements are not known in many instances, is due to the many complex interrelationships between minerals. Most minerals have many physiological functions and there are usually other minerals that can substitute for at least some of their functions. Also, excesses of some minerals can tie up or prevent utilization of other minerals.

TYPES OF MINERALS

Minerals are normally classified into two broad categories: (1.) macro minerals and (2.) micro or trace minerals. Macro minerals are those which are required in relatively large amounts; such as sodium (Na), calcium (Ca), phosphorous (P), potassium (K), and magnesium (Mg). The micro or trace minerals are required in very small amounts. Indeed, the requirements for trace minerals are so small that they are usually expressed in parts per million. Examples of trace minerals that animals are known to have a need for are: cobalt (Co), copper (Cu), zinc (Zn), manganese (Mn), iron (Fe), iodine (I), and selenium (Se).

MACRO MINERALS

CALCIUM AND PHOSPHOROUS

Calcium (Ca) and phosphorous (P) are far-and-away the most important minerals from a practical animal feeding standpoint. Typically animals will need to be supplemented with one or the other of these two minerals.

Grains are good sources of P, but are almost devoid of Ca. Therefore, animals on high grain rations must usually be fed supplemental Ca.

Grasses and most other forages (with the exception of some legumes) often contain adequate calcium, but are low in phosphorous. Therefore grazing animals, or animals fed high forage diets, must usually be supplemented with phosphorous.

Deficiency Symptoms of Calcium. The classic deficiency symptoms of Ca is the condition known as rickets.

---

* This is true for temperate climates only. In tropical climates clinical deficiency diseases are common.

Rickets occur as a deficiency of Ca during the early growth stages of a young animal. The condition results in a buckling of the knee joints, giving the animal a bow-legged look.

In adult animals a Ca deficiency results in a condition known as osteomalacia. Calcium (as well as P and Mg) is stored in the bone. It can be pulled out of the bone as required and used for physiological purposes. In the case of a deficiency (in an adult animal), Ca can be pulled out to the point of severely weakening the bone. As a result, stresses that would ordinarily cause no injury will result in bone fractures.

Poultry, laying hens in particular, have an extraordinary requirement for Ca. This is because eggshells are almost pure calcium carbonate. If inadequate Ca is not provided laying hens, osteomalacia will occur, while eggshells become thinner and more brittle.

Deficiency Symptoms of Phosphorous. As mentioned in the Ca section, some of the deficiency symptoms of P are the same as Ca, that is, rickets and osteomalacia. Rickets and osteomalacia, as caused by P, occur with some frequency in grazing animals. This is because in many areas forages are relatively low in P. Therefore range animals must often be supplemented with P, or fed concentrates (which contain relatively high levels of P).

In cattle and sheep, one of the first signs of P deficiency is what is known as pica or depraved appetite. The animal will instinctively chew on foreign objects such as wood, bones, etc., in an attempt to obtain P. (In a high grain feeding situation, chewing on wooden fences, etc. is more of an indication of Potassium deficiency; eating dirt can be an indication of a deficiency of a number of minerals, including salt.)

In swine and poultry, deficiency problems with P typically do not occur. The reason is that rations for these animals typically contain substantial grain and grain by-products, which are relatively high in P.

CALCIUM:PHOSPHOROUS RATIOS

In monogastric species, excesses of either Ca or P can tie up and prevent utilization of the other mineral, and therefore Ca:P ratios can become relatively important. In swine and poultry, ratios between 1.1:1 and 1.5:1 are usually considered ideal. Ratios of 2:1 are normally considered a good average for all species, although ratios of over 2.5:1 have caused problems in poultry.

With ruminants a somewhat wider range of ratios can be tolerated. Ratios as wide as 1.1:1 to 4:1 are generally considered satisfactory. The only exception would be cattle or sheep fed a ration high in grain sorghum. In that case, if the Ca:P ratio is less than 1.5:1, urinary calculi (kidney stones) will occur. Maintaining a 2:1 ratio on high grain rations will almost always mean that supplementary Ca will have to be provided.

MAGNESIUM (Mg)

Magnesium is the third most abundant mineral found in the body. Along with Ca and P, Mg is a constituent of bone and can be mobilized from bone. Although the mechanisms are not completely understood, Mg is also known to be involved in muscular control. Apparently there are a number of interrelationships between Mg and Ca. It has also been reported that K and P can reduce Mg absorption.

As a practical matter, most feeds contain an adequate level of Mg. The only time Mg becomes a real problem is with lactating cows or ewes grazing small grain pastures. At certain times, for a yet unexplained reason, the Mg in these grasses may become unavailable. The Mg requirement for lactating animals is quite high, and without adequate Mg, tetany and death will result. Therefore, cows grazing improved pastures must be supplemented with Mg.

SALT (SODIUM AND CHLORIDE)

Common salt, or sodium chloride is the most frequently supplemented mineral. In many cases, because salt is so inexpensive it is often over-supplemented.

Whenever supplementing salt, only the sodium (Na) requirement need be considered. While animals do have a requirement for chlorine (Cl), it is much lower than the requirement for Na, and is nearly always provided in a normal diet.

Clinical cases of salt deficiency rarely occur. Ruminants have a greater need than swine or poultry; but since ruminants usually consume substantial amounts of forage, which is high in potassium, clear cut deficiencies are usually abated. This is because potassium can substitute for sodium in many body functions.

Horses have the greatest need for sodium because they have well developed sweat glands. For example, horses need 4 to 5 times as much salt as cattle (85 grams vs. 17.5). The actual amount required, however, will depend on how warm the climate is and how hard the horses are worked (how much they sweat).

When deficiencies of sodium do occur, it is usually in the form of what is known as a subclinical deficiency, that is, the animal shows no specific symptom, but just doesn't grow or perform as well.

When supplementing salt, be sure and consider the amount of sodium occurring naturally in the feed. Some feeds, such as alfalfa, contain quite high levels and may even provide enough to satisfy the requirement.

POTASSIUM

As mentioned in the Na section, potassium (K) can substitute for a substantial amount of Na. However, there is at least one role in which Na cannot substitute. Potassium is required for normal heartbeat. Severe deficiencies will not allow the heart to contract and death will result.

As a practical matter, however, in livestock clinical deficiencies do not occur as most feeds are relatively good sources of K. Grasses contain about 1% K and legumes can run as high as 2.5%. Molasses is probably the richest source, it can run as high as 4%. Grains run about .4%. The requirement for K varies with the species, but typically runs between .3 - .5%.

SULFUR

Sulfur (S) is a required mineral normally present in most feeds at adequate levels. A major exception is when ruminants are fed non-protein nitrogen (NPN). In that case, additional S must be fed.

The reason is that there are 3 essential amino acids that contain S. Therefore, without additional S, the rumen microorganisms cannot form these amino acids. Whenever NPN is fed, S should be added at a ratio of 1:15, sulfur to nitrogen.

Deficiencies of S produce nonspecific symptoms that typify many different conditions. The animals will gradually go off feed, slowly emaciate, and if the condition continues, death will occur. But again, because S is contained in reasonable quantities in most feeds, clinical deficiencies do not ordinarily occur.

Toxicity problems, however, are relatively common, as excesses of S are often found in groundwater and by-product feeds. For example, sulfurous acid is used in some of the processes of the corn milling industry. If errors occur in processing, by-products such as corn gluten feed can be extraordinarily high in S. In some countries S is added to molasses to prevent it from being distilled into potable alcohol.

Toxicity Symptoms. High levels of sulfates will cause profuse diarrhea. Sulfates have a low digestibility, and therefore toxicity symptoms are generally limited to diarrhea and subsequent poor performance.

If trace minerals are marginal, or sulfates are particularly high, a variety of other problems may occur. Sulfates can combine with several trace minerals, and limit their absorption. The trace mineral most commonly involved is Cu as Cu readily combines with S. Indeed, feeding sulfates has been used as a method to prevent Cu toxicity.

If the S excess is exceedingly high, death can result. Relatively common in cattle, a brain disorder occurs, and the animal dies in a stupor.

THE MICRO OR TRACE MINERALS

As mentioned, the requirements for trace minerals are so low, that rather than computing them into rations on a percentage basis, trace minerals are typically computed on a parts per million basis. Indeed, two of the trace minerals (Co, and Se) are required at only about one tenth part per million.

As a practical matter, classic deficiency symptoms of trace minerals do not ordinarily occur, except in localized areas where the soil is known to be deficient. More commonly, a non-specific reduction in performance occurs when trace minerals are marginally deficient. As a general rule, humid, wet climates with sandy and/or acid type soils are most likely to be deficient. Areas such as the southeastern U.S. and most tropical areas of the world would fit into that category. Also, within temperate areas there are also often specific regions that are deficient in specific trace minerals. For example in the U.S., portions of the Northwest are known to be deficient in selenium and much of the Midwest is deficient in iodine.

Because the levels of trace minerals required are so small, supplementation is relatively inexpensive. (In most cases, the carrier used to deliver the trace minerals will cost more than the minerals themselves.) For that reason, if the livestock operation is in an area where general or specific deficiencies may occur, then it would be wise to supplement.

Toxicity with trace minerals. Whenever supplementing with trace minerals it should be realized that most of them can be extremely toxic if excessive amounts are accidently fed. Probably the most toxic trace mineral is selenium. Required at from .1 to .3 parts per million (ppm); if added at 5 ppm, toxicity and death loss can occur in most species.

Copper is another potentially toxic trace mineral, and indeed, there are a great number of livestock poisonings attributed to copper each year. Most poisonings occur from livestock accidentally gaining access to copper sulphate, a common disinfectant. However, some by-product feeds are quite high in copper and some species, such as sheep, are particularly susceptible to Cu toxicity. For example, swine are often fed high levels of Cu, but if sheep are accidentally fed the same feed, death loss can occur.

Discussion of the functions of the various trace minerals could fill an entire text since most are interrelated. However, the following generalization can be made:

Cobalt - A mineral extremely important in ruminant animals. Cobalt is central to the synthesis of vit. B12 and must be present if rumen microorganisms are to snythesize B12. Without cobalt, ruminants become lethargic and ematiated. It is required in exceedingly small amounts (about .1 part per million).

Copper - Required for a number of body enzyme systems as well as the coloration of hair. Indeed deficiency symptoms include a dullness of the hair coat. Copper is also involved in hemoglobin formation, and thus anemia may be another sign of deficiency. Excess sulfur, or the presence of molybdenum can tie up copper and thereby increase the requirement. Copper is highly toxic when fed in excess.

Iodine - Known as the anti-goiter factor, Iodine is required for the formation of thyroxin, the hormone produced by the thyroid gland. If deficient, the thyroid swells up to a grotesque size, creating what is known as a goiter. Iodine deficient soils are prevalent through much of the Midwestern United States.

Iron - The primary function of iron is in the formation of hemoglobin. The function of hemoglobin is to transport oxygen in the blood; therefore without adequate iron, anemia develops.

Manganese - Manganese is required in a great number of enzyme systems. Severe deficiencies do not ordinarily occur in temperate areas, although reduced growth may sometimes exist as a non-clinical problem.

Selenium - The most controversial of all trace minerals. For many years the addition of selenium to feeds was illegal. The reason was that it can be highly toxic and even carcinogenic when included at excessive levels. However, selenium is a required mineral, and when deficient will create a form of muscular distrophy (known as white muscle disease). What must be always kept in mind, however, is that the margin for error in selenium is

Figure 17-2. An apparent selenium deficiency. Selenium deficiency results in a form of muscular distropy.

quite small. As explained earlier in the text just 5 parts per million can be toxic (requirement for most species runs .1 to .3 parts per million.

Zinc - Zinc is required in many different enzyme systems, but is best known for its effect upon skin quality. Swine are apparently more susceptible to Zinc deficiencies than any other species. The classic deficiency symptom is known as parakaratosis, which is a series of skin eruptions and lesions.

VITAMINS

The term vitamin actually means "vital for life", and indeed, vitamins are vital for life. Even so, the very existence of vitamins wasn't even discovered until the turn of the century. Up until that time, deficiency diseases were often thought to be due to toxic substances found in feeds.

The discovery of vitamins revolutionized animal agriculture. Without the incorporation of vitamins in animal feeds, modern intensive methods of production could never have been developed. For example, without vit. D (the sunshine vitamin) it would be impossible to raise poultry and swine in the climate controlled confinement buildings so common today.

VITAMIN CLASSIFICATIONS

Vitamins are divided into two different categories: 1. fat soluble and, 2. water-soluble. The fat-soluble vitamins include vitamins A, D, E, and K. The water-soluble vitamins are comprised of vitamin C, and all the B vitamins; Thiamine, Riboflavin, Pantothenic Acid, Biotin, Pyroxidine (6), Folacin, and Cobalamin (B12).

The fat-soluble vitamins can be stored in the body fat and in some cases the liver. As a result, there isn't a pressing need for a daily intake of them. When daily intake is inadequate, the stored reserves can be used. Likewise, when more than enough is consumed on a given day, part of the excess is stored for future use.

When vitamins are not consumed in sufficient amounts, deficiency symptoms occur. In most situations, adequate vitamin nutrition will reverse deficiency symptoms. However, reduction in gain, growth, reproductive efficiency, etc. may already have been experienced. Therefore, it is necessary to predict vitamin requirements and supplement if the feed, forage, ration, etc. will be deficient.

The necessity of supplementing vitamins depends greatly upon the species and the ration being fed. While all livestock species have physiological requirements for all the vitamins, as a practical matter, not all the vitamins need to be supplemented. This is because animals can often synthesize various vitamins and thus don't have a dietary need. For example, farm animals do not ordinarily need dietary vit. C as they can synthesize it. Ruminants can synthesize all of the B vitamins and do not ordinarily need a dietary source.

For ruminants, as a practical matter, only vit. A is a major concern. Even then, if green leafy forage is fed or grazed, vit. A supplementation is probably not required. The reason is that green leafy forage contains carotene, a compound that the body can synthesize into vit. A. Indeed, carotene is often referred to as the vit. A precursor.

Monogastric animals, such as swine and poultry, can utilize carotene, but as a practical matter it must normally be supplemented with synthetic vit. A because of the type of rations they are usually fed. Monogastric animals require a dietary source of all the B vitamins, and generally need to be supplemented. In addition, in confinement facilities where the animals are kept indoors, vit. D is a necessity.

VITAMIN A

Deficiency Symptoms. In almost all species the first symptom is night blindness, which is followed by a drying up of the tear gland. This causes the surface of the eye to dry up. Ultimately the cornea may rupture, leaving it open to infection and blindness.

By the time the tear gland of the eye begins to dry up, the epithelial tissues of the body also begin to become dry (and scaly). However, the most serious conditions occur in the mucous membranes, particularly in the lungs. Possibly as a result, animals suffering vitamin A deficiencies are more susceptible to respiratory infections. Greatly impaired reproduction is another sign of chronic vit. A deficiency.

Toxicity Symptoms. It should be noted that excesses of vit. A can be toxic. Miscalculations with injectable vit. A, or when formulating feeds, can cause death loss.

Sources of vit. A. As mentioned previously, one of the most important sources of vit. A is the carotene contained in green plants. Also as discussed, carotene is not vit. A per se but a precursor. It is a compound that can be converted into vit. A within the animal's body. Green forages usually contain relatively high levels of carotene, whereas grains and most other concentrates are extremely low in carotene. Therefore, animals grazing green pastures or otherwise fed rations high in green forages do not usually need vit. A supplementation. Animals fed rations high in concentrates, however, usually do require vit. A supplementation.

It is highly important to realize that as plants mature and turn yellow the carotene level declines rapidly. Dry mature grasses and plants are essentially devoid of carotene.

Animals are able to store vitamin A in large amounts in the liver, and mature animals in good condition can go 3-4 months on deficient diets without supplemental vitamin A. This mechanism allows grazing animals to get through the worst of winter when the grass is dormant. If bad weather keeps the grass dormant longer than usual, grazing animals may exhibit clinical signs of deficiency. Stress can rapidly use up vitamin A stores, thereby effectively reducing the time required for deficiencies to occur. It is therefore a good idea to supplement vitamin A any time grazing animals do not have access to green grass.

Carotene is relatively unstable and is readily denatured by heat and light. Hay and silages therefore usually have much lower levels of carotene than the green feeds they are made from.

VITAMIN D

Vit. D is involved with the absorption of the minerals Ca and P. Without vit. D, the absorption of these minerals is less than optimal, and, in young animals, rickets may occur. In poultry, particularly laying hens, vit. D is extremely important because of all the calcium deposited in eggshells. Known as the "sunshine" vitamin, animals exposed to sunshine usually do not require supplementary vit. D. Ultraviolet rays striking the skin's surface cause the body to manufacture its own vit. D.

Crops cured in the sun, especially hay, are reasonably good sources of vit. D for all livestock except poultry. This type of vitamin D, known as vit. D_2, has a much lower value for poultry. Vit. D_3, the vit. D found in irradiated yeast, cod and shark liver oil, is the other form of vit. D, which is much more available for poultry.

Whenever adding vit. D to feeds extreme care must be exercised, as vit. D can be very toxic in excessive quantities. In humans, for example, just 5 times the requirement can be lethal.

As a practical matter it should seldom be necessary to add vit. D by itself to a feed. Most vitamin manufacturers routinely add vit. D to vit. A supplements, so that if vit. A is added adequate vit. D will already be present.

VITAMIN E

Vitamin E is a fat-soluble vitamin, but is somewhat different than other fat-soluble vitamins in that it is not stored in the body in large amounts. Thus, when fed in excess, it is not as toxic as other fat-soluble vitamins. Also, injection is not as effective a means of delivering this vitamin. Originally vit. E was found to be involved in reproduction, and was first known as the anti-sterility factor. Indeed, the scientific name for vit. E is tocopheral, which means "birth of life" in Greek. Vitamin E is required for spermatogenesis, and a deficiency will result in reduced fertility and eventually testicular degeneration. In the female, a deficiency of vitamin E will result is the inability of the fertilized ovum to implant in the uterine wall.

In farms animals, vit. E has practical significance in preventing what is known as white muscle disease, which is a form of muscular dystrophy. Sometimes called nutritional muscular dystrophy, the muscles develop a characteristic white streaking which refers to the other name, white muscle disease. The animals become stiff and movement is difficult. Later, the animal will not be able to support itself on the pasterns and will "knuckle over". Ultimately, when the heart muscle is affected, the condition can be fatal.

The trace mineral selenium is also involved with vit. E. There are specific areas which are known to have recurring problems with white muscle disease, and animals in these areas should receive supplemental vit. E and Se.

VITAMIN K

Vitamin K is a fat-soluble vitamin that is of practical significance only in poultry. All animals require vit. K, but as a general rule, bacteria in the large intestine are able to synthesize adequate quantities (except poultry).

Vitamin K is required for proper blood clotting. Without adequate vitamin K, wounds heal slowly, and internal hemorrhaging is possible. There are certain molds that are capable of functioning as vit. K antagonists, which result in internal hemorrhaging. Dicumarol is the name of the most well-known antagonist, which is produced by a mold that grows on sweet clover. It is a very powerful compound and is capable of causing death. Indeed, commercially produced dicumarol is commonly used as a rat poison. Therefore, one must be very cautious about feeding sweet clover hay. If any has spoiled, do not feed it.

Aside from antagonists, about the only time vit. K deficiencies might be seen is when high levels of certain antibiotics are used. That is, there are some antibiotics that function against the bacteria that synthesize vit. K. In poultry, the antibiotic sulfaquinoxaline commonly used for coccidiosis control has caused problems, and in humans, the drug neomycin has been implicated.

Sources of vit. K. Aside from internal synthesis, vitamin K is found in leafy green forages. Alfalfa is a particularly rich source. Indeed, before synthetic vit. K became available, a small amount of alfalfa meal was usually included as a supplement in poultry rations.

As discussed previously, poultry are the only species where vit. K supplementation becomes a necessity. Ruminants, which are normally fed at least some forage which is high in vit. K, can synthesize vit. K as well.

THE WATER-SOLUBLE VITAMINS

The water-soluble vitamins include vitamin C, and the B vitamins. Unlike the fat-soluble vitamins, the water-soluble vitamins cannot be stored in the body in significant amounts. As a result, there is a daily need for these vitamins. However, all farm animal species can synthesize vitamin C, and ruminants can synthesize the B vitamins.

VITAMIN C

Only humans, primates, and guinea pigs have a requirement for dietary vit. C. All other animals can synthesize it. Therefore, as a practical matter, vit. C is not a consideration in formulating rations.

THE B VITAMINS

Originally, all the B vitamins were believed to be one factor. When it was discovered that there was more than just one compound, the succeeding vitamins became known as the B vitamin complex.

As a general rule, most of the B vitamins are involved in energy metabolism. While some of the B vitamins have specific deficiency symptoms, most B vitamins will result in lethargy and generalized poor growth when deficient.

Whole grains and green leafy forages are reasonably good sources of most of the B vitamins. The one major exception is vit. B12, which is only found in animal proteins. As a general practice, monogastric diets are routinely fortified with B vitamins even though whole grains and animal proteins may be in the diet. Poultry especially, often show increased growth rates over the B vitamin levels found in normal diets. In addition, poultry have exceptionally high requirements for some of the individual B vitamins.

As discussed previously, B vitamins are important in relation to monogastric animals but not ruminants, under most circumstances. Ruminant microorganisms are capable of synthesizing adequate quantities of all the B vitamins.

In relation to ruminants, B vitamins become of practical importance only in baby calves and lambs (before they have functioning rumens). The only other time B vitamins may be of practical significance is in sick or diseased animals that may not be ruminating properly.

SUMMARY AND OVERVIEW

Feeds are normally broken down into two broad categories: concentrates and roughages. Concentrates would include feeds high in starch, protein, fat or sugar. Examples would be grains, oil seed meals, tallow, and molasses. The term concentrate simply refers to the fact that these types of feeds are more "concentrated" in nutrient content than roughages.

The term roughages typically refers to feeds high in fiber. The most common examples would be hay and silage. In addition, there are a broad array of by-product feeds that would also be considered roughages. Examples would be corn cobs, straw, cotton by-products, soy hulls, peanut hulls, etc., etc.

There can be enormous variation in the digestibility between roughages. In addition, there can be enormous variation within the same type of roughage depending upon the stage of maturity at which it was harvested. For example, lush, green immature grass can run 15% protein or more and have a total digestible nutrients (TDN) value of 65% or more. But when yellow and mature, the protein level can be only 5% and the TDN only about 45%. Therefore, when utilizing roughages one must consider the stage of maturity as well as the type.

Feeds are often bought and sold on a crude protein basis. It is important to realize that the term crude protein actually refers to a laboratory procedure and gives no indication of the actual digestibility of the protein. Crude protein is simply a measure of the nitrogen content of a feed; and it is "assumed" that the amount of nitrogen can be extrapolated to represent the amount of protein in the feed. As a general rule that is true, but if the feed has been damaged (rained-on hay, etc.) the digestibility of the protein will be greatly reduced.

Non-protein nitrogen (NPN) refers to urea and similar compounds that contain nitrogen but are not true proteins. They can, however, be used as a partial substitute for true proteins in the diets of ruminant animals. But caution is in order as the total amount that can be used is limited. If amounts used are excessive, animal performance will be reduced and/or toxicity and death can occur.

Carbohydrates are the main source of energy in animal feeds. Cellulose is not normally thought of as a carbohydrate, but technically it is. The reason has to do with the fact that cellulose is not digested nearly as quickly as other carbohydrates, such as starches and sugars. Likewise, monogastric animals cannot digest cellulose (fiber) to any major degree. Only ruminants can utilize cellulose as a source of energy.

Minerals are broken down into two major categories: 1. macro minerals and, 2. trace or micro minerals. Macro minerals are required at fairly high levels (usually from .1 to .3% of the diet). Trace or micro minerals are required in exceedingly small amounts; so small that requirements are usually reported on a parts per million basis. Whenever supplementing minerals, especially trace minerals, it should be kept in mind that while problems may occur if feeds are deficient, most minerals can be toxic if excessive levels are fed.

Vitamins are divided into two broad categories; fat-soluble and water-soluble. Fat soluble vitamins include vit. A, D and E. Water-soluble vitamins include vit. C and the B complex vitamins. As a general rule, water-soluble vitamins cannot be stored in the body and therefore a daily requirement occurs. Ruminants, however, can synthesize all the water-soluble vitamins and do not ordinarily have a dietary requirement. Monogastrics are usually supplemented with the B complex vitamins. Only man, guinea pigs, and primates require a dietary source of vit. C.

Green forages contain carotene which can be synthesized into vit. A. Therefore when animals are on green pasture or otherwise fed lush green forages, supplementary vit. A does not ordinarily need to be supplemented.

Adult animals can store vit. A in the liver for 60 to 90 days. After that period, if green forages are not in the diet, vit. A must usually be supplemented. Moreover, vit. A is the most important vitamin in practical animal nutrition and deficiencies often occur. Vit. D is required for all livestock which are kept indoors and do not otherwise have access to sunshine (sunshine can be used to synthesize vit. D in the body).

CHAPTER 17A   FEED PROCESSING FOR LIVESTOCK

(excerpted from Modern, Practical Feeds, Feeding
and Animal Nutrition by D. Porter Price Ph.D.)

With the exception of pasture, nearly all animal feeds are processed in one way or another. Indeed, feed processing is a major industry servicing animal agriculture. Understanding the effect of processing upon feeds is vital to the practical application of most types of livestock feeding.

Although feeds are processed for a variety of reasons, they generally fall either under the heading of increased digestibility, or changes in physical form. Changes in physical form can have an effect on the ability of the animal to utilize the feed, or simply be for the purposes of improving the mechanical handling of the feed, or both.

PELLETING

Pelleting, one of the most common forms of feed processing, generally falls under the category of improved mechanical handling. However, in some instances pelleting is also used to increase the acceptability or utilization of a feed.

Pelleting is accomplished by grinding the feed to an extremely small particle size, and then forcing it under steam and pressure through small openings in what is known as a pelleting die. The openings generally vary in diameter from 1/8 inch to ¾ inch. As the feed emerges through the holes in the die, a rotary knife cuts the feed into the desired length of pellets. As a general rule most pellets are cut to have a length of about 4 to 7 times the diameter.

The one-eighth inch pellets are commonly used for baby pig, calf, rabbit, and poultry feeds. One-quarter to 3/8" pellets are commonly used for adult pigs, cattle, and horses. The larger ¾ inch pellets are commonly referred to as "cubes". These extra large pellets are commonly used for feeding cattle on pasture. Poultry feeds are often further processed by breaking up the pellets into smaller pieces, known as crumbles.

Pelleting allows a complete ration to be blended together, and ensures that separation will not occur. Pelleting also greatly facilitates the handling of feed through automated machinery. Thus, intensive poultry and swine confinement operations often utilize fully pelleted rations.

Increased utilization of roughages due to pelleting. With respect to high fiber roughages, pelleting can actually increase the utilization (for ruminants). The grinding increases the surface area and allows rumen bacteria to digest the fiber more rapidly. This in turn allows the animal to eat more of the feed (roughage). Thus, fiber does not build up in the digestive tract and reduce consumption as much.

Although it is the fine grinding that actually allows the increased rate of digestion of high fiber roughages, the pelleting is also important. For without pelleting, the feed would be very bulky, dusty, and unappealing to animals. Consumption would decline, and therefore the utilization of the feed would not be as great.

Other common methods of roughage processing. In addition to pelleting, roughages are also commonly chopped, or ground. (Technically, the ensiling of roughages would be considered a processing method, but the ensilation process is covered in Chapter 19, Forage Harvesting.)

Chopping, sometimes called slicing, is normally done with specialized equipment designed specifically for the task. Grinding is usually done in hammer mills.

Chopping/slicing is normally used with hay and is done with a series of rotary knives. The idea is to cut the hay up into lengths of about 2 to 4 inches. Chopping/slicing is usually done whenever fine particles are not desired but hay is to be included in a mixed ration. Dairy rations, in particular, most commonly utilize chopped/sliced hay. Grinding is usually done whenever dust and fine particles are not a problem, and/or other roughages besides hay are utilized (peanut hulls, corn cobs, etc.).

DRY GRINDING AND ROLLING OF GRAINS

Dry grinding and rolling are most commonly practiced with feed grains to break the seed coat. This makes grains more digestible to livestock. That is, the seed coat is a barrier to digestion. When grain is fed to most livestock species in the whole form, much of it is not chewed and will be swallowed whole. Therefore, in most animal feeding situations some sort of grinding or rolling of grains is necessary.

While dry grinding and rolling are often considered together, they can be quite different. In some cases they can actually have different applications.

In dry rolling, the grain is passed through a pair of corrugated rollers. The corrugations (see illustration) grip the grain, and pull it through the roller. The pressure exerted by the rollers causes the grain to crack or break up into pieces. The size of the pieces depends upon how closely the rollers are set, and how fast the grain is fed into the rollers.

Setting the rolls close together causes the pieces to be smaller since there will be a smaller space the grain will have to crumble to fit. Setting the rollers fairly far apart causes the grain to only be broken up into a few larger pieces. (Quite often this is referred to as "cracking" grain.)

Grinding grain refers to putting grain through what is

known as a hammer-mill. A hammer-mill is a device that pulverizes grain by means of a series of rotating metal bars. These bars or "hammers" pulverize the grain by breaking it down against a series of holes or openings in a circular metal plate, known as a screen. The size of the holes in the screen determines the ultimate size of the particles.

In theory, both methods are supposed to be capable of being able to produce similar sized end products. In actual practice, rolling is usually a much better method if coarse "cracking" (large particle sizes) are desired*, and grinding is usually more feasible if very small, fine particle size is desired.**

STEAM ROLLING OF GRAINS

Steam rolling of grains (not to be confused with "steam-flaking") is a process by which grain is softened with steam before being passed through a roller mill. Instead of crumbling into pieces (as in dry rolling), the grain emerges in one piece. The kernel is flattened, and the corrugations crush the seed coat, giving the grain a distinctive textured look.

The primary purpose of steam rolling is to avoid the dust and fines associated with dry rolled grains. In some cases, as in dairy feeds, steam rolling or "texturizing" the grain results in better acceptance by the animals. Steam rolling also produces a more pleasing appearance and is therefore commonly used in "consumer" type products such as horse feeds.

STEAM-FLAKING

Steam-flaking, often confused with steam rolling, is undertaken for a different reason. Whereby steam rolling is undertaken for the purpose of texturing grain, steam-flaking is undertaken to increase the digestibility.

---

* Hammer mills typically produce more fines and dust than roller mills, which is not desired when coarse cracking of grains is called for.

**Trying to set rolls very close together to obtain very small particle size can be done, but inevitably the corrugations will come into contact with one another, and greatly reduce roll life.

Figure 17A-1 A&B. Steam flaking mill. Notice the massive springs (below) used to put pressure on the rolls. This pressure, plus extra time in the steam cabinet, is what separates steam flaking from steam rolling.

The difference between the two processes lies in the amount of time exposed to live steam, (15 to 30 minutes vs. 5 or 10) and the amount of pressure exerted by the rolls. An ordinary roller mill is incapable of exerting enough force to truly flake grain. A flaking mill consists of rolls that are much larger and heavier than those used in ordinary roller mills. In addition, hydraulic rams or very heavy springs are used to maintain pressure on the grain passed between the rolls.

The response in terms of digestibility depends upon the animal species, and the type of grain. As a general rule, only ruminants show a digestibility response to steam-flaked grains. Of all the feed grains, sorghum by far shows the greatest response.

This is because the starch granules in sorghum are surrounded by a tough protein matrix. This matrix is a barrier to digestion, and is broken down during the pressure exerted upon the grain.

While the steam-flaking of grains has been done to a limited degree in dairies,* by far the greatest application has been in the beef cattle and sheep feedlot industry. For feedlot purposes, steam-flaked grain can be fully 10% more digestible than dry-rolled or ground sorghum. In addition, consumption is often greater than dry processed and so animal performance often shows improvement in excess of 10%. Corn does not show the response to steam-flaking that grain sorghum does. Generally about a 2% to 4% response (over dry-rolling) is all that can be expected. Wheat and barley apparently respond very little to steam-flaking. Consumption is usually a little better, but steam rolling will suffice as well or better than steam-flaking.

PRESSURE-FLAKING

Pressure-flaking is very similar to steam-flaking, except that a pressurized steam chamber is used. By placing the grain under pressure with steam, time in the steam chamber can be greatly reduced. Due to the expense and danger involved with pressure-flaking, it is not a popular processing method.

POPPING

Popping is a processing method developed several years after steam-flaking. In true popping the grain is heated until its internal moisture causes the kernel to rupture. As with steam-flaking, the primary application of popping has been with cattle and sheep feedlots. For that use, the grain is usually heated to 450-550 F to produce about a 30% pop. The grain is then run through a roller mill, and when the pressure from the rolls is exerted on the grain, the unpopped kernels usually go ahead and pop. The reason only an initial 30% pop is strived for is simply to save energy. Rolling is required to reduce the bulk density of the feed. Without rolling, a popped ration would be too fluffy and would reduce intake. Popping reduces moisture in the grain, and therefore 3-6% water is usually added back to the grain after processing.

With popping, sorghum is again the grain that responds the most. When done correctly, popped grain sorghum will perform as well as steam-flaked.

Popping is more energy efficient than steam-flaking. This is because dry heat is applied directly to the grain, rather than in the form of steam. Water requires over twice as much energy to change its temperature, and thus the creation of steam uses more energy than applying the heat directly to the grain.

The main disadvantage to poppers are that the equipment is very specialized, expensive, and potentially dangerous. More skilled or technical help in operating and maintaining the machinery is required. Very serious explosions have occurred with popping equipment.

HIGH-MOISTURE GRAIN (Early Harvested Grain)

High-moisture grains are used extensively in the feeding of ruminants. High-moisture grain is defined as grain harvested at about 22-30% moisture and either ground and ensiled, treated with organic acid preparations as a preservative, or stored in oxygen limiting structures.

The feeding value of high-moisture grain sorghum is substantially better than dry-processed milo. Research data varies, but in general, high-moisture sorghum approaches the value of steam-flaked or popped sorghum

---

* Steam-flaking is contraindicated in most dairy situations as it changes the acetate:propionate ratio in the rumen, and lowers butterfat. However, in areas where grain sorghum is substantially cheaper than other feed grains, steam-flaking has sometimes been used.

Figure 17A-2. High moisture grain being ground for ensilation.

(on a dry matter basis). The reason for the increased value is that when grain sorghum is harvested early, it is immature. Being immature the protective protein structure surrounding the starch has not had time to form.

High-moisture corn apparently has about the same energy value as dried corn on a dry matter basis. It is difficult to compare high-moisture grain to dry grain because high-moisture grain has a number of feeding idiosyncrasies. To begin with, in most feeding situations, high-moisture grain should not make up the entire grain portion of a ration. At least 20-30% dry grain should be added. Without additional dry grain, consumption tends to decline.

High-moisture grain rations must be fed at least twice a day (preferably three times a day), and the animals must be allowed to clean up between feedings. High-moisture grain molds rapidly, and this will reduce consumption and performance if left in the feed bunk.

RECONSTITUTED GRAIN

Reconstitution means adding water back to dry grain to bring a total moisture up to 24-30%. The only grain that responds significantly to reconstitution is sorghum. When done correctly, reconstituted milo, like high-moisture milo, approaches the value of steam-flaked or popped milo.

To reconstitute sorghum it must be stored whole, and in some sort of oxygen-limiting silo. The usual treatment is to add water and store the grain for at least 14 days. The grain is removed and ground or rolled just before feeding. Grinding dry sorghum, adding water, and ensiling in a pit (as commonly done with early harvested (high-moisture) sorghum, does not produce the response obtained when sorghum is stored whole and then ground.

This is because reconstitution is actually the germination process set into motion. When the seed soaks up moisture, the germ secretes an enzyme which breaks down the protein matrix that surrounds the starch molecules. This is to allow the seed to fully utilize the stored energy (starch). But being in an oxygen-limiting silo, full germination (sprouting) does not occur.

Figure 17A-4. An oxygen limiting silo being used for the reconsitution of grain.

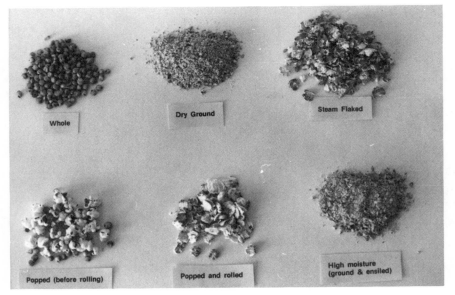

Figure 17A-3. The various end products for the different methods of processing for grain sorghum.

TEMPERING

The practice of tempering (soaking) grain in water before feeding has been carried out for many years. A very common practice with barley, tempering sometimes increases performance (over dry grains) due to increased palatability and intake; but it does not measurably improve digestibility.

TOASTING

The term "toasting" is usually used in reference to the heating of legume seeds, soybeans in particular. Soybean contains what is known as a trypsin inhibitor (a compound that ties-up or otherwise renders the amino acid trypsin indigestible). Heating or toasting soybeans deactivates the trypsin inhibitor and thereby greatly increases the value of soybean meal for monogastric animals.

For ruminant animals toasting also has value, in that toasting decreases the degradability of soybean meal in the rumen. This allows more to escape rumen digestion and increases the value of soybean meal protein.

OTHER FORMS OF HEAT TREATMENT

Aside from the toasting of soybean meal, heat is used for various reasons in the processing of an array of feeds.

DEHYDRATION

Dehydration is a very common use of heat, and is used for a great number of feeds. Brewers and distillers grains, corn gluten, meat meal, blood meal, alfalfa meal, whey, and milk are a few of the more common feeds processed by dehydration.

As a general rule, dehydration is used as a means of preserving the feeds, and/or reducing the weight in order that they can be shipped long distances. Dehydration is expensive, and is often used only if the feed cannot be sold locally in its wet form.

Most of the feeds dehydrated are protein concentrates and one caution applies to all of them. The caution is that excessive heat can denature proteins and make them indigestible. Known as the Browning reaction, excessive heat turns most proteins a dark color, which is indicative of reduced digestibility. In years past this was a major problem with dehydrated milk, although in recent years most milk plants have improved their methodology. Excessive heat remains a problem in the alfalfa meal and distillers feed industries. In the case of alfalfa, excessive heat also makes the product quite bitter, and can cause feed refusal and sorting in several animal species.

ROASTING

Roasting generally refers to the heating of corn grain, which reportedly increases its feeding value. However, roasting has never caught on in the commercial feed industry as any improvement is quite marginal.

SUMMARY

As a general rule, feeds are usually processed either to improve the digestibility or to change the physical form. Whenever the physical form is changed, it is usual-

Figure 17A-5. An alfalfa dehydration plant as discussed in the text, the dehydration of alfalfa, as well as the heating of soybean meal, meat meal, and other protein concentrates must be carefully controlled. If excess heat is used, the digestibility of the protein will be greatly reduced.

Figure 17A-6. The aftermath of a popping machine explosion. As explained in the text, the popping of grain is energetically more efficient than steam flaking. However, the machinery is complicated, and if not maintained properly, very serious explosions can result.

ly for the purposes of improving the handling characteristics. Pelleting would be the primary example of this.

Feed grains are usually ground or rolled in order to break the seed coat and increase the digestibility. Feed grains are also sometimes subjected to various methods of heat processing. Steam rolling, flaking, and popping are the most common.

Other feeds such as soybean meal are also subjected to heat. In the case of soybean meal, heat is used to deactivate an amino acid antagonist which prevents the utilization of trypsin (an amino acid). Other feeds such as whey, brewers grains and skim milk powder are dried with heat.

Whenever heat is used, care must be excercised to be certain the heat is not excessive. Excessive heat denatures the protein and reduces the digestibility.

CHAPTER 18 BASIC ANIMAL DIGESTION

Excerpted from *Modern, Practical Feeds, Feeding and Animal Nutrition*,
by D. Porter Price, Ph.D.

Among farm animals there are two basic types of digestive systems. Swine and poultry have what is known as the "monogastric" type of system, and cattle, sheep and goats have what is known as the "ruminant" system.

The monogastric system is the type that we humans have, and if the word is broken down, the meaning is easy to understand. Mono means one, and gastric means stomach. Therefore humans, or animals with the monogastric system have one stomach.

The ruminant animal, on the other hand, is said to have more than one stomach. Actually it is more correct to say that ruminants have more than one compartment. Indeed, as is stated in most textbooks, ruminants have stomachs with four compartments. While it is true that ruminants do have four separate compartments, it is more important to recognize that ruminants have two separate digestive systems; i.e. there are four stomach compartments, but together they comprise two different types of digestive systems.

Ruminants have all the digestive organs contained in the typical monogastric system (stomach, intestines, and secretory organs), plus a "rumen". The rumen is a large fermentation vat that is connected in front of the monogastric system. The rumen gives the animal the capability to digest roughages (hay, grass, silage, etc.) which cannot be digested by monogastric animals. Thus the ruminant has the capability to digest all the feeds monogastrics can digest, plus fibrous feeds (roughages) that monogastrics cannot digest.

The Monogastric System. As stated in the introduction, in the monogastric system there is but one stomach. Attached to the stomach are the intestines. The small intestine is first, followed by the large intestine. The end of the large intestine is known as the colon, where fecal matter collects until excretion. There are also a number of glands that secrete digestive fluids into the tract. These glands consist of the pancreas, the gall bladder, and the salivary glands.

Contained within the stomach are a series of strong acids and enzymes. Hydrochloric acid, the most powerful, is present in an extremely concentrated form. To protect the stomach from these acids and enzymes, the stomach is lined with a mucous membrane. Indeed, if it weren't for the mucous membrane, the stomach wall would be digested by its own secretions.*

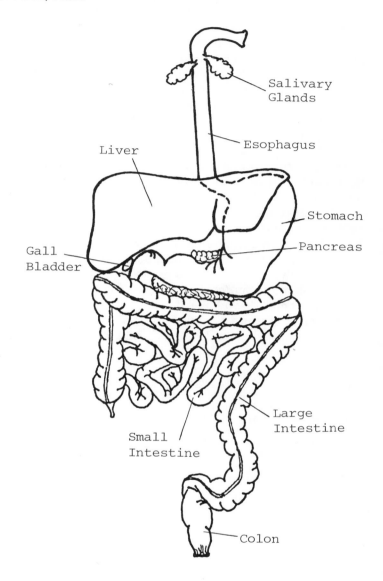

Figure 18-1. The Monogastric Digestive System. This would be analogous to the system found in swine, dogs, and is similar to poultry (see also Fig. 23-2).

"Pepsin" the principal enzyme secreted by the stomach is used to break down protein. Actual absorption does not take place in the stomach, but takes place after passage into the small intestine.

The upper part of the small intestine that connects to the stomach is known as the "duodenum". One of the most important parts of the digestive tract, the "duodenum" is the site of secretion of a number of enzymes. Some of the enzymes are secreted by the

---

* In humans, this is why alcoholics tend to have more problems with stomach ulcers. Alcohol tends to dissolve the mucous membrane and leaves the stomach more vulnerable to its own secretions.

duodenum itself, and others come from specialized organs separate from the intestine.*

As stated, the stomach secretes pepsin** a proteolytic enzyme (an enzyme that breaks down protein). Because of the action of pepsin, proteins enter the small intestine broken down into short chains of amino acids known as peptides. Additional enzymes secreted in the small intestine, known as peptidases, break the peptide chains down into their individual amino acid constituents. These individual amino acids are then capable of being absorbed.

Another very important enzyme is secreted in the small intestine (although not produced there). Known as "amylase", this enzyme is used for breaking down starch (the scientific name for starch is amylose). Also known as pancreatic amylase, amylase is synthesized in the pancreas and secreted into the small intestine via a duct. (Some amylase is also secreted by the salivary glands, but the amounts in most species are quite low.) Lipase, another enzyme produced in the pancreas is also secreted into the small intestine. The purpose of lipase is to break down lipids (fats).

The liver produces a digestive secretion that also aids in the digestion of fats. Known as bile, it is actually secreted from the gall bladder, which is a storage organ attached to the liver.

Bile is not an enzyme, in that it does not break down the chemical structure of fats per se. Rather, it is an emulsifying agent that causes fat to disperse into small globules. This allows the true enzyme, lipase, to attack the fats more readily. Without bile, fat digestion would be greatly retarded, and even moderate amounts of fat in the diet would create substantial intestinal disorders.

As mentioned earlier, most absorption occurs in the small intestine. To increase absorption, the small intestine is lined with hundreds of thousands of small finger like protrusions known as "villi". Through the use of these villi, the surface area available for absorption is increased approximately 600 times.

Undigested material passes into the large intestine. Except for water, sodium, and calcium, there is very little absorption in the large intestine. However, in species such as the horse and rabbit, the large intestine does play a major role in digestion. In that case the large intestine is known as a cecum, and is used for bacterial fermentation, much like a rumen.

THE RUMINANT SYSTEM

As mentioned previously, the ruminant system has all the organs contained in the monogastric system, plus a rumen. The rumen is situated in front of what would otherwise be considered a monogastric system, which radically changes digestion patterns. Specifically, it doesn't change the way the true stomach (abomasum), intestines and secretory organs function, so much as it changes the feed (chyme) that reaches those organs.

---

* "Sucrase", "Maltase" and "Lactase" are three of the enzymes secreted in the small intestine. The function of these enzymes is easy to understand, since enzymes are often named for the compound they break down. The letters "ase" are usually just added to the end of the name of the compound. Thus, sucrase, maltase and lactase break down the sugars sucrose (table sugar), maltose (a sugar often contained in grain), and lactose (milk sugar).

** The name of pepsin is an exception to the naming of enzymes.

Figure 18-2. Photograph of a preserved rumen from a feedlot steer.

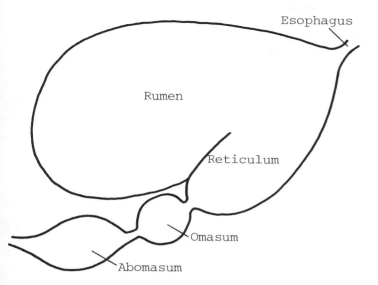

Figure 18-3. Diagram of the Ruminant Digestive Tract.

Ruminant Protein Digestion. In ruminant protein digestion, rumen microorganisms break down the ingested proteins for their own use. Proteins are made up of amino acids, which simply described, are organic molecules with nitrogen attached. The nitrogen is indispensable, for without nitrogen, there is no protein.

The rumen microorganisms break the nitrogen away from the rest of the protein molecule, and resynthesize it into protein for their own use. They use the protein to reproduce by division. Eventually the microorganisms are moved from the rumen into the abomasum. They are then digested by the digestive juices in the abomasum or the small intestine and used as a source of protein themselves (see Figure 18-4). In essence, whenever a ruminant eats a proteinaceous feedstuff, it is actually feeding the microorganisms contained within the rumen, which are in turn used as a feedstuff themselves.

Since the rumen microorganisms break down most of the protein the animal eats, and then resynthesizes it, the amino acid balance is of little concern under most circumstances. This is why non-protein nitrogen compounds such as urea may be used for part of the required protein. In essence, ruminants have a nitrogen requirement, not a protein requirement per se.* With monogastric animals the amino acid content of the proteins eaten must be balanced, or the animal can not effectively utilize it.

---

*In certain special situations there are circumstances where amino acid balance is of importance. Amino acid balance is of particular importance in very young calves, and in some feedlot, and high producing dairy rations.

Energy Digestion in the Rumen. Starch and cellulose are the principal forms of energy digested in the rumen. Starch is the primary form of energy found in grains, and cellulose the primary form of energy found in roughages. Animal digestive systems do not have the enzymes required to break down cellulose. Only microorganisms have the necessary enzymes to break down cellulose. Thus, ruminants can eat and utilize cellulose since the rumen microorganisms digest it for them.

This gives ruminants an advantage over non-ruminants since they can eat and digest a much broader array of feeds. However, when eating grains or other concentrate feeds, ruminants are at a disadvantage. The reason is that the rumen microbes waste a great deal of the feed energy, by giving it off as heat and gas.

The gas given off is methane, which has the same chemical formula as natural gas used for home heating. This is an energy loss, which occurs as a by-product of rumen digestion.

An even greater loss is incurred by the energy that must be used by the microorganisms to conduct the reactions that liberate the gas. This is known as the Heat Increment or Fermentative Heat Loss. The chemical bonds that hold cellulose together are particularly strong, and therefore considerably more energy is required to break the bonds of cellulose (as compared to starch). However, the loss from the fermentation of either starch or cellulose is substantial.

The combined energy loss from gaseous and fermentative energy losses will vary with the type of feed used. As a general rule, it may be said that approximately 25% to 40% of all energy digested in the rumen is lost. This energy loss is the primary reason that ruminants are poor converters of grain; the primary reason why hogs and other monogastric animals are so much more efficient. That is, this is why feedlot cattle will convert grain rations to meat at a ratio of about between 7 and 8 to 1, whereas hogs will convert at a ratio of about 3.5 to 4 to 1. Poultry can convert at up to 2.5:1 to 2.0:1. The hog and poultry, of course, do not have a rumen, and therefore the starch from the grain they eat is digested and absorbed directly; there are essentially no fermentative energy losses.

However, while the ruminant is not as efficient as monogastric animals at converting grain, it should be pointed out that in terms of meat or milk produced, only a portion of the feed utilized comes from concentrates. Feedlot cattle, for example are brought to the feedlot for grain feeding only after they have spent a considerable amount of time on pasture. In most cases only 30-40% of the final weight of a feedlot steer can be attributed to grain feeding.

Ruminant Digestion of Fat. Most feedstuffs fed to ruminants do not contain appreciable amounts of oil or fats. However, ruminants can utilize small quantities of

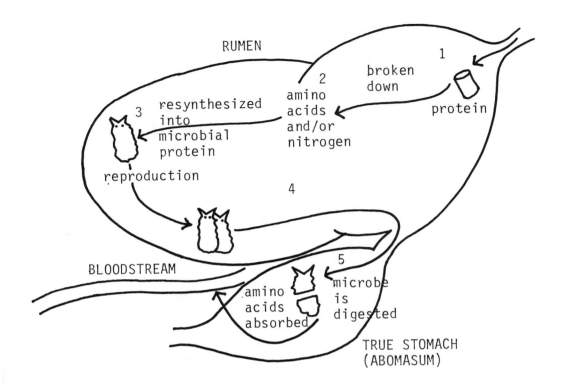

Figure 18-4. Schematic of Ruminant Protein Digestion.

1. Protein is eaten and ingested by the animal.
2. The rumen microorganisms break the protein down into its constituent amino acids, and/or nitrogen compounds.
3. The rumen then resynthesize the amino acids or nitrogen compounds into proteins for their own use.
4. With these proteins the rumen microorganisms are able to grow and reproduce.
5. Ultimately some of the microorganisms are transported to the abomasum or true stomach, are digested, and used as source of protein.

fat, and are often fed tallow or feed grade fats in dairy, and feedlot rations. The textbook energy value for fat is given as 2.25 times the energy value of carbohydrates. For use in grain rations for cattle, the energy value of fat vs. carbohydrates will be somewhat greater.

The actual mechanism is not clearly understood, but fat is known to inhibit the production of methane gas. As pointed out in the section on energy digestion, methane gas is an energy loss, and so the decrease in production of methane is an energy saving. Fat will therefore show a gain response (in feedlot grain rations) in excess of what can be attributed to the energy it contributes to the ration. However, at levels higher than 4% of the ration, fat will usually reduce intake, and can cause cattle to go off feed.

In high roughage rations, fat apparently depresses the utilization of fiber. Again, the actual mechanism is not understood, except that cellulose-digesting bacteria are somehow inhibited by the presence of fat.

In dairy rations, the use of fat is sometimes used to increase the energy density of the feed, as well as to increase milk fat percentage. However, not all rations react the same to fat addition.

Utilization of Low Quality Roughages. The fiber content of roughages limits the amount of roughage a ruminant can consume. The higher the fiber content the lower the intake. This is primarily due to two factors:

(1.) Cellulose, the most common carbohydrate in roughages, is difficult for the rumen microorganisms to break down. As a result, cellulose is broken down much more slowly than other carbohydrates. Since it is broken down relatively slowly, passage of cellulose through the rumen is also slow.

(2.) Lignin, a structural component in fiber, is almost completely indigestible. Even the rumen microorganisms cannot break down lignin to any degree. As the fiber content of the roughage goes up, so does the lignin content.

Cellulose and lignin therefore accumulate in the rumen. This causes the animal to feel full, which eventually reduces intake.

As the fiber content of roughages increases, the protein level usually decreases. Particularly serious is the fact that as the total protein decreases, the amount of digestible protein decreases more rapidly. In low quality digestible roughages, not only is there not enough protein to meet the needs of the animal, but often there is not even enough protein to maintain an active rumen microbial population.

From 3% to 5% of what is reported as crude protein in roughage is really just nitrogen complexed with the lignin, or indigestible fraction of the fiber. Straw and cornstalks will show crude protein values of 3% to 5%, and many grass hays will have only 5% to 7% protein. In essence, straw and cornstalks will have 0% to 2% digestible protein. Under most circumstances that is not enough protein to maintain a healthy population of rumen microorganisms. As a result then, the rumen microorganism population decreases.

As the microorganism population decreases, the rate of digestion decreases at a greater rate. Microorganisms are responsible for breaking down roughages, and therefore as the numbers of microorganisms decrease, so does the rate of roughage digestion. "The less bugs there are to work on the fiber, the slower it is going to be utilized." The slower it is utilized, the more the fiber builds up in the rumen, and the fuller the animal feels. Intake therefore decreases rapidly after the first few days. For example, hungry cattle can eat up to 3% of their bodyweight in poor quality grass or hay for two or three days. Over a longer period, however, intake would be closer to 1.25-1.75% of their bodyweight.

Feeding Concentrates to Ruminants. Special care must be exercised when feeding grain and other concentrates to ruminants. That is, ruminants that have been consuming primarily roughages must be gradually adjusted to high concentrate rations. Normally it takes about 3 weeks to bring ruminants (cattle and sheep) up to a high grain ration (about 80-90% concentrate).

Actually, it is not the animal that needs to become adjusted, so much as it is the microbial population in the rumen. When starch is introduced into the rumen, it is broken down rather quickly by the microorganisms. Lactic acid is given off as a by-product, and in animals that have been consuming mainly roughages, there will be a very low population of bacteria that can utilize the lactic acid. Thus, lactic acid can increase very rapidly in cattle abruptly moved to rations high in concentrates. The pH in the rumen of cattle eating roughages will be about 6.5; when abruptly moved to high concentrate rations, it can go as low as 3.5. When this happens, the animal suffers what is known as lactic acidosis. Acid enters the bloodstream and severely upsets the animal's metabolism. In acute cases the animal may stagger, fall, go into convulsions and die. In less severe cases, the symptoms of founder may appear. In founder, the animal's feet become extremely tender, making it very difficult for them to walk. This is because the hoof becomes engorged with blood, and through a mechanism not thoroughly understood, the 3rd phalanx bone of the foot rotates downward, creating severe pain and pressure upon the sole of the foot. Over a period of time the hoof will grow at an extraordinarily rapid rate from the toe, with distinctive annular rings running the full length of the hoof.

Treatment consists of taking the animal off feed and drenching the rumen with mineral oil, which causes the

Feedstuff	Ration 1	2	3	4
Grain	48	60	72	83
Dry roughage	45	33	21	10
Ration conditioner (molasses, fat, etc.)	3	3	3	3
Protein supplement	4	4	4	4
	100%	100%	100%	100%
Days fed	7-10	5-7	5-7	finish

Table 18-1. A typical ration series for bringing ruminant animals up onto high concentrate rations. If not allowed to adjust to grain or other concentrates slowly, a condition known as acidosis (described in text) will develop. Acidosis can be fatal.

digestive system to be purged, and to a certain extent, prevents absorption of nutrients (lactic acid). In acute cases the animal may be intravenously injected with sodium bicarbonate.* Of course, the best solution to the problem is prevention; newly arrived cattle brought up on grain type rations gradually.

As a general rule, acidosis and founder is a problem with cattle and sheep feedlots, although dairy cattle suddenly given access to concentrates could also be affected. As mentioned, the solution is just to move the animals on to grain or other concentrates slowly. To do this, most feedlots utilize a series of approximately 4 rations; a top high concentrate ration, and 3 intermediate rations. In addition, most feedlots will feed the 1st or starter ration over hay for 2 or 3 days (increased roughage). A typical ration series would look something like Table 18-1.

SUMMARY

Among farm animals there are two basic types of digestive systems; 1. monogastric, and; 2. ruminant. Monogastric means one stomach, and typifies the type of system possessed by swine and poultry.

The monogastric system consists of the esophagus leading to the stomach. Connected to the stomach is the small intestine, which is, in turn, connected to the large intestine.

A very important part of the digestive system is the accessory organs that secrete fluids that aid in the digestive process. One of the most important is the pancreas which secretes the enzyme amylase which is used in the breakdown of starch. The liver secretes bile, which is an emulsifier used in the digestion of fats. Bile collects in the gall bladder which is a storage organ. From there it is secreted into the small intestine. The salivary glands are also secretory organs. In some species small quantities of amylase are secreted, and in ruminants buffers are secreted.

The ruminant animal (cattle, sheep, and goats) have all the organs contained in the monogastric system, plus a rumen. The rumen is sometimes called a stomach, but actually it is nothing more than a large fermentation vat. A true stomach secretes acid and pepsin (an enzyme used in the digestion of protein). The ruminant animal has a true stomach, which is known as the abomasum. The rumen itself, however, secretes nothing. Rather, the rumen is filled with billions of bacteria and other microorganisms that actually conduct the digestion.

The microorganisms in the rumen have the capability to digest fiber, which gives the ruminant animal the capability to utilize roughages as a source of food. The microorganims break the fiber down into compounds the animal can absorb and utilize as a form of energy.

The microorganisms are also capable of forming proteins from non-protein nitrogen compounds. That is, nitrogen is a vitally important constituent of protein. However, not all nitrogenous compounds are protein. But rumen microorganisms can take many of these non-protein nitrogen compounds and synthesize them into true proteins. This allows substitution of less expensive non-protein nitrogen compounds for a portion of the protein in the ruminant diet.

Like monogastrics, ruminants can utilize grain and other starches for energy. But whenever ruminants are to be fed grains or other high energy concentrates, they must be allowed to adjust to them over a period of time. If ruminants are abruptly put on high energy diets, founder and acidosis may occur.

---

* If this procedure is undertaken, extreme care must be exercised. Because if an excess of bicarbonate is injected, the pH of the animal's blood may go too far the other way. That is, go from acid to basic. Alkalosis, can be fatal just as can acidosis.

CHAPTER 19 HAY AND SILAGE MAKING
(Forage Preservation)
Excerpted and condensed from *Modern, Practical Feeds, and Feeding and Animal Nutrition* by D. Porter Price, Ph.D.

THE PROPER HARVESTING OF FORAGES

A major consideration in many forms of animal feeding is the proper harvesting of forages. It is important for livestock managers to be able to visually appraise the nutritional value of a particular forage. Likewise, it is highly important for farm managers to know how to preserve forages for maximum nutritive value. The purpose of this chapter is to provide information on both these topics.

HAY AND HAYMAKING

The making of hay is nearly as old as the domestication of livestock. Early man learned that by cutting and drying grass in the summer, it would be preserved for feeding to livestock in the winter. If cut while green, much of the nutritive value would be retained, whereas if left to mature and turn yellow, the feeding value would deteriorate greatly.

Today, the primary motive for hay making is the same. What has changed is the equipment and special considerations made for that equipment. For when we discuss haymaking, we must also discuss the equipment, or packaging systems to be used. That is, once hay is made, it is either baled, cubed, or pressed into large movable "stacks".

Basic haymaking. The basic idea in hay making is to cut

Figure 19-1 A&B. Forage that has been cut and windrowed (to dry) is being picked up by a round baler.

forage (usually a grass or a legume) at a green, immature stage and then dry it in the sun. The actual stage when it is cut will vary with the individual crop, but will normally be a stage that allows near maximum plant growth (yield), but without allowing the plant to approach maturity. That is, as plants mature the nutritional value declines rapidly.

After cutting, the forage is left in the field for drying. Then, when sufficiently dry, the forage (which is now considered "hay"), is picked up.

In times past, this process was relatively simple. That was because hay was hand stacked or otherwise put into loose bundles. In that case, the moisture was not necessarily critical; anything in the area of 25% moisture or less would suffice.

Today, however, the moisture level is critical. For most hay packaging systems, 25% moisture would be excessive. Excessive moisture in tightly packed bales or hydraulically pressed "stacks" will cause bacterial fermentation to develop. This fermentation causes heating which leads to degradation of the nutrient value of the hay. The bacteria use the readily available carbohydrates and sugars to make the heat, and the heat itself destroys the digestibility of the protein. Thus, excess moisture depletes the energy value of the hay, while denaturing the protein. But in addition to the potential for destroying the feed value, there is also the danger of spontaneous combustion.

Figure 19-2. A conventional square baler.

Figure 19-3. Rained-on bales stacked in the field for drying. If hay is rained on, or otherwise too wet, it must be dried before being stacked in the barn, otherwise the moisture allows microbial growth. These microbes produce heat which can destroy much of the nutrient value of the hay, and even cause spontaneous combustion.

Spontaneous combustion. Whenever there is excessive moisture in tightly packed hay, spontaneous combustion is a very real danger. Indeed, millions of dollars of hay and hay storage facilities are lost every year to spontaneous combustion.

Figure 19-4. Spontaneous combustion. Spontaneous combustion can cause an enormous roaring fire, or a smaller smoldering fire that burns for days, or even weeks. Putting out a sponanteous combustion fire is extremely difficult with ordinary equipment. As discussed in the text, injecting CO_2 is about the only effective means.

For that reason, if it is suspected that packaged hay is excessively wet, it should never be stored indoors, or even stacked together in large quantities. If the hay is already stacked, then extreme caution is in order. Ideally, the stack should be opened up to allow the hay to dry. Unfortunately however, opening up an already hot or fermenting stack can cause it to burst into flame. In that case, a thermometer should be probed into the stack. If the temperature reading is over 175°F, fire should be anticipated. In that case, CO2 gas should be pumped into the stack. The CO2 (available in high pressure cylinders) cools the hay as well as displaces oxygen, and thus reduces the danger of fire.

Correct Moisture. As discussed, excessive moisture can cause heating. But if hay is put up too dry (particularly alfalfa) many of the leaves will shatter (fall off). The leaves are the most nutritious part of hay, and therefore this is a situation to be avoided also.

The correct moisture will depend upon the actual forage crop, but will usually vary in the vicinity of 16-20%. In recent years a number of hay preservatives have been marketed as a means of allowing hay to be picked up at a higher moisture. Experience with these products as been highly variable.

Hay preservatives. Success or failure with hay preservatives not only depends upon obtaining a viable product from a reputable manufacturer, but also a delivery system that allows even and continuous distribution throughout the hay. The decision to pick up hay at a higher moisture level can have serious, indeed even life threatening consequences (due to spontaneous combustion), and therefore is not to be taken lightly. Before attempting to use one of these products, it should be investigated thoroughly, including the details and idiosyncrasies of application. (Many of these products contain organic acids which are corrosive to machinery and are potentially dangerous.)

Hay conditioners. A very common practice to speed up the drying of hay in the field is to use what is known as *conditioning* equipment. In years past, the conditioner was a separate piece of equipment, but today it is usually included with the hay cutting equipment.

Actually, conditioning equipment is simply a set of rollers designed to squeeze the hay. By squeezing the hay, much of the juice is removed which greatly reduces the time required for drying.

A variation of conditioning is known as *crimping*. Crimping is simply the addition of corrugations on the conditioner rollers. The corrugations bend or "crimp" the stalk at periodic intervals which, in some applications, is superior to conditioning alone (squeezing).

The disadvantage to conditioning (or crimping) is that some of the leaves will be lost. That is, the extra mechanical handling of the hay does cause a portion of the leaves to be shattered and lost. In most climates however, the potential loss from conditioning is deemed

Figure 19-5. The "big package" hay systems, are very popular with livestock producers, and are very efficient.

Figure 19-6. The covering or sheltering of hay is of greater importance with round bales since a greater portion of the hay is exposed to weathering.

to more than offset the potential loss from rain.

Rained-on hay. Rain is the one single most confounding factor in hay-making. Each year hundreds of millions of dollars of hay is ruined by rain.

The worst damage occurs if rain comes after hay has been baled, but not yet picked up and stacked. That is, each individual bale gets wet. In that case, all that can be done is to leave the bales in the field until they dry (to avoid spontaneous combustion). In so doing, however, they will go through a "heat", and much of the nutritive value will be lost.

If the rain occurs before the hay is baled, then the situation is not quite as bad. In that case the hay will have to lay in the field a longer time before it is baled; that is, before it is dry enough to be baled. The main nutritive loss will be vitamins A and E, due to bleaching. The physical falling of the rain, however, may cause shattering of the leaves. The amount of shattering depends on the intensity of the rain and how dry the hay was when the rain occurred.

Other shattering will occur if the hay must be turned

to prevent heating. The heating may not be as severe as if the hay were already baled, but there still may be some.

The most common error, in respect to maturity, occurs when grain crops such as oats, sorghum, etc. are to be cut for hay or silage. The perception is often that harvesting should be delayed until a good seed head (grain) has developed. It is often perceived that allowing the grain to mature will increase the value of the hay or silage. Nothing could be further from the truth.

Allowing grain to mature means that the plant itself will mature. Maturity of the plant means a conversion of the tender, digestible fiber, sugars, and soluble carbohydrates into coarse, indigestible fiber. Specifically, the digestibility of the vegetative parts of the plant will decline by as much as 50%.

What we must also keep in mind is that grain in most forage crops is utilized very poorly by livestock. When grains are allowed to mature, a hard, impenetrable seed coat develops. Fed in an unprocessed form (as in harvested forage), very little grain actually gets utilized. Most simply passes through the digestive tract in the whole (undigested) form.

We must, therefore, cut forage when it is green and immature. For grain-type crops, that means prior to the mid-dough stage of maturity.

How to Judge Hay Quality. Judging hay quality is relatively easy. While laboratory analysis is often relied upon, visual appraisal can be a good estimate of overall value. Indeed, at times, laboratory analysis can be very misleading, and visual appraisal is required to detect the error.

This is because the two most common laboratory analyses used in forage testing, *crude protein* and *crude fiber*, are unaffected by heat damage. Crude protein is merely an analysis for total protein. The crude protein procedure is not able to detect changes in the digestibility of protein.

The eye, however can detect heat damage. As forages undergo heat damage, the color darkens. The color can go from bright green to light brown to dark brown, and ultimately, to a near-black color. *The severity of heat damage normally follows the darkness of the color.*

Good quality hay should be a bright green color. Any color other than green indicates a problem.

A yellow color can indicate sun bleaching from excess drying (usually due to being rained on in the field). A yellow color can also indicate that the plant was too mature before being cut. To determine which is the case, examine the stem. If the stem is hard and woody, that indicates maturity. If the stem is softer and more flexible, chances are the yellow color is due to bleaching.

A yellow color due to bleaching is preferable to a brown color. Brown indicates heating, which indicates damage to protein and a loss of soluble carbohydrates. A yellow color due to maturity may not necessarily be preferable to a brown (heating) color. The onset of maturity in most grasses, cereal hays, and sorghum-type hays (haygrazer, sudax, etc.) can drop protein digestibility and carbohydrate availability to the same low levels as immature hay destroyed by heating.

A very dark brown or black color indicates the protein is almost totally destroyed and indigestible, and all soluble carbohydrates and sugars have been burned. All vitamin A and E will have also been destroyed. This type of hay cannot be fed to livestock without protein, vitamin, and energy supplementation. Otherwise, emaciation and vitamin deficiency symptoms will appear.

After color look for leaves, especially with legumes such as alfalfa. Leaves have many times more nutritive value than the stem. Thus, the greater number of leaves, the higher the digestibility and protein level. As discussed, this is why hay-making methods that avoid shattering of the leaves are the most desirable.

In conclusion, visual appraisal can determine the overall digestibility and protein level of most hays. If the hay is a bright green color with plenty of leaves and the stalks are not woody or coarse, then it is a good hay.

The eye, however, cannot detect the mineral or nitrate level. A laboratory must be depended upon for this. *Never purchase hay based solely on a laboratory analysis as common laboratory procedures cannot detect heat damage, whereas the eye can.*

SILAGE AND SILAGE MAKING

The ensiling process. The ensiling process relies upon what are known as anaerobic bacteria (bacteria that function in the absence of oxygen) to produce acid which preserves the forage. This acid prevents the growth of other bacteria, yeasts, molds, or other fungi.

The silage making process begins when the forage is cut. Ideally, forage should be cut when it is green and immature. This is important for two reasons: 1. to ensure the forage has a high nutritive value, and 2. to ensure enough moisture is present to ensile properly.

Most forages ensile best at a moisture of 62 to 68% (32 to 38% dry matter). For most forages, an absolute minimum of moisture would be in the range of 56%, and an absolute maximum of 74%.

If moisture exceeds 74%, run-off will occur. This run-off should be avoided for two potential reasons: 1. a good deal of the soluble sugars and carbohydrates will be contained in the liquid run-off, and thereby will be lost, or 2. the liquid fraction will cause undesirable fermentation products to be formed. Specifically, excessively wet silage creates butyric acid as a fermentation by-product, which has a foul odor and is unpalatable to livestock.

Silage that is too dry is difficult to pack. That is, a re-

Figure 19-7. A small on-farm pit silo.

Figure 19-8. The bagging of silage is popular in heavy rainfall areas where "pit" silos are not feasible.

quirement of the ensiling process is that silage be tightly packed to exclude oxygen. Light, fluffy, low moisture silage makes packing difficult or impossible. The presence of oxygen will allow aerobic (oxygen-using) organisms, such as molds and yeasts, to grow and ruin the silage. These organisms not only burn up most of the available energy, but in the case of molds can also leave toxic residues. Known as mycotoxins, their effect can range from mild retardation of animal performance to death loss. In the absence of oxygen, however, these organisms cannot grow. Also, the absence of oxygen allows favorable anaerobic bacteria to produce favorable fermentation by-products that preserve the silage.

The most important of these bacteria are the lactic acid producers. Indeed, silage quality is often measured by the amount of lactic acid present. Other important acids produced by anaerobic bacteria are propionic, acetic, and valeric. These acids are identical to the fermentation products produced in the rumen and are important sources of energy to ruminant animals. Together with lactic acid, in silage they take on the added significance as preservatives. These acids bring the pH of silage down to the range of about 5.2-4.5. This effectively stops further bacterial growth, and, as long as ox-

Figure 19-9. A "classic" concrete stave silo.

such as the cellulose portion of the fiber, can ultimately be converted to acids. The problem is that the time required for this allows the undesirable bacteria, yeasts, and molds more time to grow and damage the silage.

However, in all but the most extreme cases, the silage will not be ruined; rather, the digestibility of the protein and energy will be reduced. The delay in time that it takes to drop the pH to a level whereby bacterial growth ceases, allows more of the available energy to be burned up. This not only causes more of the energy to be lost, but it increases the temperature of the silage, which denatures the protein.

As mentioned, alfalfa is one of the most difficult crops to ensile. The reason is that alfalfa not only has limited amounts of soluble carbohydrates (like grasses), but also has a very high calcium content. The calcium acts as a buffer, which further delays the drop in pH.

ygen is not permitted to enter, the growth of yeasts and molds is prevented.

Problem silage crops. There are a number of silage crops that are notoriously difficult to ensile. Most of the grasses would fit in that category, but the most difficult of all is alfalfa.

The primary problem with grass, alfalfa, and similar legumes, is a shortage of soluble carbohydrates and sugars. That is, in order for anaerobic bacteria to produce the acid fermentation products that preserve silage, a suitable fermentation substrate is required. In silage crops such as corn or sorghum, that is not a problem. These plants contain large quantities of soluble carbohydrates and sugars that can be rapidly fermented by bacteria. Grasses and legumes, however, contain much lower quantities.

Given enough time, the bacteria can drop the pH in grass and alfalfa low enough to shut off further bacterial, yeast, and mold growth. Less soluble carbohydrates,

Figure 19-10. Oxygen limiting silos. These types of silos are very expensive, and are often used in the reconstitution of grain (discussed in Chap. 17A).

The ironic thing about overheated alfalfa silage, is that animals seem to crave it. Cattle, in particular, will eat overheated alfalfa silage as well or better than well-preserved silage. Thus, how well animals eat silages is not an indication of their value.

The best indication as to the relative value of silages is their color. Good quality, well-preserved grass or alfalfa silage will be bright green in color. Burned up, overheated silage will be a brown, tobacco color; while colors in-between indicate lesser damage.

Silage additives. When ensiling difficult crops, it is highly advisable to use some sort of additive. One of the oldest, and still one of the best, is molasses, or finely ground grain. By adding molasses or grain, the bacteria are ensured a rapidly available source of carbohydrates. Thus, grass and alfalfa silages ferment as quickly as corn or sorghum silages.

In more recent times, commercially prepared bacterial cultures have been marketed as silage additives. Unfortunately, generalizations concerning these products cannot be made. Some of them do appear to be beneficial, and others appear to have no value at all. Preparing cultures of bacteria, and then preserving them in a manner so as to ensure their viability, is a highly technical operation. Therefore, whether a product works or not depends to a great degree on the sophistication and ethical policies of the company.

COMPARING SILAGE AND HAY

The making of silage is nearly as old as the making of hay. Indeed, ancient man probably developed silage making as a means of avoiding rain damage to hay. And today, in many areas, silage making is still undertaken for this very purpose.

But there are other advantages to silage making as well. Probably the most important, but least understood, is the fact that with many crops a greater amount of the soluble sugars and other carbohydrates will be saved. That is, with most hay crops, during the drying process a good deal of the juice (which contains the sugars and soluble carbohydrates) is lost. Since silage is not dried, much of the sugars and soluble carbohydrates are preserved (converted to organic acids). Also, with crops that have fragile leaves (such as alfalfa) the loss due to shattering is greatly reduced (because leaves are much more liable to be shattered when dry).

But silage making has many disadvantages as well. Probably the top of the list is the fact that silage is not a salable product. Once it is put up, it cannot be moved, and thus cannot ordinarily be sold as a cash crop. Hay, of course, can be moved and sold.

Some crops, particularly grasses and legumes, can be difficult to ensile. If not managed properly, a great deal of the nutritive value can be lost.

Silages are also less palatable to livestock than hay. Although with most species, and in most situations, this is not a problem, with very young or freshly weaned calves or lambs it can be. (Once animals begin eating silage, there is no problem. However it normally takes 2 or 3 days before most animals will eat the same amount of dry matter as they will hay.)

SUMMARY AND OVERVIEW

The proper harvesting of forages is dependent upon two basic concepts: 1. cutting at the correct maturity and, 2. proper storage. Most farm managers understand proper storage techniques. Far and away, most errors arise out of failure to harvest at the proper maturity. The most common error occurs whenever cereal crops are used as forage. In many cases it is thought best to allow the grain head to fully develop. In so doing, however, the plant is allowed to mature which drastically reduces overall digestibility. Grain crops should be harvested no later than the mid-dough stage of maturity. Forages should be cut when green and immature.

Once a forage crop has been cut, storage at the proper moisture level is the next most critical item. For hay crops, that generally means something in the area of 14-18% moisture. The actual amount will depend upon the type of hay system used. Allowing the crop to dry excessively will mean increased shattering of the leaves, which is the most nutritious part of most hay crops (especially legumes such as alfalfa). Putting hay up too wet means the potential for heating, subsequent nutritive loss, and even spontaneous combustion.

Silage crops normally should be in the range of 60 to 70% moisture. The absolute limits for that range would be 56 to 74% moisture. Putting up silage crops too wet will mean run-off and poor fermentation end products (butyric acid). Putting silage up too dry will make it difficult to pack and lead to oxidation and spoiling.

To ensile properly, the crop must have a good supply of available carbohydrates. With corn, sorghum, and most small grain silages, this is normally not a problem (when cut at the correct maturity). With grass crops and legumes such as alfalfa, this is usually a problem. Therefore, some sort of silage additive is required. Molasses and finely ground grain make excellent additives. If commercially prepared additives are used, be very careful as there are enormous differences between products.

CHAPTER 20 RATION FORMULATION

Condensed from *Modern, Practical Feeds,
Feeding and Animal Nutrition*
by senior author D. Porter Price, Ph.D.

ADJUSTING FOR MOISTURE

One of the first, and most confusing aspects of ration formulation is adjusting and compensating for variations in moisture. As discussed in Chap 17, all feeds contain some moisture. Dry feeds, such as grain or hay, usually contain from 8 to 15% moisture. Wet feeds such as silages normally contain 60-70% moisture. Many by-product feeds such as cull vegetables, corn cannery waste, etc., can have over 90% moisture.

Also as discussed in Chap. 17, laboratory analyses or evaluations of feeds are usually given on an *as fed basis*, a *dry matter* (100% dry) basis, or on an *air dry* basis. "As fed" means with the full amount of moisture; "dry matter" means with all the water removed; and "air dry" generally means with 10% moisture. Taking a sample of corn with 85% dry matter (15% moisture) and 10.5% crude protein on a dry matter basis, the analyses on an as fed, dry and air dry basis would vary as shown:

Corn

Reporting basis	Crude protein
As fed	8.9%
Dry matter	10.5%
Air dry	9.45%

A very striking difference is seen with high-moisture feed such as silages. A 35% dry matter corn silage (65% moisture) with 8% crude protein on a dry matter basis would thus appear:

Corn Silage

Reporting basis	Crude protein
As fed	2.8%
Dry matter	8%
Air dry	7.2%

Therefore, when comparing feeds, it is necessary to know the moisture and the basis by which chemical analyses are reported. To adjust as fed analyses to dry matter, simply divide the figure by the fraction of dry matter; for example, in the case of corn silage, 2.8% crude protein would be divided by .35 dry matter (2.8 / .35 = 8.0%). To obtain the "air dry" basis, multiply the 100% dry figure by .9 (90% dry matter).

THE PEARSON'S SQUARE

The Pearson's Square is the oldest and simplest method of balancing rations. When using two feeds, it will tell exactly the ratio of each feed required to meet the desired specification. For example, assume it is desired to feed a corn ration using alfalfa hay as the source of additional protein. If the ration is to contain 11.5% crude protein (CP), and the corn has a CP value of 9.5%, and the hay 15%, the following arithmetic steps would be used:

Setting Up the Square

Set up the square by placing the desired CP level in the middle, and the CP values of the two feeds in two corners on one side:

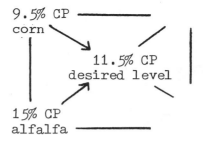

The next step would be to subtract the CP of the feeds across the desired CP level (one of the feeds must be higher than the desired level, and one must be lower).

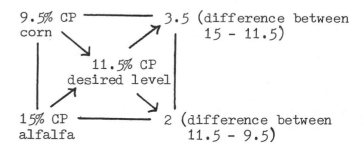

Now add the total of the differences to get the total parts of the two feeds to go into the ration:

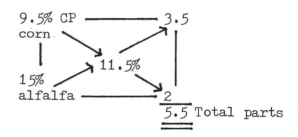

To get the percentage of alfalfa to go into the ration, divide the difference from the subtraction of the desired level from the corn, by the total. That is, divide the number diagonally across from the feed by the total. For alfalfa, the percentage in the ration would be:

$$\frac{2}{5.5} = .36 \text{ or } 36\%$$

For the amount of corn, the percentage in the ration would be:

$$\frac{3.5}{5.5} = .64 \text{ or } 64\%$$

To check and see that 36% for alfalfa and 64% for corn are right, multiply the percentage times the CP value as a fraction, and add the results:

	Ration		(Fraction)		
Alfalfa hay	36%	x	.15	=	5.4%
Corn	64	x	.095	=	6.08
					11.48%

As a practical matter, in most rations the ratio of roughage to concentrate is as important as the protein level. Thus, the way most rations are actually balanced using the Pearson's Square is to first decide what the grain/roughage level is to be, and then to use a third feed as the protein concentrate. In this case, let us assume that a 90% corn 10% alfalfa hay mix is desired, and soybean meal is to be used as the additional protein source. The first step then, would be to determine the protein content of the 90% corn 10% hay mix.

Feed	%	x	CP Fraction	=	CP in Ration
Corn	90		.095		8.55
Alfalfa	10		.15		1.5
					10.05% CP

The Pearson's Square would then be used like this:

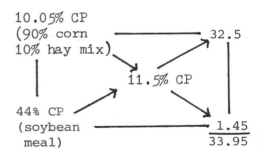

$$\% \text{ corn/alfalfa mix} = \frac{32.5}{33.95} = 95.7\%$$

$$\% \text{ soybean meal} = \frac{1.45}{33.95} = 4.3\%$$

The only problem in using this method is that the final ration will not contain 90% corn and 10% hay. Rather it will have something less than those values. The ration would actually be:

	Original %	x	Adjusted fraction	=	Actual % in ration
Corn	90		.957		86.1%
Alfalfa	10		.957		9.57
Soybean meal	4.3		-		4.3
					100.0%

BALANCING FOR TWO OR MORE NUTRIENTS

The Pearson's Square can be used to balance for any number of nutrients. For example, let's assume we want to balance for both TDN and crude protein (say 11.5% CP and 75% TDN). Once again let's assume we are again going to use corn and alfalfa as the main ingredients, with soybean meal as the protein supplement.

In this case we would balance for TDN first, and then if additional protein is needed, we will balance a second time utilizing soybean meal as the supplement. Corn has an "as fed" TDN of about 80% and alfalfa about 55%. Our first square would therefore look like this:

80% TDN Corn

75% TDN desired level

55% TDN Alfalfa

In this ration the amount of corn would be:
  20/25 = .80 or 80%

The amount of alfalfa would be:
  5/25 = .20 or 20%

In order to determine how much soybean meal to add, we would determine the protein contained in this mixture of the main ingredients. To do that, we would multiply the decimal fraction of the protein in each feed, times the percentage of that feed to go in the ration:

Corn .095 CP x 80% = 7.6% CP
Alfalfa .15 x 20 = 3.0 CP
                  10.6% CP

Thus if we want 11.5% CP in the ration, we will need to add .9% CP from soybean meal. Again, we can use the Pearson's Square:

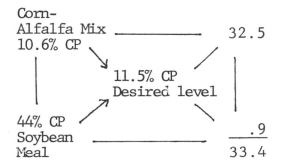

% corn/alfalfa mix = $\frac{32.5}{33.4}$ = 97.3%

% soybean meal = $\frac{.9}{33.4}$ = 2.7%

  Corn 80% x .973 = 77.8%
  Alfalfa 20% x .973 = 19.5
  plus soybean meal at 2.7
                      100.0%

If we want, we can carry this procedure even further. That is, we can balance for calcium, phosphorous, vitamin A etc., by using Pearson's Squares and adjusting with additional calcium, phosphorous, and vitamin A supplements. However, after a while, the process gets rather tedious and time consuming.

As a practical matter, in real life ration formulation, most professional nutritionists decide what the main ingredients will be, and then leave a set percentage within the ration for a supplement. They then calculate what is supplied by the main ingredients. From there, they can determine what will have to be contained in the supplement.

To the novice, the calculations may seem complex and difficult to understand, but in reality, it is only simple arithmetic. With a little practice most anyone can learn to calculate rations and supplements.*

Example: Let's assume that we decide we will leave 5% of a ration for supplement. Assume further that we decide the rest of the ration will be 85% grain sorghum, 10% timothy hay. We also have determined that we want the ration to contain 11.0% crude protein, .45% calcium, and .28% phosphorous. To determine what the supplement will consist of, we would calculate the nutrition supplied by the ration. To do that, we must first determine what the nutrient content of the feeds are. Let's assume that by laboratory analysis we have learned that the feeds contain the following nutrient levels:

Nutrient (as fed basis)

Feed	CP	Ca	P
Grain sorghum	10.0%	.04%	.22%
Timothy hay	8.5	.24	.18

Therefore, to determine the nutrients in the ration, we would multiply the decimal fraction of the feed contained in the ration, times the percent nutrient in the feed:

Determination of Crude Protein

Sorghum .85 x 10% CP = 8.5
Hay   .10 x 8.5% CP = .85
                        9.35% CP

Determination of Calcium

Sorghum .85 x .04% Ca = .034
Hay   .10 x .24% Ca = .024
                        .058% Ca

---

* **Special Note on Computers** - It's important to realize that the calculation of rations is just the mechanics of formulation. Anyone or anything (computers) that can understand arithmetic can calculate rations. The skill or experience required to formulate rations involves knowledge of the animal to be fed and the feedstuffs to be used.

This may seem elementary, but it is easy to lose sight of with respect to computers. A computer-formulated ration is often deemed to be the last word in animal nutrition. In fact, however, a computer is nothing more than a programmed adding machine. It is the knowledge of the person that designed the program that determines whether the ration calculated by the computer is the best ration, or even a suitable ration, for the animals to be fed.

The plain truth of the matter is that the vast majority of the computer programs available for livestock nutrition do nothing but calculate the numbers. Interactions between feedstuffs (known in scientific circles as the "Associative Effects of Feedstuffs"), special physical (texture) requirements of a ration, special management procedures required for certain feeds and/or combinations of feeds, as well as a myriad of other details that arise with individual feeds under specialized situations, are not covered by any software program commercially available at the time of this writing. The fact is that computers are a tool. They are not a substitute for knowledge or experience.

Determination of Phosphorous

Sorghum .85 x .22% P = .187
Hay   .10 x .18% P = .018
                        .205% P

Now that we know the nutrient levels contained in the ration, we can subtract those levels from the desired levels. This will tell us how much is lacking in the ration.

	Crude Protein	Calcium	Phosphorous
Desired level	11.00%	.45%	.28%
Amount in ration	− 9.35	−.058	−.205
Amount lacking in ration	1.65%	.392%	.075%

We must now determine how much of each nutrient must be contained in the supplement. To do this, simply divide the amount lacking in the ration, by the decimal fraction of how much supplement will be contained in the ration. That is, divide by .05 (5%).

*Thus the amount of protein required will be 1.65/.05 = 33%. The amount of calcium required will be .392%/.05 = 7.84%. The amount of phosphorous required will be .075%/.05 = 1.5%.

To determine what the supplement will consist of, divide the percentage level of each nutrient required by the percentage level contained in the supplementary source. For example if we use 44% CP soybean meal, we would divide the required level 33% by 44%. Thus 33%/44% = .75 or 75%. We now know the supplement will have to be 75% soybean meal.

The next nutrient to consider is phosphorous. The reason we should consider phosphorous before calcium, is that most supplementary sources of phosphorous, also contain calcium. For example, the most commonly utilized source of phosphorous is dicalcium phosphate which usually consists of 18.5% P plus 20% Ca.

We must also take into account that soybean meal also contains Ca and P. The amount of Ca and P added by the soybean meal will reduce the amount of other sources of Ca and P needed.

Soybean meal contains .6% P. Thus, this amount of P added by the soybean meal to the supplement will be .75 x .6% = .45%. We will therefore subtract .45% from the total amount required in the supplement (1.5% - .45% = 1.05%). We now only need to add 1.05% instead of 1.5% If we use dicalcium phosphate as the source, which contains 18.5% P, then we will divide 1.05% / 18.5% = .056 or 5.6% to be added to the supplement.

To determine calcium to be added, we must now go back and calculate the calcium added by the soybean meal and the dicalcium phosphate. They are:

Ingredient In Supplement	Decimal Fraction In Ingredient	Percent Ca In Ingredient	Percent Ca added In Supplement
Soybean meal	.75	x .36%	= .27%
Dicalcium phosphate	.056	x 20.00	= 1.13
		Total	1.40%

We now subtract the total added by the soybean meal and dicalcium phosphate from the required level (7.84% - 1.40% = 6.44%).

Limestone (calcium carbonate) is the most commonly used source of supplementary calcium. Limestone normally runs 38% calcium. To determine how much limestone to add to the supplement, divide the required level by the level contained in the supplementary source: 6.44 / 38% = .169 or 16.9%. This is the level (16.9%) of limestone we will add to our supplement.

We now know the levels of the major ingredients to go into our supplement;

Ingredient	Percent
Soybean Meal	75.00%
Dicalcium phosphate	5.60
Limestone (calcium carbonate)	16.90
Total	97.50%

The total ingredients only come to 97.50%. The remainder is what is commonly referred to as "filler" or "carrier" although the terms are not totally synonymous. Fillers are usually low cost ingredients simply used to make a formulation come out even (to 100%). The term "carrier" implies an ingredient used to dilute highly concentrated compounds used in small quantities, such as minerals and vitamins.

LEAST COST RATIONS

The least cost ration concept has been around for a long time but has been exploited much more since the availability of computers. The idea is that each available feedstuff is evaluated on a price per nutrient basis, and only the most economical feeds are included in the ration. The idea is basically sound, however, there are a number of associative interrelationships between feeds that must also be considered.

While computers are often used to develop least cost rations, simple arithmetic can often be just as quick. Only when numerous by-product feeds are available is the computer a really valuable tool. For the most part, most livestock feeders usually have no more than 10 or 12 feeds available and their use can be worked out with a simple calculator rather quickly. For example, the typical livestock operator may have the following feeds available at the following prices:

- Corn - $89/ton
- Grain sorghum - $78/ton
- Barley - $85/ton
- Alfalfa hay - $70/ton
- Corn silage - $25/ton
- Wheat straw - $40/ton
- Molasses - $90/ton
- Feed grade fat - $360/ton

To determine the price per unit of energy of these feeds, simply divide the dry matter content and the energy value into the price. Using TDN* values from the Feed Composition Table in the appendix, the feeds would appear as shown in Table 20-1.

Of the grains available, grain sorghum appears to be the best buy at $111/ton of TDN. The feeder should realize, however, that the 78% TDN reported for milo is an intermediate value. As discussed in Chap. 17, grain sorghum tends to be quite variable. In addition, in ruminant rations the method of processing will affect the digestibility. Therefore, the species to be fed will make a difference. In the case of feedlot cattle, if steam-flaking, popping, or reconstitution is available, the 78% TDN value would be low and thus sorghum would be the grain of choice. If dry grinding or rolling were the only processing methods available, the 78% TDN figure would be high, so corn would be the best choice.

For the roughages, corn silage is (and usually is) the cheapest on an energy basis. Cottonseed hulls are less expensive on an energy basis, but the protein provided by the alfalfa would also need to be taken into consideration.

Of the ration conditioners, fat appears to be the more economical choice. Again, the other ration ingredients must be considered. Fat usually should not be used as more than 4-7% of the ration for most species or reduction of intake will occur.

Special Note on Computers. What the author has tried to do is explain how basic least cost values are obtained (arithmetically), while pointing out some of the extraneous factors that must be considered. In a good computer program, these factors are put in as "constraints"; e.g. only 3% fat: no more than 6% molasses or 4% molasses and 3% fat, etc.

The computer may not, however, have all the "constraints" necessary to cope with an individual feeding situation. The feeder should therefore consider a computer-formulated ration very carefully before using it.

PREDICTING ANIMAL PERFORMANCE
WITH CALCULATED RATIONS

For each major livestock species, tables of data are available that can be used to predict animal gain or milk production. In the U.S., the National Research Council's data is the most widely used, although other countries also have published tables.

Before learning to use these tables, it should be pointed out that the values were obtained from feeding

---

* TDN has been chosen for simplicity. Net energy figures would be more accurate, but confusing to the first time reader. The concept, however, can later be applied to net energy values.

Table 20-1. PRICE PER TDN* VALUE OF SELECTED FEEDS.

Feedstuff	Price	−	Dry matter (fraction)	=	Price/unit dry matter	−	TDN (fraction)	=	Price/unit TDN
Corn	$89/ton		.855		$104/ton		.91		$114.38/ton
Milo	78/ton		.90		88.6/ton		.78		111.11/ton
Barley	85/ton		.89		95.5/ton		.83		115.06/ton
Alfalfa hay (28% fiber)	70/ton		.90		77.7/ton		.53		146.60/ton
Corn silage	25/ton		.35		71.42/ton		.65		109.89/ton
Cottonseed hulls	46/ton		.90		51.11/ton		.41		124.66/ton
Molasses (beet)	90/ton		.685		131.38/ton		.70		187.69/ton
Feed grade fat	360/ton		.995		361/ton		2.25		160.00/ton

Figure 20-1. Whenever using animal performance prediction tables (or computer calculations) it must be remembered that the data used (usually National Research Council data) was developed from university trials with good quality animals, held in excellent facilities, usually with near ideal weather conditions. Animal performance under actual farm conditions, might be quite different.

trials conducted by universities. The animals utilized were of good genetic quality and the conditions under which the animals were kept and fed could be considered ideal. Therefore, one should use caution in predicting animal performance based purely on calculated values.

ESTIMATING CONSUMPTION

The first step in projecting performance of farm animals is to estimate consumption. In some instances, such as with dairy cows in which we will be limiting how much feed they receive, this does not present too much of a problem. We know how much they will receive and we can calculate their nutrient intake accordingly.

In other situations, however, such as in finishing beef cattle or swine, estimating intake is more difficult. It is also crucially important. Indeed, overestimating intake can make gain projections unrealistically high, and underestimating intake will make projections pessimistically low.

DILUTION OF THE MAINTENANCE REQUIREMENT

The reason accuracy in estimating feed intake is so important is due to what is known as dilution of the maintenance requirement. That is, there is a certain amount of energy (feed) that every animal must consume in order to maintain bodyweight. That level of energy is known as the **maintenance requirement**, and is the foundation for all methods of calculating animal gain. What is so important about estimating intake correctly is that everything the animal eats in excess of his maintenance will go toward gain. For example, a 990 lb steer has a maintenance requirement for 7.9 lb of TDN. If a ration is fed that is 78% TDN, then the steer must eat 10.1 lbs. of the ration, just to meet its maintenance requirement (7.9 lb / .78 = 10.1 lb). Everything eaten over 10.1 lb will go toward weight gain. Using data from NRC, look at how performance and cost of gain varies with intake (ration cost @ $.06/lb):

TABLE 20-2. THE EFFECT OF INTAKE VARIATION ON GAIN RESPONSE.

Intake Ration*	20 lb	21 lb	22 lb	23 lb	24 lb
lb TDN	15.6	16.4	17.2	17.9	18.7
Less lb TDN for maintenance	-10.1	-10.1	-10.1	-10.1	-10.1
lb TDN left for gain	5.5	6.3	7.1	7.8	8.6
Daily gain	1.8 lb	2.25 lb	2.5 lb	2.95 lb	3.15 lb
Conversion**	11.1:1	9.3:1	8.1:1	7.8:1	7.6:1
Cost of Gain***	66.6¢/lb	56¢/lb	52.8¢/lb	46.8¢/lb	45.7¢/lb

*78% TDN ration
**Intake ÷ gain
*** @ 6¢/lb ration cost x conversion

Figure 20-2. Severely foundered hooves. As explained in Chap. 18, whenever ruminants (cattle, sheep, and goats) are fed high grain rations, they must be introduced to the grain gradually. Otherwise, acidosis and founder can occur.

PREDICTING PERFORMANCE UTILIZING THE CALIFORNIA NET ENERGY SYSTEM

Predicting (calculating) performance for cattle and swine utilizing the older TDN system is relatively easy and straightforward. As illustrated in the previous example (on the effect of intake), it is just a matter of calculating the TDN in a ration, multiplying the decimal fraction of TDN times intake, and then comparing total intake of TDN versus tabular data on animal response.

The newer system of Net Energy, developed at the Univ. of California, has become the standard for gain predictions with cattle. Often referred to as the California Net Energy System, it is somewhat more complex than the TDN system. More complex, since it differentiates between the ability of a feedstuff to simply maintain an animal, and the ability of the feedstuff to supply energy for growth or gain. Thus, the system places two separate energy values on each feedstuff; NEm, Net Energy for Maintenance; and NEg, Net Energy for Gain.

As with the TDN system, the first step in estimating performance is to determine how much energy is in the ration. In this case, two energy figures, NEm and NEg, have to be calculated. The format for calculating the two figures is essentially the same as the format used to calculate the TDN in the previous section.

Feedstuff	NE_m (Mcals/cwt)		Fraction in ration		Dry matter fraction		NE_m added to ration
Barley	96.6	x	.85	x	.89	=	73.1
Alfalfa hay (24% fiber)	56.6	x	.12	x	.88	=	6.0
Molasses (cane)	83.3	x	.03	x	.77	=	1.9
							81.0

Feedstuff	NE_g (Mcals/cwt)		Fraction in ration		Dry matter fraction		NE_g added to ration
Barley	64.4	x	.85	x	.89	=	48.7
Alfalfa hay (24% fiber)	26.6	x	.12	x	.88	=	2.8
Molasses (cane)	53.3	x	.03	x	.77	=	1.2
							52.7

Thus, this ration contains .81 megcal/lb. NEm, and .527 megcal/lb. NEg. (Figures in the table and calculations express energy as megacalories per cwt.) To calculate gain it must first be determined how much of the ration the animal needs to satisfy its maintenance requirement. To do this, take the average weight of the animal and look at the Appendix Table for cattle to see what the maintenance requirement of that weight of animal is. For example, if dealing with steers going from 650 lbs. to 1,100 lbs., the maintenance requirement for 880 lb. steers (average weight) would be about 6.95 megacalories of NEm. The ration contains .81 megcal/lb. NEm.

To determine how many pounds of ration must be consumed to meet the maintenance requirement, divide the maintenance requirement by the NEm in the ration (6.95 / .81). Thus, the cattle must consume 8.6 lbs. of the ration to meet their maintenance needs. If consumption is 22 lbs., then there will be 13.4 lbs. of ration left over for gain.

Multiplying 13.4 lbs. of ration times the NEg contained in the ration determines how much NEg will be available for gain. Thus, there are 7.1 megacalories of NEg available for gain (13.4 ration x .527 megacalories/lb).

Looking at the Appendix Table on Cattle Gain, it is shown that 880 lb. steers receiving 7.1 megacalories of NEg (left over after maintenance) will gain at a rate of between 2.8 and 2.9 lbs. In comparison with the TDN calculations done in the previous section, there is a discrepancy of about .25 lb/day. This discrepancy can be explained by the fact that the California Net Energy System assumes that the cattle have been implanted with a growth stimulating hormone. Implants normally increase gain by about 10%, which would therefore increase gain about .25 lb.

In general, the California System is more accurate for prediction of cattle feedlot performance. However, the author would again like to point out that given less than ideal weather, animals of below average genetics, failure to allow for associative factors in rations, etc., the California System (as well as the TDN System) can significantly overestimate performance.

CALCULATING MILK PRODUCTION

Calculating rations for milk production for a dairy cow is somewhat different than calculating weight gain for beef cattle or swine. The main difference is that the roughage and the concentrate portion of the ration are usually considered separately. Typically, a set amount of roughage (usually 1-2% of bodyweight) is allotted to all cows, and the amount of concentrate is varied according to the production capability of the cow. Thus, a cow genetically capable of giving 80 lbs. of milk, would be allotted more concentrate than a cow capable of milking only 50 lbs.

For example, let us assume that all cows are to be fed 18 lbs. of good quality alfalfa. Let us also assume that a concentrate mix is available that contains .87 Megacalories of NEL (Net Energy for Lactation) per lb (1.9 megcals/kilogram).

The first step is to determine how much energy (NEL) the cows are receiving from the hay. Looking up the value of NEL for hay in the Appendix, we see that good quality alfalfa runs about .63 megcal/lb (1.4 Megcal/kg) of NEL. However, this value is on a "dry matter" basis. Hay runs about 90% "dry matter" so we multiply the energy value by .9 to get the energy values on an "as fed" basis. Thus the alfalfa has an NEL value of .57 Megcal/lb. (1.3/kg).

Therefore, at 18 lbs. per head, the cows are obtaining 10.26 Megcal of NEL just from their hay. Next step is to look up the maintenance requirement for the cows. To do this, we must know what the cow weighs. For this purpose, we will assume the cows weigh 1300 lbs. When we look up the requirements, we find that a 1300 lb cow (600 kg) needs 10.3 Megcals of energy. Thus, at 10.26 Megcals of energy from the hay, the cows are essentially meeting their maintenance requirements with the hay.

Now we need to look up the energy requirement for 3.5% fat milk. Looking at the Appendix for Dairy Cows, we find that .69 Megcal NEL is required per **kilogram** of milk. Dividing by 2.2 (there are 2.2 lbs. in a kilogram), we find that .31 Megcal are required per pound of 3.5% fat milk.

A cow giving 40 lbs. of milk would therefore need 40 x .31 Megcals, which equals 12.4 megcals. The ration contains .87 Megcals/lb so we divide what is needed by what the concentrate mix contains (as a decimal fraction) to determine how much of the ration is needed; 12.4 / .87 = 14.2 lbs of concentrate.

A cow giving 60 lb of milk would need 60 x .31 Megcals = 18.6 Megcals NEL. In terms of concentrate mix, she would need 21.3 lbs. (18.6 / .87 = 21.3).

OVERVIEW

Formulating rations can be done by hand as well as by computer. While it may take longer to formulate by hand, accuracy can be very similar. Actually, most computer software programs utilize the same arithmetic steps used in hand calculation. When large numbers of feeds are considered, the computer is a great time saver. However, it is important for students to learn to hand calculate rations, if for no other reason than to be able to understand the limitations of computer formulated rations.

BOOK IV LIVESTOCK PRODUCTION

CHAPTER 21   PRINCIPLES OF FARM ANIMAL REPRODUCTION

An understanding of the reproductive functions of animals is vitally important to those involved in the breeding of livestock. Ignorance can result in very poor reproductive performance, and indeed, is a problem that holds back animal agriculture in many underdeveloped countries. For example, the U.S. has only 20% of the world's cattle population, yet produces 80% of the world's beef. Part of this is due to more rapid feeding and finishing of animals, but a great deal is also due to reproductive performance. In the U.S. 80 to 85% calf crops are considered average. In many parts of the world 45 to 50% would be average, and in some places even as low as 30%.

REPRODUCTIVE TRACT OF THE FEMALE

The female reproductive tract of all domestic livestock is basically the same (cow, ewe, mare, and sow). The ovaries, or female gonads, are attached on both sides of the body cavity, just ahead of the pelvis, above and behind the uterus. The ovaries produce eggs or ova which are carried down the oviduct to the uterus.

If sperm cells are present and fertilization occurs, the fertilized ovum (embryo) will develop in the uterus. This can take place in either horn (side) of the uterus. The cervix, which is located between the vagina and the uterus, plays an immensely important role in reproduction. Resembling and feeling like a chicken or turkey neck, it is composed of annular rings of cartilaginous tissue. During pregancy the cervix closes tightly and secretes a mucous plug to effectively seal off the uterus from outside bacterial contamination. During estrus, the cervix opens to allow the passage of sperm cells, and completely dilates during parturition to allow passage of the fetus.

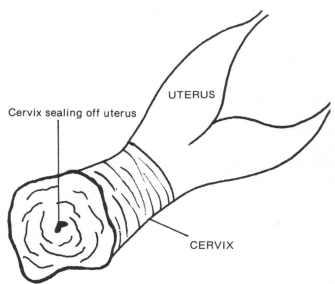

Figure 21-1A. CERVIX SEALING OFF THE UTERUS DURING PREGNANCY.

Attached to the posterior of the cervix is the vagina, which is a tube of elastic tissue. During parturition the vagina greatly dilates to afford passage of the fetus. Below the vagina lies the bladder which attaches to the vagina via the urethra at about the mid point of the ventral (bottom) side of the vagina. During urination, the urine travels through the posterior half of the vagina to the vulva, which is the external female genitalia.

THE OVARIAN CYCLE

Mature, nonpregnant females go through regular ovarian cycles. In the cow, mare, and sow this normally takes an average of 21 days to complete. The ewe will cycle every 17 days on average. Understanding this cy-

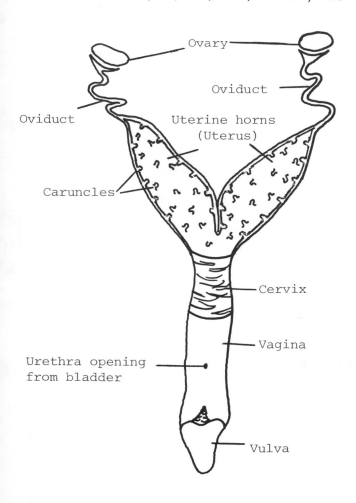

Figure 21-1. REPRODUCTIVE TRACT OF THE FEMALE.

cle is a basic requirement to those engaged in animal breeding.

The cycle begins with the secretion of gonadotropic hormones from the pituitary gland, which initiates the production of ova or eggs in the ovaries. The ova develop in a fluid-filled compartment known as a follicle.

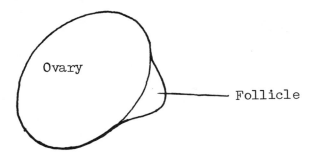

Figure 21-1B. FOLLICLE ON OVARY.

Estrogen, a female hormone, is produced within the follicle. Just prior to rupture of the follicle, large amounts of estrogen are released into the bloodstream which causes the animal to display signs of being in estrus or "heat".

During estrus, the cervix will dilate to allow passage of sperm and mucous will be secreted in both the cervix and the vagina to facilitate copulation. This mucous also serves as a medium for the sperm cells to swim through. A pheromone (hormone-like odor) is given off, which will attract the males and "tell" them that the female will stand for a service. The amount of time the female will be in a "standing heat" depends on the species.

At some point; before or after estrus, the follicle will rupture and the ova will pass down the oviduct. If the female has been served and sperm cells are present in the oviduct, fertilization will occur. After a period of a few days the ova will pass through the oviduct into the uterus and, if fertilized, will attach itself to the uterine wall and a fetus will develop.

Regardless of whether or not the egg has been fertilized, the ruptured follicle on the ovary will regress to form a glandular structure known as a corpus luteum. The corpus luteum will then secrete a hormone known as progesterone, which will cause the cervix to close and otherwise reverse all signs of heat. Under the influence of progesterone, the female will become calm and will refuse to be serviced.

If fertilization occurred, the corpus luteum will remain and continue to secrete progesterone until the time of parturition. If fertilization did not occur, the corpus luteum will recede in a few days and new follicles will develop.

REPRODUCTIVE ANATOMY OF THE MALE

The reproductive anatomy of male farm animals is quite similar to most mammals. Sperm cells are produced in the testes, and stored in the epididymis (tubular like storage vessel attached to the testes). The epididymis is connected to the urethra by a tube known as the vas deferens.* Just before reaching the urethra, the vas deferens opens up into a much larger tube, known as the ampulla. During copulation, sperm cells are transferred from the epididymis to the ampulla. Simultaneously the prostrate and the bulbourethral glands secrete accessory fluid to cleanse and lubricate

---

* In a vasectomy, the vas deferens is severed so that sperm cells cannot reach the urethra. Vasectomized bulls are sometimes used for heat detection in artificial insemination programs.

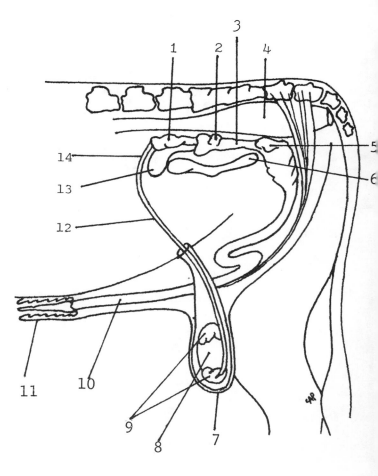

Figure 21-2. REPRODUCTIVE TRACT OF A BULL.

1. Vesicular gland
2. Prostate
3. Urethra
4. Rectum
5. Bulbourethral glands
6. Pelvis
7. Scrotum
8. Testis
9. Epididymis
10. Penis
11. Sheath
12. Vas deferens
13. Bladder
14. Ampulla

the urethra. When ejaculation occurs the vesicular glands secrete accessory fluid to add volume to the semen (semen is considered to be the sperm cells plus the accessory fluids).

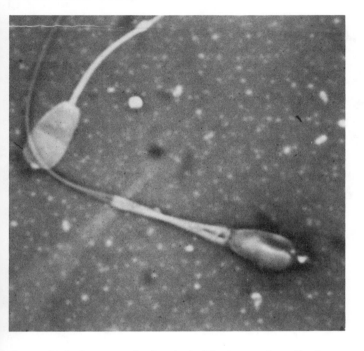

Figure 21-3. Sperm cells (bovine). (Photo courtesy Dr. Larry Rice, Okla. State Univ.)

The amount of sperm cells produced in the testes will vary with the amount of sexual activity. If the male has been sexually inactive for a period of time, sperm production will be greatly reduced. If the period of inactivity is extensive, sperm viability will deteriorate. Because of this, the first ejaculate of the breeding season will often produce very poor quality semen; i.e. sperm cell motility will be greatly reduced. Normal sperm are very active "swimmers", but sperm cells from the first ejaculate will usually be sluggish or totally inactive. Fertility testing males involves viewing sperm cells under a microscope, and motility is one of the most important criteria. Obviously, if males are to be fertility tested at the beginning of the breeding season, two ejaculations (some time apart) may be necessary.

The amount of sperm produced will vary greatly, depending upon the species and the individual. While the total amount of sperm cells produced per ejaculation are tremendous, in relative terms, bulls and rams produce the most, while boars and stallions produce much less. Bulls will normally produce between 5 and 15 billion per ejaculation. (It should be pointed out, however, that for any individual, the bottom end of the range can be 0.)

METHODS OF MATING

Basically, there are three methods of mating: 1. hand mating; 2. pasture mating; and 3. artificial insemination. Most swine and horse farms perform hand mating. A portion of the purebred beef cattle industry also uses hand mating, along with a small percentage of the dairies. Pasture mating is typically utilized by sheep and commercial beef cattle operations. Artificial insemination, to a certain extent, is used by all types of farm animal breeders, but far and away is used the most by the dairy industry. Over 80% of all dairy cattle are bred artificially.

Recently, a new and very expensive form of mating,

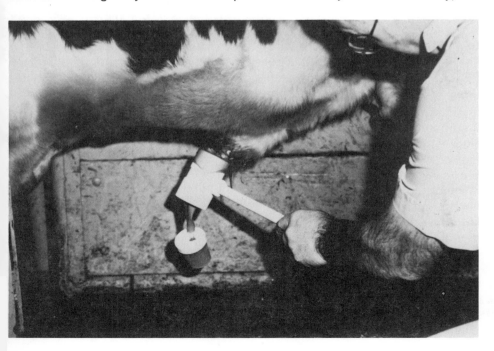

Figure 21-4. Collecting semen from a bull for fertility testing. In this case an electro-ejaculator is being used to create ejaculation. (Photo Courtesy Dr. Larry Rice, Okla. State Univ.)

known as embryo transfer has come into use. Although not widespread it has found a place in the dairy industry, and to a lesser extent in the beef industry.

HAND MATING

Hand mating consists of keeping the male separate from the females. When a female is found to be in heat, she is taken to the male and bred. The exact time the female is taken depends upon the species.

Sperm cells will normally remain viable within the reproductive tract for 12 to 24 hours. Therefore, it's important that the service be within 12 to 24 hours of ovulation. In some species, such as the cow and the ewe, that presents no real problem. Breeding can simply be done toward the end of estrus (estrus normally lasts about 18 hours). In the mare and sow, however, the length of estrus is much longer and more variable. Therefore, more than one mating is often practiced; one mating soon after the animal is first noticed to be in heat, and a second service two or three days later.

PASTURE MATING

As stated previously, pasture mating is primarily used by the commercial beef cattle and sheep industry. The advantage to pasture mating is the obvious reduction in labor. After the bulls or rams are put in the pasture, all that is required are occasional riders to see that the bulls or rams are actually with the cows or ewes, and that the cows or ewes are actually being settled.

If toward the end of the breeding season it is noted that cows and/or ewes are coming back into heat, the situation must be investigated promptly. Much has been said concerning fertility testing of males, and certainly infertility in one or more bulls or rams could cause such a problem. More commonly however, nutritional deficiencies are the culprit (discussed later in the chapter). The worst possible situation would be a venereal disease outbreak. But regardless of the cause, the situation must be detected early.

The number of cows or ewes a bull or ram can service under pasture breeding will depend greatly upon the types of pastures. In irrigated or improved pastures, where there are no physical obstructions (mountains, deep ravines, dense brush, etc.) and the cows are concentrated together, a mature bull can be expected to service up to a maximum of 40 cows. A ram can be expected to service about 80 ewes. Under range conditions, 20 to 25 cows per bull and 35 to 40 ewes per ram is much more common. Where forage is extremely sparse and the stocking rate very light, cow to bull ratios of 15:1 and 20 to 25 ewes per ram may be appropriate. (These ratios assume that the sheep are unherded.)

With respect to cattle, in each ratio given, it is assumed that the bulls are three years or over in age. Two year olds should be turned out only if they have proven themselves to be aggressive breeders. Even then it is best to reduce the number of cows they are to serve. It is also best to place two year olds with the replacement heifers as sometimes they are hesitant to breed mature cows, or may not be tall enough to effectively mate; i.e. deposit semen at the cervix. Likewise, it is best not to put two year old bulls with mature bulls as they are often intimidated by the older bulls. Yearling bulls cannot be depended upon to successfully conduct pasture mating.

Rams can be relied on to pasture mate at two years of age. However, other than age the same general rules for bulls apply. Young rams should be given smaller flocks to service and should not be placed in the same pasture with older rams. Likewise it is best to place them with ewe-lambs, rather than mature ewes.

ARTIFICIAL INSEMINATION

Artificial insemination (A.I.) is the method of mating primarily used by the dairy industry. Other animal breeders occasionally utilize A.I., but it is the dairy industry that far and away utilizes A.I. to the greatest extent.

There are two reasons for this. First of all, dairy cows are individually handled and observed each day. Likewise, the cows are confined, which means that dairying physically lends itself to A.I.

But the biggest reason has to do with economics. In dairying, there are cows that are up to three times more productive than average cows. There are cows that will average 40-45 lbs. of milk per days and there are cows that will produce 120 lbs.

Nowhere in animal agriculture are there differences like that. In sheep, swine, and beef cattle, there are individuals that may put on meat at up to a 50% greater rate, but nothing on the order of the 100 to 200% increase in productivity that is so common in dairy cattle.

Bulls that are capable of siring such highly productive daughters quite obviously are extremely valuable ... more expensive than the average dairyman could afford. But as mentioned earlier, a bull produces up to 15 billion sperm in one ejaculation. However, if sperm cells are placed inside the uterus (as in A.I.), only 10 to 15 million are needed. Therefore, one bull can be used to breed up to 1000 cows per service, and the cost per service is greatly reduced.

Therefore, only a very few bulls are needed for breeding purposes. In that way, only the very best bulls can be used, and genetic progress can be very rapid. Indeed, in 20 years of A.I., the dairy industry has been able to double milk production, using half the cows.

To a lesser extent, A.I. is used in beef cattle. One of the main advantages to cattle in general is the fact that bull semen can be frozen and kept for a long period of time. Thus, bull semen can be shipped long distances, which totally eliminates the need to keep male breeding animals nearby.

Boar and ram semen is much more difficult to freeze or otherwise store for any appreciable length of time. For this reason, commercially prepared boar and ram semen has generally not been available. This has seriously curtailed the A.I. of swine and sheep as collecting, extending, and storing semen requires a great deal of expertise and technical capability.

A.I. of horses has generally been curtailed by stringent registration regulations by breed associations. Most horse breed associations require that the stallion be present on the same farm as the mare, which nullifies most of the advantages of A.I. This, plus relatively low conception rates in mares, has made A.I. in horses relatively uncommon.

The Artificial Insemination of Cattle. As mentioned, artificial insemination in cattle, especially dairy cattle, is a relatively common practice. Therefore, a description of the practical aspects of the technique in cattle is provided.

The Technique in General. As mentioned previously, the reason A.I. has become so practical with respect to cattle is that bull semen can be frozen.

The semen that the stockman buys comes to him frozen and stored in liquid nitrogen. Normally it will be in the form of an ampule or straw. One ejaculation can be used to breed hundreds of cows, and therefore the ampule or straw contains only a fraction of an ejaculate mixed with an extender. This is why semen from bulls that would cost enormous sums of money can be purchased for nominal fees.

The semen is delivered and kept in an insulated tank (containing liquid nitrogen). The first step in A.I. (after finding a cow in heat) is to remove the semen from the tank and thaw it. The company merchandizing the semen will provide detailed instructions on how to thaw the semen, and these instructions should be followed closely as the extenders in the semen may vary with the processor. Generally though, ampules are placed in ice water, and straws allowed to thaw in the air (after placement in an insemination catheter).

After thawing, ampules are cut and/or broken and the semen siphoned up into the breeding catheter. This is either accomplished by a small rubber bulb (poly bulb), or a small syringe attached by a short piece of flexible tubing.

The cow to be bred is placed in a simple chute with a pipe or post placed low behind her hind legs (to reduce danger from kicking, as well as to provide restraint). Squeeze chutes can be used, but tend to excite the animals.

The breeder then places an obstetrical sleeve on one arm, lubricates it (soapy water works quite well), and places it in the rectum of the cow. This is best done by forming a wedge with the fingers and slowly pushing the hand in as the anal sphincter relaxes.

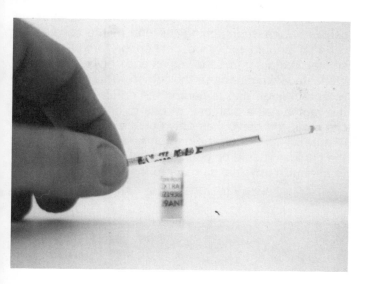

Figure 21-5. An ampule and a straw of semen, as they are usually purchased in the commercial artificial insemination industry.

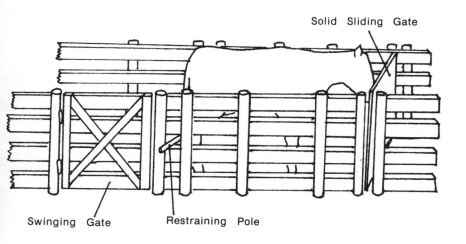

Figure 21-6. AN IDEAL BREEDING CHUTE FOR ARTIFICIAL INSEMINATION (or pregnancy checking).

Deposition of the semen should be made deep within the cervix, near the uterus, and therefore after insertion of the hand into the rectum, the operator must find and "pick up" the cervix (through the rectal wall). The cervix is the most distinctive organ to be found in the pelvic region as it has the approximate size and feel of a turkey neck. The cervix is attached to the anterior (forward) part of the vagina and is usually found "lying" on the floor of the pelvis (Figure 21-7).

When the crevix is located, the catheter is then placed into the vagina. Absolute cleanliness is essential as bacteria introduced into the vagina or cervix can create infection. Most technicians will exert a downward pressure with the arm that is in the rectum, which causes the lips of the vulva to "pop" open. The catheter can then be placed into the vagina without coming into contact with contamination on the vulva.

The catheter is then guided into the cervix with the hand holding the cervix from the rectum above. When the tip of the catheter is up next to the uterus, the semen is deposited (Figure 21-9). The walls of the vagina and cervix are fairly tough, but the uterus can be punctured quite easily. The operator should therefore use extreme caution when approaching the end of the cervix. The catheter should never be pushed or forced into the animal. During estrus the vagina and cervix will be well lubricated with mucus so the catheter should slide easily. The difficult part is finding the orifice of the cervix; the vagina has a diameter of one to one and a half inches, but the cervix has only about a ¼ inch opening (Figure 21-10). When the semen is actually deposited, the plunger should be pushed or squeezed gently so as to minimize physical damage to the sperm cells.

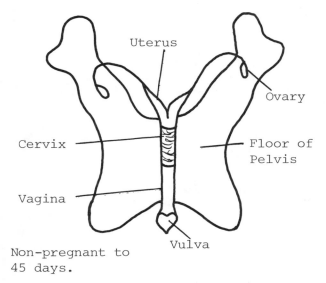

Non-pregnant to 45 days.

Figure 21-7. The Cervix. As stated in the text, the cervix has the feel of a turkey neck.

Note: As pregnancy develops the cervix will move forward and drop over the brink of the pelvis

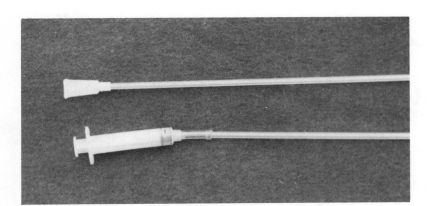

Figure 21-8. INSEMINATING PIPETTE.

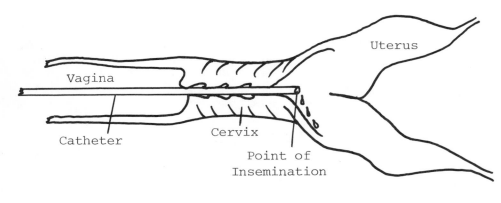

Figure 21-9. PROPER SITE OF INSEMINATION.

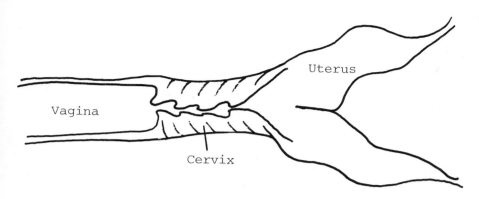

Figure 21-10. LONGITUDINAL CROSS SECTION OF VAGINA AND CERVIX. Notice the blind endings in the cervix which can make passing an insemination catheter difficult.

When to Physically A.I. A Cow in Heat. Ovulation usually occurs after the cow passes out of "standing heat". Sperm cells are generally thought to remain viable for 12-24 hours, and since cows can remain in heat up to 18 hours, many authorities recommend breeding several hours after the cow is discovered to be "in".

Heat Detection. The most difficult part of any artificial insemination program is the process of detecting heat. Difficult in that it requires the most time and labor. Throughout the breeding season, the cows must be observed carefully for 1 to 3 hours every morning and evening. The morning observation should begin just before daylight, and the evening observation should end just before dusk. Failure to do a proper job of heat detection is the most common cause of A.I. program failure.

Signs of Heat. The most well-known, and the most certain sign of estrus is the "standing heat". Cows or heifers in heat give off a pheromone (odor) that alerts the rest of the herd to their condition. Other cows or heifers will then come up and mount them as a bull would. Females that are in heat will "stand to be ridden"; females that are not in heat will not "stand". It is the heifer on the bottom . . . the one that stands, that is in heat. Occasionally a female in heat will ride, but when approached by others she will also usually stand to be ridden herself.

Other signs of estrus would include bellowing, walking (pacing) the fence (separation from the herd), going off feed (not grazing while others are grazing), or otherwise unusual behavior. The tail head may be ruffed up from being ridden by other females, and sometimes mucus can be seen on the vulva.

Estrus will generally be expressed early in the morning or late in the evening and therefore this is when observation should take place. A very effective technique is to herd the cows together and let them sniff and inspect each other. Stirring them around will usually stimulate riding activity.

Embryo Transfer. As mentioned earlier in the text, embryo transfer (ET) as a means of mating has come into existence on a limited but significant basis. It is a very expensive procedure, and is therefore only used on highly valuable animals. To date, the only commercial application has been with cattle.

In embryo transfer, a highly valued female is injected with hormones to cause what is known as a superovulation. That is, the hormones cause the cow to release several ova rather than the usual one ova. The cow is then bred by A.I. in the usual manner. After six days, these ova are taken out of the uterus and placed into much less valuable "recipient" cows. In this way calves can be developed that are totally unrelated to the recipient cows. Thus, many more calves can be produced from a highly valued donor cow than could otherwise be produced.

Estrus Synchronization. In the last decade a technique for bringing all the females in a herd into heat at one time was developed. The process involves injections with what are known as prostaglandins. These are extremely powerful hormonal-like compounds that affect the estrus cycle. (The actual details concerning injection vary with the drug manufacturer.)

To date, in the U.S., only cattle have been utilized in the commercial application of estrus synchronization. The purpose has primarily been to facilitate artificial insemination. That is, estrus synchronization eliminates the need for heat detection. Rather than having to check for heat on a daily (twice daily) basis, the entire herd can be injected and bred at the same time. This is a distinct advantage in beef herds where cattle are not usually confined and/or observed daily.

Special Supplement.
Nutritional Aspects of Reproduction

More reproductive failures are due to nutrition than any other cause. There are many mineral and vitamin deficiencies which can directly or indirectly affect reproduction. However, as a practical matter, with the possible exception of phosphorous, most reproductive failures usually result as a simple matter of malnutrition. Therefore the two nutrients most commonly deficient in general malnutrition, energy and protein, will be discussed in this section. (The effect of vitamins and minerals are discussed in Chap. 17.)

Figure 21-11 A & B. High grade "donor" Simmental cows being used in an embryo transfer operation (above), and a low grade Zebu cross cow being prepared as a recipient (below).

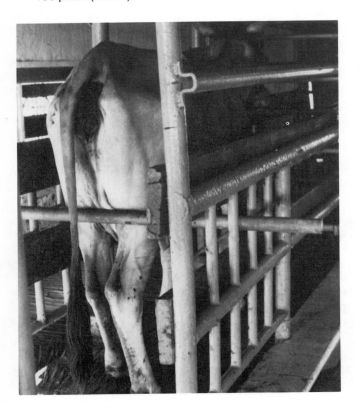

Energy. While it is difficult to generalize across all species, it can be said that food energy is the single most important item with respect to the initiation of estrus (the heat cycle). Studies with various animal species have shown that when females receive less than adequate energy the number of animals coming into heat and/or conceiving is greatly reduced. Probably the most striking example is in sheep. When ewes are supplemented with energy, not only do more animals conceive, but the percentage of twin lambs born per ewe also increases.

Timing of Energy Deprivations. The actual timing of the deprivation of energy has a major role in determining what the final result will be. As mentioned in the previous section, deprivations of energy during the breeding season results in fewer females actually conceiving.

After conception, the effect of energy depends upon the stage of pregnancy. During about the first two thirds of the period of pregnancy, most animals can withstand a moderate deprivation of energy (this would not be true for dairy cows in production) . . . if they receive adequate and/or compensatory energy during the last third of pregnancy. This is because the greatest amount of fetal growth occurs during the last "trimester" of pregnancy, and the dam can usually draw from her own energy stores (fat) to supply fetal development. As a practical matter, however, whenever general malnutrition occurs, a protein deficiency also usually occurs, and as discussed in the next section, protein deficiencies cannot be compensated.

Protein. While energy appears to have its major effect on the estrus cycle and conception, protein tends to have its major effect on gestation. When gestating females are deprived of protein, the fetus will not grow at a normal rate. Under extreme deprivation the fetus will actually be resorbed. Under less severe deprivations, the fetus will be undersized. When born, the offspring will be weaker and less resistant to disease; likewise, the number of stillbirths will be significantly greater.

Effect of Lactation. Depending upon the actual species, lactation can increase the requirements for various nutrients by 50 to 100% (dairy cow requirements can be increased 200 to 400%). This is because of the great amounts of nutrients flowing out in the milk. As a general rule, lactation has a priority for nutrients that exceeds all other bodily functions. Even body maintenance takes a secondary role to lactation. If sufficient feed is not consumed, the lactating animal will become emaciated drawing from its own body tissues in order to meet its lactation requirements. Indeed in some species, such as dairy cows and dogs, the animal may die from diversion of nutrients into the milk flow.

Figure 21-12. An embryo being passed into a donor cow. (Photo taken at the Liconsa Embryo Transfer Unit, Pachuca, Mexico).

In species such as cattle (beef and dairy) in which it is desired to breed the animal while it is lactating, nutrient requirements become critical. If the full lactation requirements for energy are not met, then conception will be reduced. This is because lactation has priority over ovulation (the estrus cycle) and only if the full lactational requirements are met, will a good conception rate occur.

OVERVIEW

Understanding the basics of reproduction is of great importance to anyone involved in the breeding of livestock. Understanding the factors affecting the estrus cycle of the female is probably more important than any other single aspect of livestock reproduction. Depending upon the species, the non-pregnant female will usually go through the estrus cycle every 18 to 21 days. Ovulation generally occurs during or just after the estrus cycle. In cattle the estrus cycle normally lasts about 18 hours; in sheep about 36 hours. Estrus in swine and horses lasts for several days. Sperm cells can normally survive in the female reproductive tract for 12 to 24 hours. This means that the timing of insemination is important. That is, insemination must coincide with ovulation. For that reason, swine and horses are usually bred more than once.

It is also vitally important to understand how nutrition affects the estrus cycle. As discussed in the chapter, more reproductive failures are due to nutritional reasons than anything else; most commonly, simple malnutrition. Certainly various venereal diseases and genetic faults can and do occur among livestock. But the majority of reproductive failures occur simply because the animals were either not given enough to eat, or were given poor quality feed. This is particularly true for grazing animals.

It therefore behooves every livestock manager to learn the nutrient requirements for the species in question; and how those requirements vary with the various stages of the reproductive cycle. The livestock manager must then determine if the feeds provided contain the required nutrients.

CHAPTER 22 PRINCIPLES OF GENETICS
(As related to animal breeding)

In order to more completely understand the breeding of livestock, one must understand a few basic principles of genetics. Only those principles which have practical application to animal breeding will be presented here.

GENES

Genetics is defined as the study or science of inheritance. Genes are the actual chemical entities that transmit the inheritance. Genes are carried on the chromosomes, which are the strands of genetic material that are carried with each individual cell at the time of its division. Chromosomes come in pairs, so that as cells divide half of the genetic material goes with the new cell, ensuring that it will be a perfect duplication of the parent cell. This is the case when cells in the body divide to form new cells, or in bacteria or plants that reproduce through *asexual methods*.

In *sexual reproduction* the reverse of this situation occurs. Instead of two chromosomes from the same source dividing, two different chromosomes from different sources come together. That is, one chromosome comes from the male, and one from the female ... and together they form a pair. Thus a cell is formed that is similar yet different than either parent.

More specifically, the cells of each animal will contain a given number of chromosome pairs; for example, swine have 38 pairs. When the animal forms gametes (ova and sperm cells), one chromosome from each pair is carried by the gamete or germ cell. Thus, sperm and ova cells from swine each contain 38 single chromosomes. When fertilization occurs, and the sperm enters the ova, the 38 single chromosomes from each germ cell will seek out its matching chromosome and form 38 chromosome pairs.

DOMINANT AND RECESSIVE GENES

Some genes display what is known as *dominant* or *recessive* action. When the chromosomes from sexual reproduction come together, there will be two genes (one from each chromosome) which control the same trait. If one of the two genes is a dominant gene, the trait it represents will be the one the offspring will express.

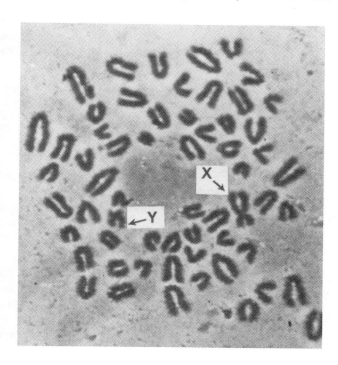

Figure 22-1. Cattle chromosomes (male) in a white blood cell, magnified approximately 1,000 X. Genes, the chemical entities that actually control heredity, are carried on the chromosomes. (From Keiffer and Cartright, 1967. Tex. Agri. Exp. Sta. Cons. PR 2483-2500.)

Figure 22-2. Asexual Cell Division.

1. Cell before division.
2. Division begins, chromosomes are paired up.
3. Chromosome pairs are separated.
4. Separate cell nuclei form.
5. Separate cells emerge.

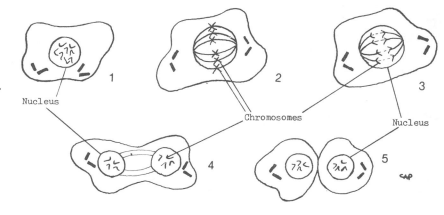

The trait represented by the recessive gene will not be expressed as long as a dominant gene is present.

The types of traits controlled by dominant and recessive genes are usually what are termed *qualitative traits*. These are traits that can readily be seen; examples would be coat color, eye color, etc. For instance, the white face of Hereford cattle, is controlled by a dominant gene. Thus, when Hereford cattle are mated with other breeds, the offspring will always have a white face. This is true, not only because the white face is dominant, but also because Hereford cattle are "homozygous" for the white face gene; i.e. they carry no other gene for face color.

There are normally only two genes that control any one qualitative trait. Thus, there are usually only two alternatives within a breed of animal. Cattle for example, usually have only two color patterns per breed, are horned or are polled, etc.

Symbolically, a capital "D" could represent the dominant gene, and a small "r" could represent the recessive gene. Assuming an equal distribution of genes, four combinations of genes could be present in any one animal.

Combination#	Genotype
1	DD
2	Dr
3	rD
4	rr

Combination #1 and #4 (DD and rr) are said to be *homozygous* because the animal carries only one gene for that trait. Combinations #2 and #3 (Dr and rD) are said to be *heterozygous* because they contain both the dominant and recessive gene. For example, mule foot in swine is a recessive gene. The more common cloven hoof is the dominant gene. If an even distribution of genes is assumed (which is not true), it could be expected that 75% of all swine would be cloven hoofed and 25% would be mule-footed.

Whenever the appearance or physical attributes of an animal are referred to, they are described as the animal's *phenotype*. Phenotype refers to something that can be seen or measured. Whenever the genetic makeup of an animal is referred to, it is known as the animal's *genotype*. The genotype is something that cannot always be determined. For instance, in the case of combinations #1, #2, and #3, it cannot be determined from physical appearance if the animal is homozygous or heterozygous, since cloven hoofs are dominant. But in combination #4, it is known that the animal is recessive homozygous since mule foot cannot be expressed if a dominant gene (D) is present.

To see if an animal is homo or heterozygous for a trait, it can be bred with animals that are known to be recessive homozygous. If the animal in question is truly homozygous, then all the offspring will display the dominant trait; i.e. all the offspring will be heterozygous:

$$D_1D_2 + r_1r_2 = D_1r_1$$
$$D_1r_2$$
$$D_2r_1$$
$$D_2r_2$$

If the animal in question is heterozygous, 50% of the offspring will display the recessive trait; i.e. be recessive homozygous:

$$Dr_1 + r_2r_3 = Dr_2$$
$$Dr_3$$
$$r_1r_2$$
$$r_1r_3$$

Animals that are homozygous and are able to always pass their phenotypical traits on to their offspring are said to be prepotent or true breeders.

GENE FREQUENCY

Gene dominance plays a very important role in determining what the qualitative traits of animals will be; but there is another factor that can be of even greater significance, the concept of gene frequency. As the term implies, gene frequency deals with the frequency by which genes are encountered within a population of animals.

For instance, in sheep the white color is dominant and the black color is recessive. With an even distribution for genes, we would expect 75% of all sheep to be white and 25% to be black. In reality that is not true, because the gene frequency for black sheep is very low. Black wool is heavily discounted, therefore black sheep are almost never kept for breeding purposes. Therefore, the occurrence of a black sheep (expression of the recessive gene) is a rarity.

There are examples, however, in which the frequency for recessive genes is very high; so high, that expression of the recessive is more common than the dominant gene. Such an example would be the presence of horns in cattle. Horns are controlled by a recessive gene. In some breeds of cattle the gene frequency for horns is very high, and as a result, the incidence of horns is very high. When crossed with cattle that are polled (carry the dominant gene), the offspring will be polled.

ADDITIVE GENE ACTION

As mentioned earlier, *qualitative traits*, or those traits which can be seen (horns, type of hooves, coat color, etc.), are controlled by only one pair of genes. *Quantitative traits*, or those which must be measured (rate of gain, production per animal, etc.), are controlled by large numbers of genes. These genes are said to have an *additive* effect; i.e. collectively the genes produce the trait. Thus, if the animal or plant receives a large number of "good" genes for production, it will be very productive. If the animal or plant receives only a few "good" genes, it will be somewhat less productive.

HERITABILITY

Heritability is defined as a measure of the ability of a parent to pass on quantitative traits to its offspring; that is, those traits controlled by additive gene action. Expressed as a decimal, it is a measure of what can be expected from the offspring of parents with known records for a quantitative trait.

For example, .3 is generally accepted to be the heritability for weaning weight in cattle. In a herd that has an average weaning weight of 400 lbs., this means that if a heifer were selected that weighed 450 lbs. at weaning ... as a cow she would have calves that would wean 7½ lbs. heavier than the average; i.e.

```
  450 lb heifer wean weight
 -400 lb herd average
   50 lb differential
  x.3 heritability
 15.0 lb
  x.5 amount of genes contributed by the female
  7.5 lb increase in weaning wt. over herd ave.
```

Likewise if the heifer were bred to a bull that weaned 50 lbs. heavier than the other herd bulls, another 7.5 lb increase in weaning weight could be expected.*

HYBRID VIGOR OR HETEROSIS

Hybrid vigor or *heterosis* is defined as the phenomenon associated with the crossing of two breeds or species of animals. For a yet unexplained reason, hybrid offspring will usually be superior in traits that are otherwise lowly heritable. Fertility is one of the most important traits with a low heritability, and crossbred animals are usually more fertile than either parent breed. Likewise, crossbred offspring are typically much hardier and therefore have a higher survival rate.

The amount of heterosis to result from crossbreeding is generally accepted to be inversely proportional to the heritability of the trait in question. This explains the increase in fertility and hardiness. Quite often, however, increases in relatively heritable traits are also experienced. This can sometimes be explained by better adaptation and resistance to disease.

In some cases, however, decreases in heritable traits will be experienced. This will occur when one parent breed is vastly superior for a trait. An example would be milk production in dairy cattle. The Holstein breed is so superior that only straight-bred Holsteins are used for dairying in most parts of the world.

It is only in tropical areas, where the Holstein is not adapted, that crossbreeding will come into play. In that case, crossing with other breeds (adapted to the area) tends to increase resistance to the heat, parasites, etc.

GENETIC ASPECTS OF ANIMAL BREEDING

CROSSBREEDING

Crossbreeding may be defined as the breeding of different breeds so as to take advantage of heterosis in the offspring. Crossbreeding has found it greatest acceptance and use in areas which create environmental stresses, heat stress in particular.

Defined as *hybrid vigor*, crossbred animals are generally hardier and more able to withstand harsh environments. The classic example of this was the development of the Santa Gertrudis breed of cattle at the King Ranch. Situated along the very hot and humid Texas Gulf Coast, the environment greatly hindered the production of Shorthorn cattle on the ranch. Brahma cattle were also present, but while they are able to with-

---

* Justification for the existence of the artificial insemination industry can quickly be recognized when one contemplates the advantage to breeding to a bull with a weaning weight of 200 or more pounds greater than the herd average for bulls. Calves with weaning weights 30 lbs. greater than the herd average could then be expected just from the bull's genes; i.e.

```
  200 lb. differential
  x.3 heritability
   60 lbs.
  x.5 amount of genes supplied by bull
   30 lb. increase in weaning weight over herd average
```

Obviously bulls of this quality would be very expensive ... the cost of which could not be justified to breed 20 or 30 cows a year. Thus, artificial insemination has become a viable business in the cattle industry.

Table 22-1. Weaning weights in cattle, comparing straight breeding with cross breeding.

Breeding of Calf	Birthwt.	Weaning Wt.
Angus X Angus	68.8 lb	393.4 lb
Hereford X Hereford	73.0 lb	383.5 lb.
Angus X Hereford	71.7 lb	407.7 lb
Hereford X Angus	70.2 lb	406.3 lb

Adapted from Gray. 1978. J. Anim. Sci. 47:370.

stand tropical environments, Brahma cattle are not particularly productive and are relatively infertile. Crossing the Brahma cattle with the Shorthorn resulted in a breed that was much more fertile and productive than either of the two parent breeds. Since then, there have been a number of crossings between Brahma and English or European breeds to produce cattle better adapted to the southern and southwestern U.S., as well as Central and South America.

In the more temperate regions, crossbreeding is practiced because it usually produces a greater growth potential in animals such as cattle, sheep, and pigs. In cattle and sheep, weaning weights are not only greater, but also more calves and lambs usually survive to be weaned. Crossbred calves seem to be hardier and have a greater survivability as baby calves than calves from straight bred mothers.

Selecting the breeds to be crossed will depend upon several factors, the most important being the environment. As a general rule, the more diverse the breeds, the greater the response to crossbreeding will be. For example, a greater response would be obtained in cattle by crossing the English breeds (Hereford, Angus, Shorthorn) with one of the Continental breeds such as Charolois, or Zebu (Brahma), than if they were crossed among themselves.

If crossbred females are used as herd replacements, weaning weights are usually improved further, as milk production is usually improved. (As mentioned earlier, this is not true in dairy cattle.) In addition, crossbred cows and ewes are usually more fertile, and therefore, greater overall calf and lamb crops are usually obtained (in addition to increased survivability of the calves and lambs themselves).

OUTBREEDING

Outbreeding simply refers to the breeding of unrelated individuals. While crossbreeding is technically a form of outbreeding, the term outbreeding usually refers to matings within a breed.

INBREEDING

Inbreeding refers to the mating of related animals. Under most conditions inbreeding should be avoided as it will increase the incidences of undesirable traits such as dwarfism. These traits are usually recessive homozygous; and inbreeding increases the probability of bringing out the recessive traits.

Traits controlled by additive genes are also adversely affected. For example, scientific estimates of reduced weaning weight in beef cattle due to inbreeding are approximately 1 lb for every percent of inbreeding.

Inbreeding has a place in purebred breeding as a means of proving or disproving that an animal is or is not a carrier of an undesirable gene. By mating to close relatives, the probability of the gene expressing itself will be much greater.

LINEBREEDING

Linebreeding is a form of inbreeding in which distant relatives are bred to an outstanding relative. This practice has a place in purebred breeding as it tends to "fix" the genes and thus make the animals more homozygous. As seed stock then, the animals are more prepotent or true breeders.

This practice should only be undertaken with caution. The traits for which the genes are to be "fixed" should be examined carefully, and the ancestries of the in

Figure 22-3. In the 1930's and 40's beef cattle were subjected to a great deal of inbreeding. Dwarfism, as exemplified by this steer, was and is increased in prevalence by inbreeding.

dividuals researched thoroughly. The breeder should be convinced that the desired traits to be fixed will indeed be accomplished by the matings. If the linebreeding is to be done simply because the breeder feels the one individual is outstanding, the decision should be based on statistical evidence. Oftentimes, visual appraisal and even personal prejudice have been the only criteria used for entering into linebreeding programs. This type of decision-making only leads to reduced productivity (and potential genetic faults, such as dwarfism).

NICKING

Nicking is a term used (typically by purebred breeders) to denote matings of 2 individuals or lines of animals which produce offspring that are superior to what would ordinarily be expected from parents of that caliber. That is, matings between certain individuals or lines that sometimes result in more favorable gene combinations than if those same animals were mated with other individuals or lines with similar production records.

The phenomenon is not completely understood, but it is thought to be the result of *epistasis*, or interaction between otherwise unrelated genes. For example, it is known that in laboratory animals and some vegetables, genes for different traits can create third traits when brought together. It is therefore theorized that additive genes in animals can do the same thing (produce superior performance) when brought together in unknown combinations.

CORRELATIONS BETWEEN TRAITS

In selecting animals for breeding, the idea, of course, is to select animals that are superior for the trait desired. For example, if it is wool growth that is desired in sheep, then animals with superior wool growth should be selected. If it is litter size that is desired in swine, then pigs from large litters should be selected.

Some traits are *correlated* with other traits. Selecting for some traits will affect other traits. The effect can be either positive or negative. For example, hardiness to cold weather would, obviously, be positively correlated with wool growth in sheep. Tolerance for hot, humid weather conditions quite clearly would be negatively correlated with wool production.

Correlations between other traits are not quite as obvious. For example, milk production in dairy cattle is negatively correlated with reproductive efficiency. The reason is that high producing dairy cows cannot eat enough feed to keep up with all the nutrients that are lost in the milk. As explained in the chapter on dairy cattle, this can cause delayed breeding.

Growth, a very important trait in most livestock species, is positively correlated with birthweight. Thus, if we select for faster growing calves/lambs etc., we can expect larger birthweights. In some cases, then, if we intend to upgrade a ewe or cow herd by mating to a breed known for more rapid growth, we can expect a higher incidence of dystocia or difficult birth.

SELECTION INTENSITY

Selection intensity is a term used by geneticists in predicting response or progress. Selection intensity is a function of *selection differential*. In layman's terms, selection differential means the differences between the herd average and new animals introduced into the breeding herd, or conversely, the differences before and after culling. The greater the difference, the greater the response will be. For example, in a dairy herd, if bulls are selected from dams that give more milk than the herd average, then the herd average will increase. Selection intensity would refer to the number of higher quality bulls used as replacements, or the total number of cows culled. Obviously, the greater the number of quality bulls, and/or inferior cows culled, the higher the genetic progress will be.

Selection intensity (and progress) is also affected by the number of traits animals are selected for. *If more than one trait is selected for, then progress for any one trait will be reduced.* This concept is of vital importance and, in the past, has been the result of some serious genetic setbacks. For example in the 1930's and 40's beef cattle breeders often selected cattle for coat color, coloration patterns, and other purely cosmetic traits. As a result, progress in growth characteristics, such as weaning weight and gaining ability, were severely set back.

What we must do is decide which traits are the most important and select just for them. Cosmetic traits, or other traits not directly affecting economic return, must be ignored. For example, in dairy cattle, milk production will be the primary trait selected for. If we do not select the animals with the very highest milk production records, then we can expect less than maximum progress.

In some cases, we may need to select for more than one trait. For example, in range cattle we may need to select for both weaning weight and calving ease. As explained in the previous section, birthweight is positively correlated with growth. Therefore, if we select for one, we will have less than maximum progress for the other. What we must do is come to a compromise; but in so doing, we realize that we will not have maximum progress for either trait.

SUMMARY

Chromosomes are strands of genetic material on which the genes are located. In sexual reproduction one chromosome will come from the female, and one will come from the male.

In some cases one pair of genes will control a trait, and in other cases several or many genes will control a trait. Normally *qualitative traits* are controlled by only one pair of genes. Qualitative traits are defined as those traits which may be seen. Examples would be coat color, presence or absence of horns, etc. *Quanitative traits* are normally controlled by several pairs of genes. Quanitative traits are defined as those traits which must be measured. Examples would be growth rate, milk production, and mature size.

Additive genes are genes that affect quanitative traits. That is, no one gene controls the trait, but taken together as a whole they influence the trait. When one pair of genes control a trait, they are said to be *dominant* or *recessive* genes.

If a gene is dominant, the pressence of that gene will control the trait. If a gene is recessive, that gene will control the trait only if the corresponding gene on the other chromosome is also recessive. Many undesireable traits such as dwarfism are recessive traits, and will not be expressed unless the animal carries recessive genes on both corresponding chromosomes.

When an animal carries recessive genes on both chromosomes it is said to be *recessive homozygous*. If it carries dominant genes on both chromosomes then it is said to be *dominant homozygous*. If the animal carries a dominant gene on one chromosome and a recessive gene on the other chromosome, then it is said to be *heterozygous*.

Since dominant genes control qualitative traits, one cannot tell by looking at an animal if it does, or does not also carry the recessive gene as well. This can only be determined by breeding to animals known to have recessive genes. In that case, a percentage of the offspring will display the recessive trait (if the animal is heterozygous). If the animal is homozygous (carries only dominant genes), then none of the offspring will display the recessive gene.

Heritability refers to additive genes and quanitative traits. Specifically, heritability refers to the proportionate increase (or decrease) one may expect to be passed on for a given quanitative trait.

Heterosis is an extremely important concept, which refers to genetic response obtained by crossing different breeds or even species of livestock. With heterosis we get response in traits that ordinarily have a low heritability. The two most important traits are adaption to severe climates andd reproduction. Known as crossbreeding, this type of breeding has been immensely important in developing breeds of livestock capable of producing well in hot, humid, tropical areas.

Inbreeding refers to mating related individuals, and is normally a detrimental practice. Inbreeding usually causes a decline in quanitative traits, and often causes genetic abnormalities controlled by recessive genes to be expressed with greater frequency.

Selection intensity refers to the emphasis given toward improving a particular trait. As more superior animals for a given trait are brought into a herd or flock (or as culling is increased), then progress for that trait will be more rapid.

For maximum progress for any given trait, one must select for that trait only. If more than one trait is selected for, then less than maximum progress will be experienced.

CHAPTER 23   POULTRY PRODUCTION
by H. John Kuhl, Jr., Ph.D.
Poultry Consultant
Rialto, California

Poultry production has evolved from a back yard, sideline business into a highly specialized industry. Efficiency and technology are the driving forces behind the growth of the worldwide poultry industry. Per bird egg production from laying hens is increasing annually. Broiler and turkey meat production is second only to aquaculture in the efficiency of converting feed to meat.

The primary focus of this chapter will be on chickens (layers and broilers) and turkeys. Pheasants and ducks will be briefly discussed. We will first discuss the unique features of the reproductive and digestive systems of birds. We will also review some common management practices, diseases of poultry, and finally, the specific segments of the industry.

THE EGG

The egg is a good starting point. After all, it is common to all classes of poultry. The oviduct of the hen and egg structures are depicted in Figure 1.

OVARY

Egg formation begins in the ovary with yolk development. The hen has two ovaries; however, only the left one is functional. This single ovary contains approximately 4,000 potential egg yolks which are visible to the naked eye and about 12,000 ova of microscopic size. The ovum is enclosed in a follicle (yolk sac) and the follicle is attached to the ovary by a stalk. The yolk sac has an abundance of blood vessels in which nutrients are carried to the yolk for growth. One segment of the follicle ruptures and the yolk is released. Yolk size increases at a rate of about 4 mm per day and the follicle ruptures when the yolk is about 40 mm in diameter. Yolk size at the time of ovulation determines the overall size of the egg.

OVIDUCT

The oviduct of the hen consists of five distinct regions; the infundibulum, the magnum, the isthmus, the uterus, and the vagina. The infundibulum, or funnel, catches the yolk after it is released from the ovary. The infundibulum is also the region of fertilization. The magnum is the region of formation and deposition of the albumen or egg white protein. The egg is moved by peristaltic action to the isthmus, where both inner and outer shell membranes are formed. The uterus or shell gland holds the egg for a little less than 21 hours while the eggshell is being formed. Eggshell pigment as well as the cuticle or outer shell covering is added in the uterus. A sperm storage area exists at the junction of the uterus and vagina. In this region, sperm may survive for as long as 35 days.* The vagina is the last segment of the oviduct and serves as a lubricated passageway for the egg's travel out of the oviduct. The average total time from ovulation to egg laying is approximately 25 hours.

ABNORMAL EGGS

Occasionally, abnormal eggs will occur. Some may be due to poor nutrition and/or management and some purely to chance. In young pullets, simultaneous ovulations may occur resulting in double-yolkers. Blood spots sometimes occur as occasionally the follicle may not

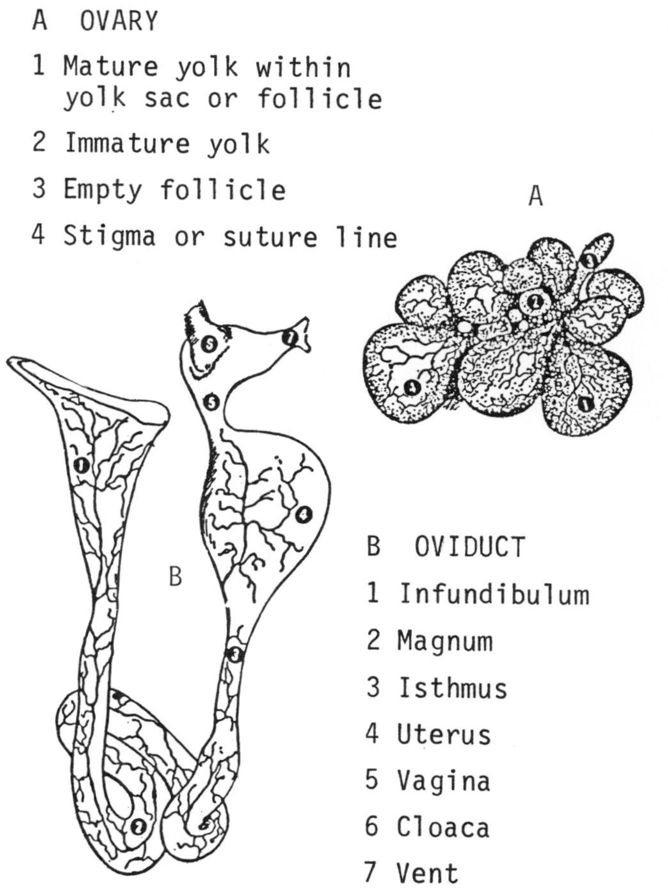

A   OVARY
1 Mature yolk within yolk sac or follicle
2 Immature yolk
3 Empty follicle
4 Stigma or suture line

B   OVIDUCT
1 Infundibulum
2 Magnum
3 Isthmus
4 Uterus
5 Vagina
6 Cloaca
7 Vent

Figure 23-1. Reproductive tract of the laying hen showing the (A) ovary and (B) oviduct. Egg Grading Manual, Handbook No. 75. U.S. Dept. of Agriculture, Consumer & Marketing Service, Washington, D.C. 20402.

---

* Contrast this with the sperm cells of most mammals which are viable for only 48 to 72 hours.

rupture along the stigma, thereby causing a bright red (blood) spot. Meat spots (usually white or gray material) are caused by the sloughing off of some type of tissue (blood clots or oviduct cells). Soft-shelled or no shelled eggs can occur because of lack of calcium in the diet, or a respiratory disease, but most often result from a violent peristaltic contraction which prematurely forces the egg out of the uterus.

THE DIGESTIVE SYSTEM

The digestive system of poultry differs from that of other monogastrics. No teeth are present, although foodstuffs can be torn or crushed in the beak. A storage area attached to the esophagus (crop) serves as a moistening chamber (in wild birds it also serves as a storage area). Starch digestion begins here. The craw attaches to the proventriculus, which is a small glandular stomach, and like most monogastric stomachs, secretes pepsin and hydrochloric acid. These secretions begin the process of protein digestion. The contents of the proventriculus are passed to the gizzard. The gizzard is a muscular stomach capable of grinding and crushing food particles as well as mixing the food with the secretions of the proventriculus. The small and large intestines are similar in function to that of other monogastrics. Two blind pouches called ceca extend from the junction of the small and large intestine. Bacterial digestion of food may occur in the ceca; no other function has been established. The cloaca is a chamber at the end of the large intestine. Ducts from the kidneys and the oviduct also enter the cloaca.

The rate of passage of food through the digestive tract is rapid, approximately 2.5 to 3 hours in most species. Excess or waste nitrogen is excreted primarily as uric acid rather than urea. Uric acid is visible as a cream-colored, pasty material in the droppings. Excretion of nitrogen in the form of uric acid requires less water than does urea, and therefore would be an adaptation for survival in wild birds.

DISEASES OF POULTRY

Some of the more common or economically significant diseases are presented in Table 1. Flocks should be observed daily and any change in habit needs investigation. Your eye may provide the first clue. Birds may appear listless or blood may be seen in the feces. Your ear may, in other cases, tell you more than your eyes. Birds may be too quiet or you may hear coughing. These observations are signs of disease. Whenever disease is suspected, a veterinarian specializing in avian disease should be contacted for diagnosis. (It is not always possible to determine the specific diseases by means of physical symptoms alone.) An accurate and timely diagnosis of a disease is the first step in treatment.

Prevention of disease is highly important. Treatment of disease is always expensive; and in some cases treatments for some diseases do not exist; therefore, the goal should be prevention.

Disease prevention has two components: 1. isolation from pathogens, and 2. immunity from their effects. Isolation is the ideal means of disease prevention, no exposure - no disease. Restricting unnecessary visitors and barring wild birds and rodents entry to buildings are the primary measures taken for isolation. Isolation is also aided by the wide separation of age groups and distance from neighboring poultry operations. Relying solely on isolation in a crowded poultry production area is not wise. Therefore, immunity (vaccination) coupled with isolation is the most practical method of disease prevention. To be effective, vaccinations must be given at the appropriate times. A typical vaccination program (Table 2) is designed to protect against those pathogens most likely found in the bird's environment at any point in its lifetime. History has shown the poultry industry that unknown types of specific pathogens can cause great losses, even in flocks successfully immunized against another type of that pathogen. Two clear ex-

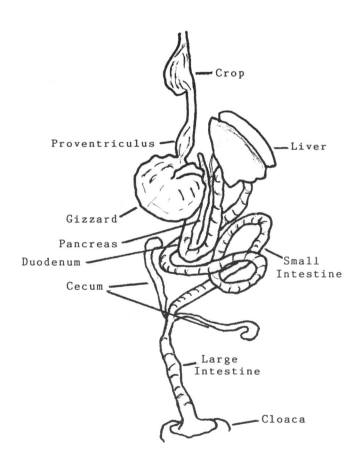

Figure 23-2. Digestive system of poultry.

TABLE 1
SOME COMMON DISEASES OF POULTRY *

Disease	Species	Signs	Treatment
Avian Influenza (AI)	ALL	Sinusitis, listlessnes, head edema, loss of egg production and fertility	None
Endemic Newcastle Disease (ND)	ALL	Watery nasal discharge, 80 percent mortality in young, severe egg production drops, egg shell thinning	None
Visceratropic Velogenic Newcastle Disease (VVND)	ALL	High mortality, bloody diarrhea, eye and nostril discharge, rough egg shell, blood spots	None
Infectious Bronchitis (IB)	Chickens	Low mortality, chirping, eye and nostril discharge, severe egg production drops, misshapen shells, thin egg albumin	None
Flow Pox (FP)	ALL	Wartlike lesions on the head and vent, distress in breathing	Vaccinate flock to stop spread
Laryngotracheitis (LT)	Chickens	Watery eyes, head and neck extended, coughing, exudate plugs in windpipe, asphyxiation	Vaccinate flock to stop spread
Mycoplasma gallisepticum (MG) (Infectious sinusitis)	ALL	Air sacculitis, yellow exudate in air sacs, foamy exudate in eyes, respiratory rattles, stunted growth, egg production drops	Tylosin, erythromycin, lincomycin, spectinomycin
Coryza (Roup)	Chickens	Swollen head and wattles, odorus nasal discharge, labored breathing, egg production drop	Erythromycin
Infectious Bursal Disease (IBD)	Chickens	Chicks only, diarrhea, 1 to 15 percent mortality, greatly impares the immune system	None
Avian Encephalomyelitis (AE) (Epidemic tremor)	Chickens Turkeys Pheasants	Chicks show dullnes, head tremors, prostration, Hens show slight egg production drop	None

TABLE 1 (Continued)
SOME COMMON DISEASES OF POULTRY

Disease	Species	Signs	Treatment
Marek's Disease (MD) (Range paralysis)	ALL	Tumors, lameness, drooped wings, incoordination, pupil mishapen.	None
Fowl Cholera (FC) (Avian pasteurellosis)	ALL	Fever, rapid death, high mortality, green-yellow diarrhea, respiratory rattle	Vaccinate flock to stop spread sulfadimethoxine
Erysipelas	Turkeys	Yellow-green diarrhea, swollen joints, swollen head parts, cyanosis of the skin, mortality up to 25 percent	Penicillin erythromycin tetracyclines
Pullorum (BWD)	Chickens Turkeys	Chicks and poults gasping, chilling, pasted vent, diarrhea, white feces, survivors are carriers	Eliminate reactor birds
Coccidiosis (Coxy)	ALL	Ruffled feathers, paleness bloody feces	Any of several coccidiostats
Histomoniasis (Blackhead) (Enterohepatitis)	Turkeys	Watery sulfur-colored droppings, drowsiness, weakness, mortality up to 50 percent	Any of several histomonastats

* Source: THE MERCK VETERINARY MANUAL, 5th ed., Merck & Co., Inc., Rahway, N.J.

amples are Newcastle Disease of the 1970's, and Avian Influenza of the 1980's. Both outbreaks resulted in the depopulation (complete destruction) of affected flocks and enormous economic loss. Flock isolation was reinforced after each of these outbreaks.

COMMON MANAGEMENT PRACTICES

INCUBATION

Artificial incubation of eggs is not a modern discovery. The Chinese and Egyptians used supplemental heat in the form of fire or solar heat for incubation thousands of years ago. In all honesty, the principals they learned still hold true today. These principals are temperature and humidity control, rotation of the egg, and the need for adequate oxygen levels. Low temperatures were found to cause uneven hatch times and high temperatures caused deformed embryos. Low humidity caused excessive water loss from the egg resulting in dehydration of the embryo. Lack of rotation resulted in the embryo growing or adhering to the shell wall. Lack of sufficient oxygen, mainly as a result of heating fires within the incubator, caused high embryo mortality.

Today, hatchery equipment is computerized to maintain temperatures at approximately 99.5 degrees F and relative humidity at about 60 percent for the first 18 days and 70 percent for the last three days. Egg rotation is automatic, and the turning angle is set between 30 and 45 degrees. The oxygen content of air in the incubator is maintained at 21%. (Prior to incubation, fertile eggs should be stored at 50 to 55 degrees F for no longer than one week.)

Eggs of different species are of different size and hatch over different time spans (as shown in Table 3). It is very seldom that one hatchery handles more than one species; except in the case of rather small mail order type hatcheries. The commercial hatchery is divided into two areas, the incubation area and the hatching area. To aid in preventing disease, commercial hatcheries are

TABLE 2 A TYPICAL PULLET VACCINATION SCHEDULE

Age	Vaccination Type	Vaccination Method
1 day	Marek's Disease	Injection -back of neck
12 days	Newcastle (B1 Strain) Bronchitis (Mass. Strain)	Drinking water, eye, nasal
19 days	Infectious Bursal Disease (Lukert's Strain)	Drinking water
5 weeks	Newcastle (LaSota Strain) Bronchitis (Holland H120)	Usually spray
6 weeks	Laryngotracheitis	Eye
10 weeks	Avian Encephaomyelitis Fowl Pox	Two prong needle-wing web
14 weeks	Newcastle (LaSota Strain) Bronchitis (Holland H120)	Spray
3 Months	Newcastle (LaSota Strain) Bronchitis (Holland H120)	Spray (repeat every 3 months for best immunity

TABLE 3 EGG WEIGHTS AND INCUBATION PERIODS

Species	Egg Weight (gms)	Incubation Period (days)
Chicken	60	20-22
Turkey	90	26-28
Duck	37.5-105	26-35
Goose	90	30-33
Pigeon	9-25	16-18
Pheasant	32	21-28
Quail	24	21-28

cleaned regularly with sanitizing agents and eggs are fumigated the day they are placed in the incubator (Figures 3 & 4). The young are removed from the hatcher (Figure 5) and the necessary vaccinations and antibiotic injections are administered. Normally they are packed in containers of 100. Young birds are very susceptible to dehydration, and therefore must be transported rapidly. This is usually accomplished by heated van or aircraft.

BROODING

Birds are warm-blooded creatures and in nature hatchlings would be brooded by the female's body heat. This is not economical on a commercial scale because of feed costs, floor space requirements, and potential for disease. Therefore, they are warmed artificially. The conventional grow house is fitted with one of several heat sources: hot water pipes in the floor, infrared heaters, coal or oil furnaces or, the most common, natural or propane gas stoves. Moisture and feces are a good breeding ground for bacteria so a layer of absorbent and insulating material such as wood shavings, peat, or fine ground corn cob is thinly spread on the floor. In pullet growing facilities, cages are sometimes used during the entire life of the bird.

Mortality is usually highest during the first week of life, a result of "starve outs" or birds not eating. Therefore, water and feed must be easily reached during the first 24 to 48 hours of life. The newly-hatched bird depends upon the egg yolk as its source of energy. However, the egg yolk is rapidly used up by movement and body functions.

Temperature at bird level is generally 90 to 95 degrees Fahrenheit for the first week. Supplemental heat is reduced as age increases. Chilled birds will always crowd together for warmth and at times may suffocate those on the bottom of the group. Twenty-three to 24 hour lighting is provided for the first week to allow unrestricted feeding and drinking.

Figures 3 & 4. Exterior (above) and interior (below) of a working incubator.

Figure 5. Interior of a working hatcher, chicks are approximately 20 minutes from removal.

LITTER MANAGEMENT

Litter can be an expensive item if it is not recycled for more than one flock. Pullet, broiler, and turkey producers have learned to reuse litter for up to 2 years. Litter depth, house ventilation, watering system maintenance, and general sanitation all play a part in determining the successful reuse of litter. Litter must be deep enough to absorb moisture from droppings and also to be turned for aeration (4 to 6 inches are normal depths). Leaking waterers obviously must be repaired or the litter near them will remain soaked. The use of a spray disinfectant and mold inhibitor will aid in disease prevention and extend the useful life of litter.

BEAK TRIMMING

Beak trimming, inaccurately referred to as debeaking, is performed on chicks and turkey poults. The purpose of beak trimming is to reduce the amount of cannibalism. Initially, chicks are beak trimmed (nip debeaked) at 7 days of age. The upper beak is trimmed from the nostrils. Beak trimming at 7 days may be all that is required for birds to be housed in light-tight layer houses. More often a touch up or final trimming is required at 10 to 12 weeks for pullets. The final trimming involves the trimming of the upper beak and the tip of the lower beak (Figure 7). Turkeys may be beak trimmed at the hatchery although some growers debeak at 2 to 3 weeks of age. The equipment used in beak trimming has been an electrically-heated blade which cauterizes as it cuts. The

Figures 6 a & b. Turkey brooder operations in Central California. The heat source, feeders and waters (above) are ringed by a nearly invisible wire. Lower photo shows poults in a pole shed at age two weeks.

Figure 7. Beak trimming was accomplished on this pullet at age 12 weeks. The primary purpose of beak trimming is to reduce cannibalism.

laser has found its way into beak trimming and should further reduce stress on the birds.

TOE CLIPPING OF TURKEYS

In addition to beak trimming, turkey growers must perform a second operation, toe clipping. In the wild, a turkey's principal weapon is its claws. Domestic turkeys are capable of doing bodily damage to each other with their claws. At the hatchery, the toes on each foot are clipped just behind the claw or nail. This prevents the claw from regenerating. Balance and movement are not impaired by this procedure. Toe clipping prevents a good deal of stress in flocks of aggressive toms and increases the carcass grade due to reduced skin damage.

VENTILATION

Ventilation in a poultry house is dependent upon the type of housing in use. Open or curtain-sided houses usually rely on wind or convective air movement to bring fresh air into the house. Temperature-controlled houses require air to be forced or drawn into the house by means of fans. A good deal of engineering is involved in designing a ventilation system for this type of housing.

Outside temperatures and season of the year make the ventilation requirement change over time. The goal is to provide fresh air. Temperature and humidity must be also considered. Air movement in hot times of the year provides a cooling effect and, generally, the cooler the better. In cold weather ventilation is more complicated. The fresh air being brought into the building is cold and should be warmed before coming into contact with the birds. Outside air is routed near the ceiling, thus heating the cold air. Air is exhausted from the building near the floor and in this way warm air is brought down from the ceiling to the bird population.

Birds release a lot of moisture into the house; 1,000 laying hens will emit 3 to 3.5 gallons of water per hour. Ventilation is a means of removing the moisture from the bird's environment. Birds also produce high nitrogen excreta, and bacteria acting upon the excreta cause a release of ammonia. If your eyes or nose become irritated by ammonia in a poultry house, the ammonia is at a level which will adversely affect bird performance (20 to 25 parts per billion). Winter ventilation of a poultry house is a balancing act, trading some temperature loss for more oxygen, less humidity and ammonia.

MANURE MANAGEMENT

This is a subject in which every poultry operation is deeply involved. Poultry production facilities are normally close to human population centers. Poultry manure can be odoriferous as well as a breeding ground for flies. If manure management is not practiced, the poultry operation will be forced to close or relocate as the human population draws near.

Since realistically it is not possible to make poultry operations desirable to live adjacent to, the goal is to handle manure in such a way so as to make it more tolerable for the neighbors. Odor from poultry manure is caused by bacterial action on wet manure. This action releases ammonia and methane. If air movement across the manure surface is provided, moisture will be removed and bacterial growth greatly impaired. Reduced moisture content of manure also reduces the number of fly eggs hatched. Fly control can be enhanced through the use of larvacides, either in the feed or sprayed on directly. Biological fly control is possible with the use of fly larva predatory wasps. In short, poultry manure management can help a poultry operation survive in close proximity to its customer base, and even generate income from the sale of dry, easy to handle organic fertilizer.

MOLTING

Molting is a natural occurrence in which birds stop laying eggs and shed feathers. Eggshell quality and rate of lay decline with age. Hens that have molted and returned to egg production produce eggs of better quality and their productive lives are extended. However, molting is not an infinite rejuvenator as each successive molt results in lower egg production rates, and eggshell

TABLE 4 EGG PRODUCTION IN SELECTED COUNTRIES *

Country	Approximate No. Layers	Per Capita Egg Consumption	Market Preference
U.K.	35,000,000	230	98% Brown
W. Germany	35,000,000	226	70% White
France	44,000,000	250	98% Brown
Italy	31,000,000	?	98% Brown
Netherlands	33,000,000	197	65% White
Belgium	10,500,000	259	80% Brown
Denmark	4,000,000	230	90% White
Sweden	7,000,000	214	95% White
Spain	36,000,000	?	70% White
Canada	25,000,000	225	95% White
U.S.	270,000,000	261	95% White
Japan	117,000,000	?	95% White

* Source: Lin Zoller 1985, Shaver tables suggest choices in world's broiler, egg industries, Feedstuffs April 29, 1985, Supplement 1.

quality declines at a faster pace.

If hens are allowed to molt naturally, the egg quality of the flock would be highly variable. An artificially induced, forced molt allows the poultry operator to manage the molt of the flock. All hens in the flock are molted at the same time and are basically synchronized in their return to production. While forced molting is a management tool used primarily with laying chickens, the practice is also applied to broiler breeding hens and turkey breeding hens.

Molt is induced through stress. Duration of lighting is increased to 24 hours approximately 1 week prior to starting the molt procedure. Feed and lights are removed to begin the stress, which leads to molt. Extra limestone or oyster shell is given just prior to or at the time of feed removal in an attempt to prevent the last eggs from becoming soft-shelled. Drinking water should never be removed. Laying chickens are generally kept out of production for 2 to 3 weeks through management of a special molt feed and lighting.

Following the period of no eggs, the hens are considered to be rested and are returned to production with a layer feed and increased length of lighting. Most production managers conclude that the first molt for laying chickens should occur between 62 and 68 weeks of age. However, molting ages are often influenced by the egg market and replacement schedules.

LAYING CHICKENS

The laying hen is an efficient converter of feed into a high quality food source for humans, the egg. A summary of laying hen numbers, egg consumption, and consumer egg color preference is shown in Table 4. The United States market is primarily white shelled with some regional preference for browns. Shell color is a genetic trait and has nothing to do with egg flavor. Due to genetic selection and better nutrition, egg production in the 1980's has increased to flocks that can produce up to 230 eggs per hen at 60 weeks of age. Hen efficiency

TABLE 5 A PULLET FEEDING PROGRAM *

	Starter 4 weeks	Grower 8 weeks	Developer 1 12 weeks	Developer 2 18 weeks
Ingredient	Pounds of ingredient per ton of feed			
Sorghum (Milo)	864.3	894.5	937.7	963
Corn	500	500	500	500
Soybean Meal 47.5%	448	345	288	218
Meat & Bone Meal 50%	165	153	145	138
Wheat Millrun	---	80	100	150
Limestone	10	15	18	20
Salt	5.5	6	5	5
Chick Starter Premix	5	5	-	-
Layer Premix	-	-	5	5
MHA	2.2	1.5	1.3	1

Nutrient	Unit	Amount per Pound of Diet			
Energy, Metab.	(Kcal/Lb)	1378	1368	1373	1368
Protein	(%)	21	19	17.9	16.5
Lysine	(%)	1.1	.94	.85	.75
Methionine	(%)	.42	.36	.33	.3
Meth & Cystine	(%)	.8	.7	.65	.6
Calcium	(%)	1.1	1.12	1.12	1.15
Phosphorus Avail.	(%)	.52	.49	.47	.44

* Source: A program designed by the author.

and productivity both hinge on genetic background and the treatment received during the first 18 to 20 weeks of life. Pullets and replacement layers are terms which describe female chickens being grown for future egg production. Factors in the pullet's life can be broken down into three "program" areas. The first program is the vaccination program discussed earlier. Two other areas are the nutrition program and the lighting program.

PULLET NUTRITION

The objective of a nutrition program for pullets is to provide the nutrients necessary to grow a bird to its genetic potential. Each breeder of layer chicks provides recommendations for nutrient levels in each formula for the grow-out period. A target weight chart which depicts the desired weight for each week of age is supplied for each strain. Skeletal development is of major importance for sustained future egg production. In addition to supporting the musculature of the bird, the skeleton also serves as a reserve source of calcium for eggshells. That is, calcium can be removed from the skeleton and deposited in the eggs. Sample pullet diets are shown in Table 5.

LAYER NUTRITION

Upon nearing sexual maturity, the ova begin to increase in size and the oviduct starts to become functional. Accordingly, the nutrient requirements of the pullet begin to increase. She is now a laying hen, and her energy needs increase, causing an increase in feed consumption. Each eggshell she produces will contain about 2.2 grams of calcium. Utilization efficiency of calcium is about 60 percent in young hens. Therefore, she must consume at least 3.75 grams of calcium to replace calcium lost in shell formation.

Eggs also contain fats, proteins, vitamins, minerals and xanthophyll (a natural yolk pigment from corn and alfalfa). If any nutrient is deficient, then smaller, fewer, or no eggs will be produced.

The actual nutrient requirement of a laying hen changes with age, production level, egg size and environmental temperature. Feeding programs should address all of these variables in order to achieve the best production performance at the lowest feed cost. Table 6 attempts to demonstrate the effects of temperature in two sample layer diets. Under the conditions stated, a change in environmental temperature of 15 degrees F.

TABLE 6 EFFECT OF TEMPERATURE ON THE FEEDING OF LAYING HENS*

Ingredient	65 degrees F.	80 degrees F.
	Pounds of Ingredient Per Ton	
Sorghum (Milo)	1692	1421
Corn	1195	1315
Soybean Meal 47.5%	515	600
Meat & Bone Meal 50%	280	315
Limestone	295	325
Premix	5	5
Salt	7	7
Liquid MHA	6	7

Nutrient	Unit	Amount Per Pound of Diet	
Energy, Metab.	(Kcal/Lb)	1320	1298
Protein	(%)	16.2	17.3
Lysine	(%)	.77	.84
Methionine	(%)	.38	.42
Meth & Cystine	(%)	.67	.73
Calcium	(%)	3.56	3.93
Phosphorus Avail.	(%)	.44	.49
Xanthophyll	(Mg/Lb)	3.00	3.30

* Source: A program designed by the author. This example is based on the following: hen age - 36 weeks; egg production - 90%; body weight - 3.7 lbs.; and egg weight - 59.6 gms.

(65 degrees increasing to 80 degrees) results in a decreased daily energy need of about 33 Kcal. of metabolizable energy per hen (325 Kcal decreasing to 292 Kcal.) Hens at 65 degrees would consume the first diet at the rate of 25 pounds per hundred birds per day. At 80 degrees, 22.5 pounds of the second diet per hundred birds would be consumed.

LIGHTING

Day length, the number of hours of natural or artificial light, is an aspect of management requiring close attention. Among other aspects of production, maturity is stimulated by increasing day length. To prevent premature egg production, day length is not increased until 17 or 18 weeks of age. The age of transfer into egg production is dependent on strain. Strain variations vary by as much as 3 weeks in maturity.

Two types of layer housing exist; light-tight housing and open or curtain-sided housing. Pullets grown for light-tight layer houses are grown in light-tight pullet houses. Under these conditions, day length is held at 8 to 10 hours of light per day from the 4th week of age to stimulation. Pullets grown for housing in buildings subject to natural lighting should receive constant day length equal to the number of hours of natural lighting they will receive at the time of movement to the laying house.

Once moved to the layer house, day length is normally increased gradually to 16 hrs. of light. Minimum light intensity ranges from .25 foot candle for light-tight housing to .5 foot candle for open housing.

LAYING HOUSE EQUIPMENT

There are two basic types of poultry house equipment. There is what is known as floor housing, and there is caged housing. Very few floor houses are in use for commercial layers. Cages provide better disease control, more uniform feed intake, and less egg breakage. An example of a cage operation is shown in Figure 8. Cage sizes are not standard.

For the floor operations, space standards for best production are approximately 60 square inches (348 square centimeters) per bird. Approximately 4 inches of feeder space should be provided per bird. Feeders may be chain or screw type, mechanical, or hand-fed open trough systems. Mechanical feeders reduce labor costs and can be activated by time clocks.

Watering systems are generally cup type. With cup water, the hen triggers the flow of water into the cup each time she drinks. Open trough, constant flow waterers need more attention to sanitation, and add to the problem of excess waste water.

Egg gathering is generally automated. Eggs roll out of the cages and onto belts. The belts move the eggs into an egg washing machine located in a separate building.

EGG PROCESSING

Eggs are either trucked or moved by belt from the laying house to the processing room. Various sizes of egg processing machinery are in use. Some are capable of daily washing and packing eggs from as many as 700,000 to 1,000,000 layers on a single machine (Figure 9). At most egg operations, eggs are washed, weighed,

Figure 8. Caged layers in a light and temperature controlled environment. Feeder chain is visible in the lower right corner. Egg gathering belts are visible on both sides under the feeders.

candled and graded in the same day they are produced. Most urban dwellers actually do receive "farm fresh" eggs.

Eggs are processed for two markets, the shell or table egg market, and a smaller breaker market. The breaker market is able to make use of eggs which mainly, due to shell quality, cannot make the grade as table eggs. These eggs are broken by machine and the yolk and white separated. Manufacturers of cake mixes, noodles and confectioners are users of liquid eggs from breaker processors. Egg components may also be spray or freeze dried for extended storage.

A small percentage of eggs are not fit for human consumption. These eggs may contain blood or meat spots or cracked shells which cause leaking. Pet food manufacturers find these eggs make an excellent feed ingredient.

EGG PRICING

The United States egg market is based on price per dozen. Egg size is the determining price factor for most world egg markets (Table 7). However, egg price in some countries is based on price per unit of weight.

SPENT HENS

Spent hens are old layers that age and low egg production have made no longer profitable. Spent hens are usually processed for human consumption in soups, canned chunk chicken, and as stewing chickens.

BREEDING CHICKENS

The principal differences between layer operations and breeder operations are the presence of males and open floor or slatted floor housing. Open floor (no cages) housing is essential for mating. Several of the vitamin requirements are higher for breeders in order to increase fertility and hatchability. Vitamin A, E, choline, niacin and riboflavin are provided at higher levels than normally found in layer diets.

Figure 9. A large capacity "in-line" egg processing machine. Eggs are brought from the layer house on conveyor-belts to a washer on the back right. Candling takes place in the booth at the rear of the photo. Scales located beneath the numbered panels weigh each egg individually and deposit them in the correct size category. Packaging is automatic.

TABLE 7 MINIMUM EGG WEIGHTS BY CATEGORY

United States		European Economic Community	
Size	Oz./doz.	Size	Gm/egg
Jumbo	30	#1	>70
Extra Large	27	#2	65-70
Large	24	#3	60-65
Medium	21	#4	55-60
Small	18	#5	50-55
Pee Wee	<18	#6	45-50
		#7	<45

BROILERS

Amazing. That is the only word to adequately describe the broiler chicken. In 49 to 52 days an average broiler may weigh approximately 4.4 pounds (2 kilograms) and have consumed less than 9.7 pounds of feed. Feed conversion rates approach 1.9 pounds of feed per pound of meat. Due to the short growing period for broilers, a broiler house may be home for up to 6 flocks in one year and that includes downtime for cleaning and new flock preparation. The vast majority of broilers are grown for the fryer market. Broiler roasters are grown to 10 or 12 weeks of age and are larger than fryers.

The broiler industry is located along the Eastern coastal states, the Southern states, and central California. Isolated pockets of broiler production are found in other states. The nature of processing plant costs and marketing channels has greatly concentrated the bulk of the broiler industry into a few very large companies.

Most of the broiler producing companies are completely integrated with breeder flocks, feed mills and processing plants in close proximity. However, the actual growing facilities may not be owned by the broiler company but may be contract facilities. Contract growers are paid by the number of birds they manage and bring to market in their own buildings. Incentives are incorporated in the contract which aid in improving cost efficiency. Total annual broiler production worldwide is approximately 4.5 billion head.

BROILER NUTRITION

Broiler diets vary from the high energy corn base diet in the United States, to lower energy milo or wheat based diets in other regions of the world. The type of diet fed is a reflection of the consumer's preference. In the U.S.A., preference is for highly pigmented (golden-yellow) broiler meat. Pigmentation of broiler skin is determined by the level of xanthophylls in the feed. Corn

TABLE 8 A BROILER FEEDING PROGRAM *

Feed to:	Starter 17 days	Grower 38 days	Finisher 52 days
Ingredient	Pounds of Ingredient Per Ton of Feed		
Corn	1274	1304.8	1331
Soybean Meal 47.5%	495	450	345
Meat & Bone Meal 50%	150	130	120
Fat - Blended	50	70	90
Corn Gluten Meal 60%	10	25	95
Limestone	5	5	5
Salt	5	5	6
Broiler Premix	5	5	5
Liquid MHA	4	3.2	1.0
Lysine 78.4% L	2	2	2

Nutrient	Unit	Amount Per Pound of Diet		
Metabolizable Energy	(Kcal/Lb)	1446	1484	1533
Protein	(%)	21.3	20.3	19.8
Lysine	(%)	1.2	1.12	.97
Methionine	(%)	.52	.47	.39
Methionine & Cystine	(%)	.90	.84	.75
Calcium	(%)	.94	.85	.80
Phosphoru Available	(%)	.48	.43	.40
Xanthophyll	(Mg/Lb)	7	8.1	12.6

* Source: A program designed by the author.

gluten meal and marigold meal are two sources of naturally occurring xanthophylls used in broiler feeds. Synthetic pigments are in wide use in Europe. Sample U.S. broiler diets are shown in Table 8.

LIGHTING

The most common lighting program in curtain or open-sided housing is 23 hours of continuous light per day. Various lighting programs exist for light-tight houses. Two hours of light and 1 hour of darkness is one lighting program recommended for light-tight housing. Light intensity of a .25 foot candle at the floor is adequate.

BROILER HOUSE EQUIPMENT

Broiler houses are temperature-controlled open floor operations. They may be either curtain-sided or solid-sided with artificial light. Wood shavings and rice hulls are typically used as litter material. Feeders are usually suspended pan type, with automated delivery. Watering systems of all types are in use. The most common is the open trough.

The amount of floor space required depends upon the market age of the broiler. Broilers being marketed at 7 weeks of age should have about .65 square feet of floor space, and at 8 weeks about .8 square feet.

BROILER PROCESSING

Plants are located close to consumer centers. A broiler processing plant will process large numbers (millions of broilers) per week. Broilers are packaged and shipped chilled or frozen to the retail outlet. Broilers are generally processed and marketed as whole body birds. Cut up parts have become popular and offer a choice of packaged breasts, legs, thighs or wings. Further processing of broiler meat has opened a large market in the fast-food business. Broiler meat patties in all shapes and sizes are in demand.

TURKEYS

Although not as efficient as the broiler in feed conversion, the turkey does outperform other meat animals (with the exception of fish). Commercial turkeys are almost entirely of the white feathered variety, and are found in three varieties, small, medium, and large white. Large white hen turkeys are generally marketed at 14 to 15 weeks of age and weigh a little over 14 pounds. Hen feed conversions at this age are about 2.5 pounds of feed to one pound of live body weight. Large white toms are marketed at ages ranging from 17 weeks to 28 weeks. A 19 week old tom should weigh about 26 pounds. The type of market dictates the age of marketing. Hens and light toms are usually sold as whole carcass packages. Toms of 18 to 20 weeks of age are often used for further processing. Older toms (up to 28 weeks of age and weighing over 38 pounds) are produced for institutions and restaurants.

TURKEY HOUSE REQUIREMENTS

The majority of turkeys are grown in confinement or semi-confinement housing. Open range growing after 6 weeks of age is practiced in some areas of the U.S., but efficiency favors confinement. Hens perform very well in confinement with 2.5 square feet per bird. Toms also do well in confinement but require at least 4 square feet of floor space per bird. Confinement housing allows year-round growing.

TURKEY PROCESSING

This is the key to the success of the turkey industry. Until the late 1950's and very early 60's, the bronze turkey was the primary variety. Although it was a beautiful bird to look at, it was not very attractive after processing. Dark pin feathers were difficult if not impossible to remove. The white turkey came onto the scene during the time frame when further processing of poultry meat was a novelty. Further processing of turkey caused a change from a seasonal market to a continuous market. A tremendous market grew for cut up turkey parts. A whole range of flavorings and texturing agents were and are being developed to make turkey meat into sausage and lunch meats. Most whole body turkeys are stored or shipped frozen. Figure 10 is a photograph of a modern turkey processing line in central Utah.

BREEDING TURKEYS

Artificial insemination is practiced on all turkey breeder farms. The modern turkey is unlikely to mate naturally due to body conformation of the male. Turkey semen cannot, at this time, be frozen and stored without great loss of fertility. Semen is therefore collected at each artificial insemination session. Hen turkeys can be expected to produce 85 to 90 setable eggs over her laying cycle. It is not uncommon for breeder toms to weigh 45 or 50 pounds and hens to weigh 20 to 25 pounds. Thus, a tremendous amount of cost is incurred in the feeding of turkey breeder stock.

No standard lighting program exists at this time. But increased lighting is applied about two weeks before the females are stimulated. Light intensity for hens ranges from 5 to 30 foot candles. Light intensity of this magnitude increases egg production by reducing broodiness. The brooding hen is a problem. She sits on her eggs and will not leave them. As a result, she does not eat enough to reach her potential as an egg producer.

Figure 10. One processing line in a turkey processing plant. This plant's capacity exceeds 3.5 million turkeys annually.

TABLE 9   A TURKEY FEEDING PROGRAM *

Age to Feed:	#1 3 weeks	#2 6 weeks	#3 9 weeks	#4 12 weeks	#5 15 weeks	#6 <20 weeks
Ingredient	\multicolumn{6}{c}{Pounds of Ingredient Per ton of Feed}					
Soybean Meal 47.5%	952.3	865	760	635	495	340
Corn	830	901.8	994.3	1111.8	1231.3	1374.3
Meat & Bone Meal 50%	120	120	120	120	120	120
Phosphate 18% P	42	38	36	34	30	23
Fat - Blended	40	60	75	85	110	130
Turkey Starter Premix	8	7.5	-	-	-	-
Turkey Grower Premix	-	-	7.5	7	6.5	6
Salt	4	4	4	4	5	5
MHA	3.7	3.7	3.2	3.2	2.2	1.7

Nutrient	Unit	\multicolumn{6}{c}{Amount Per Pound of Diet}					
Energy, Metab.	(Kcal/Lb)	1317	1360	1400	1440	1500	1560
Protein	(%)	29.2	27.4	25.3	22.8	20	16.9
Lysine	(%)	1.74	1.61	1.46	1.28	1.08	.85
Methionine	(%)	.62	.59	.54	.50	.42	.35
Meth. & Cystine	(%)	1.1	1.04	.97	.90	.78	.67
Calcium	(%)	1.4	1.34	1.29	1.25	1.15	1.02
Phosphorus Avail.	(%)	.8	.76	.73	.7	.65	.6

* Source: A program designed by the author.

PHEASANTS

Commercially produced ring-neck pheasants provide replacement stock for game and hunting preserves and are a specialty meat item for the home and restaurant. After leaving the brooder house at 5 to 6 weeks of age, birds grown for release are kept in wire enclosures until reaching 10 to 12 weeks of age. Natural vegetation should be provided in the enclosure. Feed is scattered in open areas of the pen.

If the pheasant is to be a food item, confinement housing is provided. In confinement, weight gain will be optimal. Feeders are suspended circular type. Waterers commonly in use are open trough. Cock pheasants are aggressive creatures and close confinement only increases the opportunity to become a combatant. Hoods (blinders) which block sight to the sides and reduce fighting are a standard item for rooster pheasants in confined housing. Market age for pheasant is between 16 and 20 weeks of age. Feathers and pelts from meat pheasants are the basic raw material for a segment of the hat band industry.

Nutrient requirements for pheasants are depicted in Table 10.

DUCKS

Commercially raised ducks provide meat and eggs for a relatively small specialty market. Three breeds account for nearly all of the ducks on commercial farms: Pekin, Khaki-Campbell and Runner. The Pekin duck is a hardy, meat-type duck. The Khaki-Campbell is kept primarily as an egg producer but is also a good meat duck. Egg production from the Khaki-Campbell is highest when they are kept in pens of less than 20 in number. The Runner is an egg producer and is of smaller body size. Ducks of all three breeds grow quickly. The male Khaki-Campbell will reach 5 pounds in 11 to 12 weeks.

No single set of nutrient requirements can apply to ducks in general due to the wide variation in body size among the breeds. Formula specifications should be set for each specific breed, housing condition and proposed market.

National Research Council nutrient recommendations are provided in Table 10 as a guideline only.

SUMMARY

Poultry production is a highly efficient and specialized form of agriculture. Technology has been employed to a greater degree in poultry production than in any other form of animal agriculture. Virtually every aspect of poultry production is analyzed and controlled. Everything from the genetics of the birds themselves to facilities is tailored or analyzed for each specific operation.

Because of the heavy use of technology applied in specific manners to individual operations, this chapter can only be considered a general guide. Indeed, textbooks in general can only be considered warehouses of information. The application of that information for a specific operation must come from professional consultants, or extension personnel in the field of poultry production.

TABLE 10
NUTRIENT REQUIREMENTS OF PHEASANTS AND DUCKS *

Nutrient	Unit	Pheasant Starter	Grower (6-20 wks)	Duck Starter & Grow
Energy, Metab.	(Kcal/Lb)	1275	1230	1320
Protein	(%)	30	16	16
Lysine	(%)	1.5	.8	.9
Meth & Cystine	(%)	1	.6	.8
Calcium	(%)	1	.7	.6
Phosphorus Total	(%)	.8	.6	.6

* Source: Nutrient Requirements of Poultry, 1977.

CHAPTER 24 DAIRY PRODUCTION

Joel J. Kemper, Ph.D.
Consulting Animal Nutritionist
Nutritional Service Associates, Inc.
Modesto, California

Introduction

Milk is one of the oldest foods known to man. Historically, ever since man domesticated the cow, sheep, goat, reindeer and other milk-producing animals, he has benefited from nature's "most perfect food". Cheese making was probably discovered by accident. The story goes that a caravan was crossing the desert with cows' milk in a dried goat's stomach, which was commonly used at that time for transporting liquids. When the travelers reached their destination the milk had turned to cheese; no doubt from a combination of the movement of the camel carrying the pouch and rennin left in the goat stomach.

Many animals throughout the world are domesticated for milk production. By far the greatest number are dairy cows and the greatest number are the Holstein-Freisen. In some countries this breed is simply referred to as Freisen. The Holstein breed represents over 80% of total cows in the United States with the Jersey and Guernsey breeds making up the balance and a small number of Ayshire, Brown Swiss and Milking Shorthorn comprising probably less than 5% of the total.

Dairy farms are typically located in clusters. The reasons for this are obvious. The regions with highly concentrated dairy farming are in close proximity to large urban population centers as well as milk processing facilities. Additionally, there are professional people allied to the industry such as large animal veterinarians, artificial insemination technicians, nutritionists, feed mills, milking and feed equipment suppliers as well as strong support from the Dairy Herd Improvement Associations (DHIA). Financial institutions that know and understand the dairy business are also vitally important, and would be found in areas in which dairying is common.

Milk Composition and Consumption

Milk provides high quality protein rich in amino acids. It is a major source of calcium and potassium and provides riboflavin (one of the B-complex vitamins), niacin, phosphorous, vitamin A, vitamin B-12, vitamin D (fortified) and many other nutrients. Chemically, milk contains approximately the following percentages: 3.1 protein 4.9 lactose, .7 mineral, with a total of 8.7 solids-non-fat. Fat adds another 3.6, bringing total solids to 12.3%. The balance of 87.7% is water.

Protein, carbohydrates, minerals and vitamins constitute the solid-non-fat (SNF) portion. There are basically five fluid milk types. They include: regular homogenized milk, low-fat milk (usually 2%), non-fat or skim milk, buttermilk (1.5% fat) and extra rich milk. In addition to fluid milk, there is dry or powdered milk. Dry milk has almost all of the water removed. It is whole milk powder when only water is removed; non-fat dry milk when both water and fat are removed.

The non-fluid milk products consist of cheese, butter, ice cream, ice milk and yogurt.

The largest increases in milk and milk product consumption have occurred in low-fat milk, cheeses, ice milk, with tremendous increases in yogurt. National per capita consumption of domestic milk and dairy products is over 580 pounds per year.

Keep in mind that no two cows or no two buckets of milk are exactly alike. The feeding, genetics, health, milking and general management of the dairy herd all have an effect on milk composition. The scope of this chapter will allow us to focus only in general on those factors which affect the milk quantity and quality. We will, however, be emphasizing the practical side of dairy science.

Milk Pricing

There are four things very unique to the dairy industry. It is the only agricultural product in which production is measured daily, sold daily, has the greatest investment per unit of product sold, as well as the most complicated pricing structure. Every state in America has dairy cows but the top 10 states are: Wisconsin, California, New York, Minnesota, Pennsylvania, Michigan, Ohio, Texas, Iowa and Washington. The total wholesale milk sales for the U.S. represent nearly 18 billion dollars. Wisconsin represents over three billion dollars and California nearly two billion dollars of this total amount.

Milk pricing will vary from region to region and state to state but price is based on the three major constituents we mentioned under composition; volume, fat and solids-non-fat (SNF). The demand for milk is increasing, but production is outpacing sales. Production topped 142 billion pounds in 1985, up 5% from 1984 and more than two billion pounds higher than the 1983 record. This massive output means that the federal government will purchase 14 billion pounds of surplus milk to support dairy prices. The total dairy program cost in 1985 was more than 2 billion dollars. According to the U.S. Department of Agriculture, milk cows on farms July, 1985 numbered over 11 million head, an increase of 240,000 (2.3%) from a year earlier.

Prices paid for milk will vary from one milk-shed (region) to another. For example, in Hawaii producers are paid essentially on volume of milk sold. Other states vary in their pricing procedure. The most common situation is to pay on a butterfat basis. This system goes back many years and assumes that the solids non-fat (protein, lactose, etc.) fluctuates with butterfat. There are probably two reasons this system was popular: high-fat dairy products were in strong demand (butter, cream, etc.); and, the test for fat was much simpler than solids non-fat.

In more recent times, a system known as component pricing has come into play. Component pricing takes into account the value of variations in solids non-fat as well as fat. Some cheese manufacturing plants are now including protein as part of their component pricing. The reason for this is additional protein results in higher yields of cheese and other high protein dairy products. Exact price determination is variable but it must be remembered that all milk prices are tied to the prices for butter, powdered milk and cheese.

For purposes of the U.S. dairy support program; milk is also divided into two categories, depending upon how the milk is used. Class I includes only fluid drinking milk and commands the highest price. Class II includes milk used in the manufacture of dairy products. Cheese, yogurt, ice cream, and even powdered milk are considered Class II. Class II milk receives the lowest support price. The purpose of this discrimination was apparently to ensure adequate fresh drinking milk for the public.

Not every dairyman is eligible for the government support prices. Although the system of who is and who is not eligible is complicated and beyond the scope of this text; the student should realize that at the time of this writing it is important. Known as "base", it is the right to sell at government support prices. Generally older, established dairies that were in operation at the commencement of the dairy program have "base". Newly

Figure 24-1. A refrigerated milk storage facility is a necessity for a modern dairy. It is also a major capital investment which adds substantial fixed cost. Because of this and other specialized equipment cost (fixed cost), it is often argued that government sponsored stability in milk pricing is a necesssity. As pointed out in the text, the negative side of government controlled pricing is usually large surpluses of milk and milk products.

established dairies would not have "base". However, an older dairy can sell its "base", which is highly valuable ($700-$800/lb.).

THE COW/THE MILK

The ruminant animal has been described in Chapter 18. All ruminants function basically the same way. However, we have quite different ways of feeding them. In the previous chapter it was pointed out that feedlot animals are given high levels of grain in an effort to achieve the optimum level of feed intake to body weight gain. In essence, we are putting the weight into the animal in the form of body tissue. In the beef animal we look at breed characteristics, e.g., type conformation and other phenotypic characteristics that help us predict the efficiency in putting on body tissues. In the dairy animal it is just the opposite. We look for dairy temperament, femininity, and that triangular conformation emphasized in livestock judging. We are not interested in body weight gain. We are interested in milk production; the most production at the least possible cost.

Ruminants derive most of their energy requirements from organic acids. These are waste or end products from the microbial fermentation process in the rumen and, to some extent, the lower gut of the ruminant. However, the main difference between a feedlot steer and a milk cow is the usage of the organic acids. They are referred to as volatile fatty acids (VFA). By far, the major VFA's produced by the rumen microflora are acetic, propionic, and butyric acids. There are other longer-chained acids but these three comprise the bulk fuel for the ruminants.

In feedlot animals (beef and sheep) we want high levels of propionic acid produced in the rumen. These are created by the breakdown of starchy feeds such as grains and grain by-products. High propionic acid in feedlot animals enhances the deposition of fat in the body tissues. In dairy cows we want low levels of propionic acid and high levels of acetic acid. These shorter-chained organic acids are the end products from the microbial breakdown in the rumen of roughages such as hay and silage. Acetic acid is the metobolic precursor of milk butterfat. High levels of acetic acid change the level of milk fat rather than fat deposition in body tissues. With few exceptions the dairyman is paid on milk fat level so it is to his advantage to have as high a milk fat test as possible.

High fat becomes concentrated by low volume of milk; but that is not the desirable way of achieving high fat test. The goal is high milk volume, fat test and solids non-fat. It is also desired to have high volume, test, and solids non-fat for as long a period during lactation as possible. This is referred to as lactation persistency.

Lactation persistency refers to the degree with which the rate of milk production is maintained as lactation advances. The normal rate of milk decline ranges from 6% to 10% per month, depending on stage of lactation. While lactation persistency is a highly heritable trait, only 10% to 20% of individual cow milk production differences or herd lactation persistency differences are due to genetics. The remaining 80% to 90% are due largely to herd management differences (2). Thus, the dairyman has control to a very great extent over the milk persistency of his herd.

At parturition, milk production commences at a relatively high rate and the amount secreted continues to increase for about 3 to 6 weeks. Higher-producing cows usually take longer than low-producing cows to achieve peak production. Many non-pregnant cows will continue to secrete milk indefinitely, but at a reduced rate. In terms of milk quality, fat percentage in milk decreases slightly during the first 2 to 3 months of lactation and then increases as total production decreases with advancing lactation. Most of the increase in solids non-fat components of milk is more likely associated with advancing stages of the current pregnancy rather than stage of lactation (1). The changes in the SNF are more apt to be reflected by mineral and protein changes, as the lactose portion is very resistant to change.

It does not suffice to feed the dairy cow principally roughages for acetic production to maintain fat test. A careful, practical balance must be maintained between roughages (hay/silages) and concentrates (grain and grain by-products/whole cottonseed/beet pulp/etc.) to produce the volume of milk, test, solids non-fat and persistency at the lowest practical cost. In addition to ration influences on milk production (pounds of milk per day) and milk quality (fat test, solids non-fat, low somatic cell and bacterial counts) the diet of the lactating cow has to be nutritionally balanced to assure proper heat cycles and reproductive function.

THE MILKING PROCESS

The capacity of the udder to hold and secrete milk has a major influence on milk secretion rate. Generally larger udders produce milk at a greater rate than smaller udders.

The majority of dairy herds in the United States are milked two times (2X) per day. There are a considerable number of dairymen who prefer to milk three time per day (3X). Three times per day will generally increase the total amount of milk produced. The average increase for 3X would probably be close to 15% with a range from 10% to 24%. There are occasionally ranges from 0% to 30%. Three times per day milking requires tighter management, as the cows are on concrete longer per day, require heavier and more frequent feeding, and places higher requirements on milking equipment maintenance. Likewise, there are higher utility and labor requirements, higher flush water requirements, and lagoon capacity to handle the flush water. Thus, not everyone is set up for it.

Whether one milks 2X or 3X, it is vital that periods between milking be equal. For 2X milking there should be a 12 hour interval. Eight hour intervals should be maintained between 3X milking. Research has shown that cows milked at 10- and 14-hour intervals produced 1% less milk per day than cows milked at even 12-hour intervals.

In a herd with an 18,000 lb. rolling herd average (lbs. milk/head/year) and with a milk blend price of $11.20 cwt, this would amount to $20.16 per cow per year, or $6,048 per year on a 300 cow operation simply by not being prompt with milking times. Depending on the interval, the percentage lost could be in excess of 1%.

HOUSING

Basically, there are three types of housing. They comprise the stanchion, corral or dry lot, and loose housing. The loose housing consists of both the loafing and free stall. In the mid-west, and particularly the Great Lake states, stanchion housing is very popular. In the Northeast where high rainfall and cold, wet weather persists for extended periods during the winter months, free-stall housing has become very popular. The free-stall is exactly what is says, animals are free to move into whatever stall they choose. In areas of intense sun and high temperatures such as Florida, California, Arizona and Texas, quite often housing will consist of only the provision of shades. But all types of housing can be seen throughout the United States. However, the tie-stall type is usually only found in the mid-west and eastern U.S., for herds of less than 100 cows.

MILKING SYSTEMS

The milking systems consist of the bucket or pipeline stanchion, which is commonly referred to as the flat barn. The parlor types include the herringbone, in which several animals are let free after milking at the same time; and the side-opening which allows individual cows to leave that unit after milking is completed. These systems vary in size from double -4 with smaller herds to double -20 units in larger operations. In very large herds the polygon and rotary parlors are becoming more common. The polygon allows milking simultaneously on three or more sides with efficiency of movement by the operator. The three-sided polygon is the most popular. The American rotary parlor is more sophisticated with up to 24 animals milked at one time on a moving carrousel. Australian rotaries being installed now in the U.S.A. and elsewhere hold up to 72 animals. The larger the milking system the more capital investment is required, but can be amortized or justified with more cows.

Whatever type parlor is used it is important that the equipment be operated and maintained in accordance with the manufacturer's directions. Correct and routine procedures, milking intervals and sanitation are essential in the milking parlor. Method of milking can play a vital role in harvesting the milk from the udder. It is more important that the greatest number of pounds of milk be removed from each cow rather than the speed at which cows are milked per hour.

MASTITIS - Mastitis is a term which implies infection in the udder. It can be caused by trauma to the udder as well as poor sanitation. Flakes in the milk, which can be observed using a strip cup before milking, signify an infection. However, subclinical infections, as evidenced by elevated somatic cell counts (SCC) can also take their toll on milk production. Both the creamery purchasing the milk and the monthly DHIA (Dairy Herd Improvement Association) will give regular reports on somatic cell counts or the California Mastitis Test (CMT). Both tests are recognized for detecting mastitis.

The following is a recognized standard practice to be followed for milking to reduce somatic cell counts and mastitis:

1. Milk three or four streams from each quarter.
2. Wash the udder and teats but do not wet the udder more than three inches above the teats.
3. Use a paper towel for washing and a new, dry paper towel for drying.
4. Milk only clean, dry teats.
5. Attach teat cups with minimal vacuum leakage and adjust throughout milking to prevent slippage.
6. Remove the machine after the cow is milked being sure to shut off vacuum first at the claw.
7. Do not overmilk or machine strip.
8. Dip or spray each teat after each milking using an effective sanitizing solution.
9. Dip treat each quarter using an approved antiseptic preparation when you dry off cows.
10. Properly maintain the milking system on a routine basis.

It is important that teats be dried with disposable paper towels. That is, a non-disposable used for several cows can spread infection. After milking it is essential that each teat be dipped in a solution especially designed for this purpose. If spray is used, be sure the teat ends are completely covered. Recognized dips include iodine, chlorine or chlorhexidine. Most of these dips are

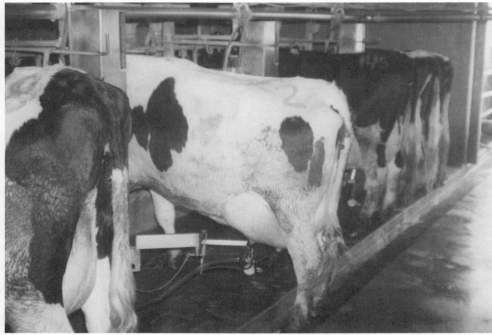

Figure 24-2. A modern milking parlor. In this type of arrangement the milker works at eye level. This is contrasted to figure 24-2b (below), the flat barn, in which the milker must bend over to wash the udder and place and remove the milking equipment.

formulated with small amounts of oil or lanolin to reduce chapping of the teats. Since mastitis is most commonly caused by bacterial entrance into the teat orifice, it is important that correct milking procedures be adhered to on a daily basis. Prevention is the key to control.

DAIRY FEEDS/NUTRIENTS

The total diet of the dairy cow may be broken down into five major divisions: roughages, concentrates, minerals, vitamins, and water.

Roughages are defined as those forages containing over 18% crude fiber. Concentrates contain less than 18% crude fiber and 60% TDN on an air dry basis (see also chapter 17).

If the roughage or concentrate is completely burned in the laboratory, the material remaining is the ash and contains all the minerals. These minerals are inorganic and critical to maintaining body functions.

Minerals are partitioned into three broad categories: major, minor, and trace minerals. These are defined by amount required, rather than importance of an individual mineral. The major minerals, i.e., calcium, phosphorous, potassium, etc. are required in such large amounts daily that they are expressed in percentages. Minor elements such as sulphur and magnesium, though just as critical to the body chemistry, are also expressed in percentages but in smaller amounts. Trace minerals such as cobalt, zinc, iron, copper, and iodine are expressed on a parts per million basis.

It is important that the correct, required dietary mineral levels be provided to the animal on a daily basis. Minerals must also be balanced for stage of lactation, reproductive cycle and balance between other minerals. Most minerals, when fed in excess of required amounts, may become an antagonist. An example of this includes sulfur and copper. Excess sulfur can tie up copper and prevent its absorption. Another example would be calcium and zinc. Excess calcium can tie up zinc and prevent its absorption.

The fourth major dairy feed nutrient is the vitamins. These are sub-divided into two major groups, viz., the fat-soluble and water-soluble vitamins. They are as the name implies, soluble in fat or water. The fat-soluble vitamins include A, D, and E. The water-soluble are the B-complex vitamins. Although they are synthesized in the rumen, some research has shown that sufficient quantities of B vitamins probably may not be formed in the gastro-intestinal tract of the cow to entirely meet its nutritional needs. (However, this area remains highly controversial.)

The fat-soluble vitamins A and E are not synthesized in the ruminant animal and need daily supplementation to handle the high demands put on the modern dairy cow. Vitamin D is the "sunlight" vitamin and is produced in the body.

We must never overlook water as our primary nutrient. All chemical reactions within the animal body are dependent upon water. It is essential that a good supply of water be available to the milking herd. A mature, heavy lactating cow will consume in excess of 50 liters (13 gallons) of water per day. Water sources in the dairy should be periodically checked for contaminants. Not only can such problems as salmonella and nitrates be transported by the drinking water, but high levels of sulphates in drinking water can devastate a herd.

FEEDING THE LACTATING COW

In feeding the lactating cows there are several decisions to be made. The first decision is whether the cows will be fed grain inside the milking barn.

Years ago this wasn't considered a decision to make. Most dairymen just assumed it was a necessity to feed during milking. Typically, the major portion of the cow's grain mix was fed to her while she was being milked. It was felt that this made heifers easier to "break" for milking, more eager to come into the barn as mature cows, made them let down their milk quicker, and otherwise made them much calmer throughout the milking process.

Today, however, a number of dairymen have gone to

Figure 24-3. A very common practice in dairying is to feed roughages and concentrates separately. However, as explained in the text, this tends to keep the rumen microbial population in a flux.

Figure 24-4. Cows receiving roughages and concentrates mixed together as a TMR (totally mixed ration). When compared to the separate feeding of roughages and concentrates, an increase in milk flow is often experienced with TMR.

what is known as Total Mixed Rations or TMR for short. In TMR feeding the cow's roughage and grain are mixed together and fed as a complete balanced ration, similar to a beef cattle feedlot type ration. The advantage to this type of feeding, in many instances, is increased efficiency.

Under the old type of system in which cows are fed their roughage outside and their grain inside, the rumen microbial population stays in a constant state of flux. That is, as explained on pg. 179, when ruminants eat only roughages the rumen stays at a neutral pH. It is neither acid nor basic, which is the optimum pH for the cellulolytic bacteria and protozoa that digest roughages.

But when the cows come into the barn and eat a large amount of grain, the rumen turns acid (see pg. 179). This acid kills off the cellulolytic bacteria and protozoa. The starch-digesting, or amylolytic bacteria can survive the acid conditions, but not the roughage-digesting bacteria. Thus, when the cows are turned out and begin eating roughages once again, the rumen microbial population must undergo a change before the roughage can be fully digested.

Because of the constant change in the pH of the rumen and subsequent changes in the rumen microbial population, production responses to total mixed rations are often experienced. Increases in milk flow of up to 10% have sometimes been experienced with a switch to TMR type feeding. (Normally it takes up to a week before the full response is realized.)

Another advantage to TMR feeding is cleanliness in the milking parlor. Specifically, when cows are not fed in the milking parlor, they are much less prone to defecate. As a contrast, when cows are fed in the parlor, the conditioned response when presented with feed is to lift the tail and defecate as eating commences.

Still another advantage to TMR feeding is that the cows leave the barn hungry. Cows fed in the barn, however, are often not hungry and lay down soon after leaving the barn. This puts the teat orifice in contact with manure and dirt before it has been fully closed, and increases the likelihood of mastitis. TMR cows, on the other hand, typically go back to their pen and begin eating. This keeps the teats clean until they have time to fully close.

GROUPING COWS

One disadvantage to TMR feeding is that it requires more pens or corrals in order to feed cows according to production. That is, it is much more efficient to feed cows as individuals or segregated groups, rather than as one big group or herd.

The primary reason has to do with genetic variability to produce milk. Some cows are capable of producing 100 lbs. of milk/day, others only 60 lbs., and still others only 40 lbs. Feeding a 40 lb. cow the same as a 100 lb. cow would just cause the cow to become overweight and increase the feeding cost. Feeding all the cows the same on an average basis, such as for 60 lbs. production, would fail to allow cows with the capability to produce more than 60 lbs., to produce what they are capable.

Therefore, it is always best to feed cows according to what they are capable of producing. Normally this is known as **grouping according to production.**

Although grain prices have declined dramatically in recent times, historically, grain and other concentrates have represented the most expensive portion of the cow's ration. The strategy has been to limit the amount of concentrate to what is just necessary to allow a cow to meet her genetic capability to produce milk. That is, all cows receive a basic allotment of roughage. From there, they are fed varying amounts of concentrates.

When the cows are fed in the barn, it is very easy to adjust the amount of concentrate to be fed. When they are fed a TMR type ration outside, it is necessary to put cows into groups, and feed different rations to each group. This, of course, requires extra pens or corrals.

Most commonly, cows are put into three groups. The groups are generally broken into 80, 60, and 40 lbs. This can easily vary 10 lbs. in either direction. In cases where farmers have pen facilities for only two groups, high producing cows are sometimes tagged and fed extra grain in the milking parlor. This essentially gives three groups, although it gets away from a total TMR system.

CHALLENGE FEEDING

The term challenge feeding means that all cows are fed in such a manner so as to give them the opportunity to be high producers. Basically the way it usually works is that as cows freshen, they are put in the high producer feeding group. This is logical since cows always produce the most milk shortly after freshening (calving). (Figure 24-5 illustrates a typical lactation curve.)

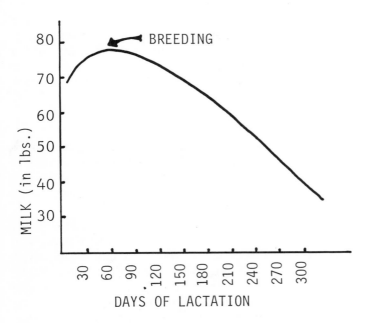

Figure 24-5. A typical dairy cow lactation curve. As can be seen, milk production normally increases during approximately the first 60 days of lactation. After that a gradual decline occurs.

As long as the cow continues to produce with the level of the high producers (e.g. 80 lbs. or more), she is left in the higher producer group. Only when her production declines is she moved to the next lower group. Again, she will stay with that group until her production declines further.

This system, "challenge feeding," is far superior to simply taking an average of her previous production and placing her in that group the whole time. Challenge feeding requires constant monitoring, but allows extra production during early lactation from many average or below average cows.

SELECTING ROUGHAGES

Roughages are the basis of any dairy cow feeding program. Roughages are the major source of fiber, which is responsible for maintaining a high fat test. It is important to realize that the quality or digestibility of fiber is highly important. Coarse, indigestible roughages retard consumption by slowing up passage through the digestive tract. This, coupled with their low energy or TDN value, magnifies their negative effect on production.

Therefore, it is important that roughages fed to lactating dairy cows be of high quality. Typically this means they must be cut when immature and succulent. Alfalfa should be cut at the pre-bloom or early bloom stage, and corn, sorghum, or cereal silages cut before passing the mid-dough stage of grain formation.

These types of roughages can also provide a good deal of energy, in addition to just fiber. Also, some roughages such as alfalfa, other legumes, and cereal silages provide a good deal of protein. Again, the key is to harvest them at an immature stage.

If harvested at an immature stage, they will be green, leafy (palatable), rich in protein, have highly digestible fiber, and contain relatively good levels of TDN. On the other hand, roughages harvested at advanced maturity will be lower in protein, less palatable, have coarse, poorly digestible fiber, and otherwise low in digestible energy.

SELECTING CONCENTRATES

For most dairymen, selecting concentrates means not only selecting the concentrates themselves, but who will process the concentrates as well. That is, some dairymen construct their own feed mill facilities, whereas others buy their concentrates already processed from a commercial feed mill.

Whether the dairyman processes his own concentrates, or purchases them already processed, depends primarily on the size of his operation and his method of feeding. Basically, however, it all boils down to cost. If he feeds his concentrates separately as a grain mix, then the type of equipment he will need to process the grain will necessarily need to be much more expensive than if he feeds a total mixed ration.

The reason has to do with fines (broken pieces of grain, grain dust, loose meal, etc.). Grain mixes fed in the milking parlor should not contain fines. It is therefore necessary to use what is described as texturized feed.

Texturized feed is basically grain that has been heated with steam, and then rolled or "crimped" through corrugated rollers. Any protein meal (soybean meal, cottonseed meal, etc.) is present in the form of a pellet. This gives the feed an appealing "texture" and prevents fines and dust in the milking parlor.

The equipment required to "texturize" feed is very expensive. Heavy duty steam rollers and boilers are required for "texturizing the grain". This is a major investment, and therefore only very large dairies can justify the cost.

Dry rolling or grinding equipment is much less expensive. If a dairy is feeding total mixed type rations, then dry rolling or grinding of grain can sometimes be feasible. That is, when fed with roughage such as silage, wet brewer's grains, or molasses to "hold" and control the fines, the dust from dry rolled or ground grains does not cause a problem.

TYPES OF CONCENTRATES TO SELECT

As a general rule, the types of concentrates to be selected depend upon price, palatability, availability, and nutritive content. (Chapter 17 explains how to determine the value of feeds in relation to nutritive content.) Throughout the U.S., corn typically rates as the number one feed grain. In some areas grain sorghum and barley are sometimes more economical, although corn is often used preferentially, due to its superior palatability.

Soybean meal is considered the premier protein meal, although cottonseed meal is used heavily in the South and Southwest. Other meals sometimes used are peanut meal, linseed meal, and in a limited number of cases, canola meal. Whole cottonseed is also a very popular supplementary protein source which also provides substantial fat in the diet.

In many areas by-products are sometimes used. Brewers grains and beet pulp are two of the more common by-products. Cull fruits and vegetables are also sometimes used, although they do not fit very well in the category of availability. That is, dairy feeds need to be available on a daily basis. It takes a period of time for the rumen to become adapted to a feed, so it is normally best not to change the cow's ration during the lactation period. Otherwise, production will usually decline during the adaptation period.

Therefore, before by-product feeds can be considered for dairy rations, the availability must be investigated. Either the supplier must be able to guarantee a continuous supply, or large enough quantities must be purchased (and properly stored) to ensure a daily supply.

FEEDING AND MANAGEMENT OF THE NON-MILKING HERD

The quality of the replacement heifers raised on the dairy will have a huge impact on the success and profitability of the operation. Time and careful attention

Figure 24-6. A large, modern well managed calf rearing barn in Japan. In raising newborn calves, cleanliness and sanitation are absolute requirements. The Japanese people are well known for their concern for cleanliness, which is quite evident in the success and management of this facility. Housing over 1,000 calves, death loss is held to under 3%. This is especially remarkable since routine prophylactic use of antibiotics is prohibited in Japan. (Manno-Chikkosan farm near Obihiro, Hokkaido, Japan).

should be devoted to the development and growth of replacement heifers.

As discussed in Chapter 17, the rumen is the primary site for digestion in the mature cow, however, the rumen of the calf does not start to function until two to eight weeks of age. It is not until approximately three months of age that the calf has the limited ability to utilize a full range of feedstuffs. Normal rumen development will take place when the calf consumes feeds other than milk replacers. Liquids normally bypass the rumen and go directly into the omasum and abomasum. The liquid is then subjected to the normal digestive process that occurs in the lower gut of the adult animal. In essence, the young calf is a monogastric animal until its rumen becomes functional. As the rumen develops, the need for milk and milk products decline as the animal becomes more dependent on conventional feed (hay and grain).

It is critical to get the newborn calf off to a good start. As soon as the calf is born, mucous should be cleared from the nostrils and the calf dried to prevent chilling (if the mother does not do this). Liberal amounts of naval dip should be applied to the cord. If it is very long it should be tied and cut off. Remove calves from their mothers 24 to 48 hours after birth. Place the calf in a clean, dry, calf-raising facility. Individual hutches are very popular.

It is important that colostrum be given the calf no later than six hours after birth. After that time, absorption of antibodies decline very rapidly. Within two hours of birth 100% of the antibodies will be absorbed for protection against disease. None are absorbed after 12 hours. The calf should be given one gallon of colostrum split between two feedings. A calf cannot be over-fed colostrum the first 24 hours. However, overfeeding can occur after 24 hours (give approximately 1 gallon per 100 lbs.). As far as the liquid portion of feed is concerned, a good rule of thumb is not to discontinue milk until the calf is eating at least 1 lb. of dry calf starter per day. Calves will normally start eating small amounts of good quality hay during the first week of life.

At two months of age the calf should be weaned. The calf should be eating a good quality and palatable calf starter, and good quality hay (green chop may also be

Figure 24-7. Hoof trimming. This is one of many important auxiliary services required by dairymen. (If it hurts a cow to walk, she will not eat properly.)

fed with the dry hay) and water should be provided free choice at all times.

A desirable growth rate of the large breed heifers (Holstein and Brown Swiss) from birth to 14 months of age is about 1.7 lbs. per day. Growth rate during this period of time will greatly affect the sexual development of the heifer. Specifically, heifers grown at lower rates will have a delayed onset of puberty. Excessive growth rate of 2 lbs. or more per day, will reduce milk production and longevity when the heifer matures. Good quality silage permits rapid growth rates, and should therefore be used with moderation. Corn silage fed heifers do not need grain after the first six months.

At breeding age of 14 months, the heifers should weigh from 750 to 800 lbs. During the last five to six weeks prior to freshening, heifers should be introduced to grain in preparation for lactation. Five to 8 lbs. of grain should be given to first lactation heifers at freshening. Salt intake will need to be restricted a month prior to calving to prevent udder edema (fluid retention of the udder). The large breed heifers calving at 24 months of age should weigh between 1100 and 1200 lbs.

The production year for the milk cow begins at parturition. Although the cow will secrete milk almost indefinitely, it has been well established that a 305 day lactation period is the most efficient. Extending the lactation beyond that time results in a steadily decreasing milk flow, and does not allow the cow to prepare for the next lactation.

By drying the cow up at 305 days, she is given a 60 day dry period (40 to 65 average) to prepare for the next lactation (parturition). That is, the dry period is the last 60 days of pregnancy. During that time the cow needs special nutritional consideration. Unlike first calf heifers, the dry cow should not receive a ration very high in energy (grain). The cow should not become fleshy during this period, or calving difficulty, and possibly ketosis (discussed later) may occur. Even high quality forages such as corn silage (which contains substantial grain) have allowed dry cows to gain excessive weight.

Keeping Ca, P, and Na, in balance is extremely important. Excess Na (salt) fed during the dry period can lead to udder edema. However, the greatest mineral problem can occur if the cow is allowed to consume too much Ca. The result can be what is known as milk fever, which occurs right after calving.

As explained in Chap. 17, one of the functions of Ca is neuromuscular control. Without sufficient Ca, muscle cells are thrown into a state of extended contraction (they cannot relax). Unless sufficient Ca is supplied, the animal's muscles literally lock down in the contracted position (this includes the heart and death can result).

When a cow freshens (calves and begins lactating), a physiologically enormous amount of calcium is put into the milk. The result can be a deficit of calcium for muscular control. When this happens, the cow will go down in a state of tetany. If Ca (in the form of calcium gluconate) is not injected (intravenously) immediately, death can result.

Calcium can be mobilized from the bones to meet a Ca shortage. However, the cow's system must be conditioned for doing this. Therefore, a dry cow should be fed a ration with the minimum requirement of Ca. If, on the other hand, she is fed a ration with excess Ca, she will

Figure 24-8. Jersey cow. The Jersey is a much smaller breed than the substantially more popular Holstein. The total milk yield is much less, but the butterfat content is much higher. Jersey milk is sometimes sold as a specialty item due to its rich butterfat content and golden color. Jerseys have an adaptation to heat and humidity, and are therefore relatively popular in the tropics. (Photo courtesy American Jersey Cattle Club)

be poorly adapted for mobilizing Ca, and will be much more likely to incur milk fever.

Dry cows should be kept in two groups, viz., far out and close up cows. After drying off, cows should be dewormed and dry cow antibiotic infused into their quarters. The far outs should be held in a separate corral or pen where they can be fed roughage. As mentioned, care must be exercised to keep Ca at a minimum level. As a practical matter, the amount of alfalfa in the diet must be limited, because of the high Ca content in alfalfa. Alfalfa may be fed, but since the calcium content may vary from approximately 0.8% to over 1.5%, it is important that alfalfa intakes be held between 8 to 10 lbs. per head per day. The balance of approximately 10 lbs. dry matter may be made up of oat hay or 10 lbs. oat hay and 30 lbs. corn or oat silage. As mentioned, however, high levels of corn silage should be avoided as it will promote weight gains and fat cows.

For the close up (springer) cows about 5 lbs. of a grain mix per head per day should be given two weeks prior to calving. Regardless of whether the cows are in the close up or far out pens, salt should be restricted to prevent udder edema. In hot weather however, cows need some salt and it is no problem to provide cows with a salt block. Cows cannot physically lick off enough salt to overload on blocks. However, loose salt commonly recommended for lactating cows should never be given dry cows.

It is important that grain be provided to the close-up cows. The purpose of the grain is two fold: 1. increase energy just prior to parturition and, 2. introduce the rumen microbial population to starch after a 45 day diet of only roughage. That is, as soon as the cow calves, it is necessary to begin feeding grain.

High producing cows need substantial amounts of grain. If not fed enough grain, the heavy loss of energy into the milk can cause the cow's blood sugar level to become dangerously low. When that occurs, a condition known as ketosis can occur.

Ketosis is created when the cow is forced to break down large amounts of body fat for use as energy. When that occurs, by-products known as ketones accumulate in the blood. The physical result is total body weakness and can result in collapse and death. Apparently it is caused by the combination of high ketones and low blood sugar.

Since it takes time for the rumen to adapt to grain feeding it is necessary to begin feeding grain to the

Figure 24-9. Jersey bull. Jersey bulls may be rather dark, as in this photo, but also a much lighter tan (although usually not as light as the Jersey cow in Fig. 24-8). Although the musculature is not as massive as other bulls, one should be aware that Jersey bulls can be extremely aggressive. Indeed, although small in stature, Jersey bulls are considered among the most dangerous of all domestic livestock, and precautions should be taken when handling or otherwise entering a pen with them. (Caution should be the order when handling all dairy bulls.) Dairy bulls are normally raised in such a manner that they have no natural fear of man. They therefore often have no inhibitions about reacting aggressively. (Photo courtesy American Jersey Cattle Club)

close up cows at least two weeks prior to calving.

In summary, it can be said that feeding the dry cows is one of the most critical aspects of dairy cow management, but is often overlooked and neglected. But no matter how close a correct dry cow program is followed, there will be an occasional cow to come down with a metabolic disease. These animals should be recorded and if they repeat at following lactations, they should be culled.

GENETIC IMPROVEMENT AND REPRODUCTIVE EFFICIENCY

Consider dairy management as a pitcher of milk on a four legged table. If each individual leg represents nutrition, milking equipment, herd health, and breeding, it is obvious that the table cannot support management if any of the legs are missing. The milk in the pitcher will be lost when the the table tips over. Genetic improvement and reproductive efficiency are both part of the breeding program and represent one critical leg of the table.

With the majority of dairy herds artificially inseminated today, the dairyman has many proven sires from which to select for his herd. There is more potential for improvement and ultimate profit when cows leaving the herd are replaced with genetically superior animals (see pg. 213).

It might appear that the desirable traits to select for would be the physical standards for the breed. One would like to breed for a good type score, including dairy character, straightness and strength of legs, udder depth, size, attachments, quartering and placement of teats.

Unfortunately, as explained on pg. 213, when we select for one trait, progress for another trait will be less than optimum. Therefore, we must select only for those traits which are the most economically important.

Since the dairyman is usually paid by component pricing, (fat, protein and SNF percentages), these are the traits that should be specifically selected for. Milk production **per se** is an obvious important trait.

Heredity defects should be selected against at all costs. Some of these include fused or missing teats, impacted molars, umbilical hernia, dwarfism and dystocia, to mention a few.

Records of performance are vital to today's dairymen. They need to understand records, how to interpret them, and how to implement the necessary changes to improve the herd. There are many ways of organizing records. The only thing that is really important is that the records be kept, and that they be understandable. They should be organized so that monthly evaluation may be made from herd summary sheets and individual cow records.

Some of the data that need careful consideration and

Figure 24-10. Ayreshire cow. Milk production of the Ayreshire is intermediate between the Holstein and the Jersey. In years past, when emphasis was placed on butterfat production, the Ayreshire (along with the Guernsey breed) was relatively popular. However, in today's market where so much emphasis has been placed on total production, the Holstein has greatly overshadowed all other breeds. (Photo courtesy of Ayreshire Breeder's Association)

can be summarized on your own form include:
1. Days in milk
2. Days open
3. Average total milk production
4. Average total fat production
5. Mature equivalent total milk production
6. Mature equivalent total fat production.

Specific items may be charted on graph paper so that a running picture of where the dairy is on a month-to-month basis will be available for the entire year. Certainly milk per day and/or fat corrected milk (FCM) need to be plotted. This can help pinpoint weaknesses or strengths in the management program.

REPRODUCTIVE EFFICIENCY

Reproductive efficiency is critical to the success of the entire operation. If more cows can be brought into production sooner by decreasing the calving interval, or reducing services per conception, this will result in higher milk production with reduced cost.

It is important to realize that time is necessary to correct reproductive problems. However, management can deteriorate in a rather short period of time if priorities are not set and followed through.

Four very important points to consider include:
1. Days dry
2. Days open
3. Conception rate
4. Calving interval

Days Dry. Short dry periods do not give cows time for rest between lactations and long dry periods increase cost.

Days open. This is one of the best indicators of her reproductive performance. Ninety to 100 days would be considered excellent.

Conception rate. Conception rate is determined by services per conception and influenced by days open. A good conception rate would be 1.5 to 1.6. Commercial dairy operations are generally less than 2.5. Registered dairy operations may be higher. Obviously, if conception rate deteriorates, it is an indication of a problem.

Calving interval. The calving interval indicates in months the lactation interval average for each cow in the herd. The average days open provides a monitor for it. Long calving intervals, i.e. 13.8+ months may indicate poor reproductive herd health and/or herd management problems. (Calving intervals on registered operations may be in excess of 14 months.)

It is important to keep in mind that success comes from mastering the fundamentals. The fundamental in the dairy business is reproductive efficiency. Reproductive efficiency encompasses herd health, nutrition, the breeding program and management.

SUMMARY

Modern dairy management is a highly technical and demanding occupation. Advances in animal genetics, reproduction, and nutrition have greatly advanced the efficiency of production.

Probably one of the most important and gratifying aspects of this technology is that it requires very little capital investment. Thus, it is available to nearly all dairymen, regardless of the size of the operation. It is, therefore, primarily a matter of becoming familiar with what is available and then applying it.

In this chapter we have discussed the general outline of this technology. Volumes can and have been written about the various disciplines (nutrition, genetics, etc.), so the purpose of this chapter has been to make the student aware of various aspects of these disciplines, and the effect they can have in a practical setting.

Moreover, the student should now realize and understand the need for grouping cows according to production; what TMR (total mixed rations) is and what its advantages are; the need for detailed ration balancing; the seriousness of mastitis and steps to take for prevention; the advantages of artificial insemination (see also chap. 21); and the need for keeping records of both milk production and reproduction performance.

Given this basic knowledge, the student should now be prepared to move on to more advanced texts and studies.

CHAPTER 25 BEEF PRODUCTION

Excerpted from *Beef Production,
Science & Economics, Application & Reality*
by senior author D. Porter Price, Ph.D.

PREFACE

Beef production is an enormously broad topic which includes classic western ranching, cow-calf farming, cow-calf confinement, intermediary "stocker" type operations, and feedlot finishing. Obviously, more than one chapter in one book would be required to present detailed information on each of these many facets of the industry. For that reason, only the most vital and/or general aspects will be described in this chapter.

WESTERN COW-CALF RANCHING

When most people think of beef production, the first image that comes to mind is the traditional ranching of the western U.S. In actual fact, however, in recent years more cattle have been produced in the Southeast, by what could be termed cow-calf farming.

However, traditional ranching is still highly important, and when we consider the contribution of Australia and Argentina, traditional ranching does constitute the majority of the beef produced worldwide. Therefore, we will begin our discussion with classic ranching, and from there proceed to more recent forms and developments in cow-calf management.

WESTERN RANCH MANAGEMENT

RANGE MANAGEMENT

Before discussing *ranch management*, one must first discuss *range management*. Indeed, range management is the most fundamental and important aspect of ranching. Unfortunately, many ranchers lose sight of this fact. It must be realized and always borne in mind that grass and forage are the commodities that the stockman actually raises . . . cattle are only the method by which the commodity is merchandised. Cattle (or other grazing animals) should always be given second priority to the forage. In actual practice, however, the reverse is all too often the case. As a result, a large percentage of the available range land has been overgrazed and is therefore, producing less than its potential. Depending upon the overall climate and available moisture, this damage takes years, or even decades to repair.

Effects of Overgrazing

Overgrazing eventually kills the individual plants by reducing the leaf area to a point where it cannot function adequately. That is, plants obtain their energy from photosynthesis, which is conducted in the leaf.

If the leaves of a plant are grazed, then a reduced amount of leaf area will be left to carry out photosyn-

Figure 25-1. Example of excessive animal grazing. Note how all that is left is unpalatable brush on the overgrazed side.

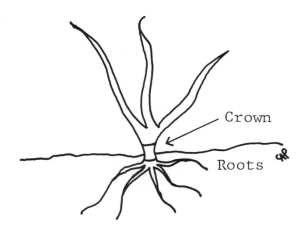

Figure 25-2. CARBOHYDRATE STORAGE AREAS IN GRASS PLANTS.

thesis. As a general rule, if more than 50% of the natural length of grass is removed, the plant will not be able to produce enough carbohydrates to meet its daily needs. The plant will then call on the reserves in the crown and roots to make up the deficit. This normally creates no problem as long as the leaf is able to grow back undisturbed.

But if the new leaf growth is continually removed (as in heavy grazing), then eventually the plant reserves will be exhausted and the plant will die.

Plant Succession

Whenever a plant dies out, a new plant species will move in to take its place. The new species doesn't actually come from another area. Usually it is a plant that is already present, but has been suppressed by the more dominant plant. That is, "survival of the fittest" is just as applicable to the plant world as it is the animal world. For a given set of environmental conditions, there will be one plant which will be the most adapted. That plant will, therefore, be able to out-compete all other plants for what nutrients are available in that particular environmental setting.

As a general rule, grasses are able to out-compete most other plant types. Normally the tallest grass that the environment will support will be dominant. If that grass is overgrazed and killed out, a shorter grass will usually take its place. If that grass is overgrazed, an even shorter grass usually takes its place. If grazing pressure continues to be severe, eventually a woody or otherwise unpalatable species will become the dominant plant.

How Overgrazing Usually Occurs

The overstocking of ranches is usually not intentional. Rather, what typically happens is that the rancher stocks the number of animals that his ranch will support during a good year. When rainfall is less than adequate, most ranchers are hesitant to cut back on their herd, hoping that rain will come soon. Typically, they will not sell any cattle until they have no available forage left. When the cattle get to the point of starvation, most ranchers will either sell or begin providing extra feed (hay, etc.). This further stresses the range simply by keeping the cattle on it longer. The amount of feed provided is usually minimal, and therefore the cattle will continue to utilize every bit of available forage.

The real damage, however, will occur when the rains finally do come. When moisture is once again made available to the range, the plants will begin to grow again. Since the plants have been dormant, nutrients for new growth must come from reserves within the plant.

The situation is that lush re-growth is suddenly made available to a cow herd that is half starved and has not had one sprig of green grass in months, or even years. Cattle, cows in particular, can eat enormous quantities of feed for short periods when they have been deprived previously. A herd of cows under these circumstances can eat 50 to 80 lbs. of range forage for several days or even weeks. By taking and re-taking the re-growth, the plant reserves will soon be exhausted; the plant will die, and the succession process explained in the previous section will occur.

Restoration of Overgrazed Range Lands

The terrible thing about overgrazing is that some range lands can never be returned to normal . . . because overgrazing not only increases the incidence of undesirable plants, it also often initiates erosion. High quality topsoil is gone. Topsoil, of course, takes eons of time to form.

Where erosion has not been a factor, as long as invader plant species have not developed, plant succession will work in reverse. Under proper grazing management, taller and taller grasses will return. The question, and the problem, is one of time.

The amount of available moisture is the determining factor in dictating the time required to restore an overgrazed range. The greater the available moisture, the quicker the range will be restored; that is, the faster plant succession can work. The sad part about it is that the majority of areas that have been overgrazed have an arid

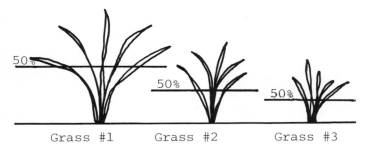

Figure 25-3. PLANT SUCCESSION DUE TO OVERGRAZING.

climate.

Total precipitation, of course, greatly influences the amount of available moisture, but the timing of precipitation is also vitally important. For moisture to be truly effective, it must come, or be available, during the growing season. In the Southwestern United States the majority of the precipitation comes as rainfall during early summer and fall. In the Northwest, most of the precipitation comes as snow during the winter, and therefore, there is far less moisture actually available for plant growth. This, coupled with the short growing season, greatly hinders range improvement. Specifically, it takes several times as long to restore the sagebrush deserts of the Great Basin than the mesquite covered deserts of the Southwest.

Range seeding, as an alternative to natural restoration, has been practiced with both success and failure. While there are a myriad of factors involved, available moisture is again the main determining factor. In relatively humid areas the main obstacle is primarily a mechanical one of removing the overstory brush and seeding the ground. If the area has adequate rainfall, the success of the reseeding is more of an economic question than a physical one; i.e., with proper technical direction, brush can be removed and grasses made to grow.

But as mentioned earlier, the real problem areas have arid climates. In arid climates, the problem is one of timing and intensity of rainfall. The most common problem being a light or moderate rain followed by drought. In that case, the seed will germinate only to wither and die due to lack of continued moisture.

Rangeland Grazing Systems

The purpose of this section will be to familiarize the reader with basic concepts concerning grazing systems. Only generalities will be covered, as specific details con-

Figure 25-4 A&B. A seed drill designed for range use.

cerning plant taxonomy, botany, ecology, and soil science are beyond the scope of this chapter. Because of the myriad of interacting details, this section has been prepared to simply help the reader recognize the need for planned grazing systems.

Continuous Grazing

Continuous grazing implies livestock left in an enclosure yearround. It takes the least amount of management effort and capital (for fencing), and, therefore, is the most popular grazing system. However, all too often it is not practiced as a system; rather, the cattle are simply "turned out".

The biggest problem associated with continuous grazing is what is known as spot grazing. As the name implies, this means that grazing animals will graze individual spots within an enclosure while leaving other areas ungrazed. The spots that are grazed and regrazed will obviously have much more immature growth on them than surrounding plants. That is, after the plants are grazed, regrowth will occur. Regrowth, of course, is very green and lush, and is much more palatable than more mature growth.

When this new growth appears, the animals will regraze that spot. In areas grazed continuously for a long time, there will be a number of areas or spots that the animals will continually return to. It is a vicious cycle. If stocking is light, the main deleterious effect will simply be that only a portion of the range will be utilized. The vegetation in the spots will be grazed and the vegetation outside of the spots will become mature and will not be utilized. (Tall, mature forage is much higher in fiber, lower in protein and digestible carbohydrates, and is therefore much less palatable than lush, green, immature forage.)

In areas with considerable rainfall, the grazing spots will be fairly evident (patches of green surrounded by yellow and brown). In more arid areas, with scattered vegetation, the spots may not be as evident. They may not appear as spots at all. Instead, one may simply find an intermixture of tall, yellow grass with short, green grass; often the tall grass will be clump or bunch grass.

At the very least, continuous grazing will result in incomplete utilization of the range. In most cases the spot grazing will result in the vegetation within the spots being grazed out. A good range taxonomist can usually tell if an area had been grazed continuously, as the more palatable plants native to that area will be missing. The most palatable plants are usually the most nutritious; and therefore, as a general rule, it can be said that continuous grazing will not only reduce the utilization of the range, but if carried over a long period of time, it will actually reduce the productivity of the land.

Seasonal Grazing

Seasonal grazing typically implies the grazing of pastures which can only be used during certain seasons of the year. Most commonly, it refers to mountain pastures which can only be used during the summer (due to ice and snow at other times).

GRAZING SYSTEMS UTILIZING MULTIPLE PASTURES FOR THE SAME HERD OF CATTLE

As the name implies, multiple pasture systems means splitting the range up into smaller plots and grazing them in some sort of alternate pattern. At some point

Figure 25-5. An example of "spot" grazing.

during the course of the grazing program, at least one of the pastures will be set aside and not grazed.

Terms Used in Multiple Pasture Systems

The terms used to describe multiple pasture systems can be confusing as they often refer to systems which are quite similar. Essentially there are two basic types: (1) rotational systems, and (2) deferred systems. Most other systems are modifications or combinations of these two basic types. The following is a description of the most commonly used systems.

Rotational Grazing - Rotational grazing commonly refers to grazing subdivided pastures alternately. The basic idea being that by concentrating a large number of animals in a smaller area (subdivided pasture) a more even utilization of forage is obtained. Rather than utilizing only the most lush or otherwise palatable forage, the animals are forced to consume a larger amount of the less palatable and more mature vegetation. There is less trampling since the animals do not have to move around as much to graze. Likewise, areas around water do not get damaged as much.

Deferred Grazing - Deferred grazing usually refers to leaving a given pasture ungrazed for a specific season or time period. The actual time period set aside for deferment will be determined by the type of vegetation contained in the pasture. For instance, if the most important plants in a pasture are propagated by seeds, the pasture would be deferred until after seed set.

Probably the most basic difference between deferred grazing and rotational grazing is that deferred grazing puts emphasis upon individual plant species, whereas rotational grazing does not. Pastures are deferred to benefit individual plants, whereas pastures are rotated to benefit all the forage.

Deferred-Rotational Grazing - Obviously this is a combination of both the deferred and the rotation systems. At some time, various pastures may be deferred and then later rotated. That is, pastures are put together in blocks or range sub-units. The pastures in one block may be entirely deferred, whereas pastures in other blocks are grazed rotationally.

Rest-Rotation Grazing - Rest-rotation grazing is a relatively complicated system used by the U. S. Forest Service. Rest is defined to be the deferment of a pasture for a full year. Each year one pasture is usually rested, while the others are grazed and rotationally deferred.

High Intensity, Low Frequency Grazing - This is an intensive plan of rotational grazing developed by Dr. Merrill at the Texas Experiment Station. With this system, the range is split into several small pastures which are grazed individually by a large number of animals for a short period; sometimes for a period as short as two weeks.

This system is well adapted to brushy areas, as it forces the animals to utilize all the available forage, including browse. This takes pressure off the more palatable plants, but may stress some of the forbs.

Short-Duration Grazing - The name given to a very intense form of grazing developed and promoted by Mr. Allan Savory of Zimbabwe, South Africa. In this system, as many as 16 pastures are used in a wagon wheel type arrangement, with the watering facility in the center. Among university and government range scientists this arrangement is considered controversial. In the author's opinion, if managed properly this type of system can be very beneficial. Under poor management, however, it can be very destructive.

CONCLUSIONS

The purpose of this section has been to point out the importance of range management. The author sincerely hopes that the reader will recognize and remember that range forage is the commodity that the rancher actually produces, and that livestock are merely the method of marketing the forage. Clearly, ranching decisions should always give priority to the vegetation.

TYPE OF OPERATION-STOCKER VS. MOTHER COW

Ranching consists of either stocker or mother cow operations, and sometimes a combination of the two. The most common type of operation is the mother cow. This simply means that the rancher runs a breeding herd of cows and sell calves produced from the herd. Stocker operations consist of taking weaned calves and running them on pasture to be sold as yearlings. Starting weights for the calves will normally run from 300 to 500 lbs., and ending weights will usually run from about 550 to 800 lbs. Steers are most commonly run, and hence the term stock-steer is often used as a term meant to be synonymous with "stocker operations".

What type of enterprise a ranch should become involved in depends upon a number of factors. The most basic factor is the quality of the range itself. In order to have a successful stocker operation, the range must be capable of producing gains of at least 1 lb./hd/day. Breeding cows need only to be maintained (do not need to gain weight), and therefore do not require nearly as good a quality of range. Since the majority of the range land in the Western U.S. and Australia is somewhat desolate, it is easy to see why cow-calf operations are the most common type of ranch enterprise. There are, of course, periods of time during the year when nearly all ranches can produce weight gains on their animals (spring and/or fall), and this is when the rancher usually plans to have his cows calve. Stocker operations must provide for weight gains over a much longer time period.

SPRING VS. FALL CALVING

Feed requirements for cows essentially double as soon as they begin lactating. Grasses and forages typically have their highest nutritive value in the spring, and so, by no small coincidence, most wild animals bear their young in the spring. In addition, the milder weather is also more conducive to offspring survival.

For these reasons, cow herds have traditionally been bred to calve in the spring. The calves are kept on the cow through the summer, and when the forages begin to decline in nutritive value in the fall, the calves are weaned. For the physical process of raising cattle, this plan works best for most western ranching areas. Marketing problems occur, however, when all the calves come off in the fall, thereby glutting the market.

To avoid what was typically a low calf market in the fall, many producers in the southwestern part of the U.S. went to fall calving. This allowed them to market their calves in the spring or summer and thereby helped to even out the calf market. In actual practice, fall calving works just about as well as spring calving, and in some areas works better. In areas such as South Texas, and the lowlands of New Mexico, Arizona, and California, fall calving lets the calves avoid the flies and 100 degree F temperatures of the summer. Winter grasses aren't any poorer in quality than the typically burned up summer grasses, and therefore fall calving can sometimes be more advantageous.

COW-CALF FARMING

This type of cow-calf production generally applies to the type of production commonly practiced in the Midwestern U.S., most of the Southeast (except Florida), and Europe. Most commonly, conventional farming is the primary enterprise with cow-calf production being secondary.

There are a few operations that run cows as the main enterprise, but these are usually purebred operations as land and other production costs in these areas are usually much greater than the return from what commercial cattle will afford.

Cows are used to graze wooded or hilly areas that cannot be farmed, and to clean up crop residues. Marginal fields are often seeded to pasture grass for additional grazing. Cows are also sometimes grazed on winter wheat fields, or other small grain pastures.

Calves from the cow herds are either sold, or kept for use as stockers for further grazing, or as growing cattle for silage feeding programs. As yearlings, the cattle are often further held for fattening programs.

THE GRAZING & FEEDING CYCLE

The actual grazing and feeding cycles of these operations vary considerably, but as a general rule, begin with the cattle grazing the wooded lots or improved pastures during the spring and summer. In the Midwest, the improved pastures may consist of a number of pasture grasses such as bluegrass, fescue, or orchard grass, etc., whatever is best adapted to the area. In addition, these pastures may also be seeded with a legume such as birdsfoot trefoil or one of the clovers. Having the grass-legume mixture increases the length of time grazing is available, as the legumes usually grow well during the heat of summer when the grasses are often dormant. Likewise, during the cooler periods of the early spring and fall, the grass usually supplies most of the grazing.

After harvest in the fall, the cows are often turned into the fields to glean crop residues. After the crop residues are gone, the cow herd is usually either put on small grain pastures, or moved into drylot for winter feeding of harvested materials. Sometimes a combination of both is used. Silage is probably the most common winter feed, although cornstalks, straw, hay, and haylage are also used.

SPECIAL CONSIDERATIONS

Grazing Untillable Pastures

Essentially the same range management principles as outlined previously apply. The only difference is that these areas typically have higher rainfalls, and thus can withstand somewhat more grazing pressure than the more arid western lands.

Grazing Improved Pastures

The types of grasses used in improved pastures are usually more productive and capable of withstanding much more grazing pressure than native species. As a consequence, they usually also require more water and fertilization. Actual utilization will depend upon the variety used.

As with range management, improved pastures are usually more productive if some sort of grazing system is used; i.e. concentrating a large number of animals in a portion of the pasture or pastures for a short period of time, rather than grazing the entire pasture area continuously. If there are several pastures available, this is accomplished simply by moving the animals periodically. In the case of large pastures, this can be accomplished by confining the animals to one section of it via electric fences.

Since improved pastures are usually fertilized, the stockman should be aware of the potential for nitrate toxicity. Likewise, bloat can also be a serious problem if the pasture contains considerable legumes, or lush, green grasses. If the pastures consist of winter small grains, the stockman should be on guard for magnesium (grass) tetany.

Grass Tetany - On lush, green pastures (primarily small grain pastures) one of the biggest problems is what is known as Grass Tetany, Magnesium Tetany or sometimes Wheat Pasture Poisoning. For a yet unexplained reason, periods of rapid forage growth accompanied by cloudy, cool weather will cause the magnesium content of the grass to become unavailable to the animal. Normal magnesium levels are present by chemical analysis, but some unknown compound or compounds tie it up and make it unavailable. Magnesium is required for normal muscle contraction, and therefore when it is unavailable to the animal, a state of tetany develops. Cows nursing calves are particularly susceptible since the magnesium requirement of lactating cows is two to four times that of dry cows and stocker cattle.

The first signs of the disorder are overall nervousness and blank staring. As the condition proceeds, the skin may begin to twitch (particularly on the face). The final stage is usually violent convulsions. If treatment is not administered promptly, the animals will usually die.

Treatment consists of intravenous administration of calcium-magnesium gluconate (calcium shares many roles with magnesium in muscle contraction). Only persons with experience in veterinary medicine should give this treatment, because if given too rapidly or in too great a quantity, the heart will fibrillate and the cow will die. Inexperienced persons can give a subcutaneous injection of 200 cc of saturated magnesium sulfate solution which will increase blood magnesium levels in about 15 minutes*

Incidences of Grass Tetany can appear very quickly, and the stockman may not be physically able to treat his cattle in time. This places a special emphasis on prevention, which means getting supplementary Mg into the cattle. There are a number of methods to do this, such as spraying Mg compounds directly on the fields, adding Mg to the water, etc. The easiest and by far the most popular method is to feed a mineral mixture with an extra high level of Mg. The drawback to this method is being sure the cattle eat adequate amounts of the mineral mix. To ensure consumption, many feed companies either add or recommend adding a palatability enhancer (soybean meal, grain, etc.) to the mix. As further protection, it is the author's recommendation to feed dry hay to pasture cattle during periods of rapid forage growth. Likewise, anytime one animal appears to have Grass Tetany, the entire herd should be hayed that day.

Grazing Crop Residues

The grazing of crop residues is discussed at length later on in the stocker cattle section. The primary difference between grazing stocker cattle and grazing cows is that cows can be run on poorer quality residues, since weight gain is not necessarily required. This means that cows may be left in the fields longer than stocker cattle. As explained later on in the chapter, most fields will contain a considerable amount of unharvested crop (lost during combining, etc.). Cattle will clean up the concentrate first and then, when forced to, will eat the residue itself.

The important thing to remember is that **cattle cannot maintain their bodyweight on residue itself** (cornstalks, wheat straw, etc.). They will lose weight when all they have is residue. The typical pattern is that a farmer puts a herd of cows on a cornstalk field, etc., and removes them 60 to 90 days later. When removed, they may weigh the same as they did when placed on the fields, but **they did not maintain their weight on cornstalks**. During the first half or two-thirds of the period they gained weight by cleaning up what corn, etc. was available, and lost weight during the last part of the grazing period when the concentrate was gone.

As explained in the next section, supplementary protein must be provided for cattle fed crop residues (low quality roughages). Residues are usually low in phosphorous and trace minerals, and therefore a mineral supplement must also be provided. In the Midwestern U.S., stalkfields can be extremely low in Mg, which can create the same symptoms of tetany normally attributed to grass pastures. In problem areas, increased Mg must be supplied.

As also mentioned in the section of stocker cattle, crop residues should be tested for nitrates before turning cattle in. This is even more important for cows since they will be left in the fields longer; i.e. nitrates are contained in the residue itself and since cows will be forced to eat a greater amount of the actual residue, the danger is greatly intensified. In fields with high nitrates, the typical pattern is that cattle will do fine and appear healthy for the first few weeks. As the amount of concentrate decreases and the cattle are forced to eat larger quantities of the residue itself, nitrate toxicity will begin to appear, often quite rapidly.

SUPPLEMENTAL FEED FOR BREEDING COWS

Supplemental feed usually accounts for the largest single cash outlay in both ranching and cow-calf farming. For that reason, it is highly important that the producer understand the fundamental principles involved in supplementation.

When the grass and other forage is green and actively growing, it is relatively high in protein and energy (soluble carbohydrates). Green, lush grass can be as high as 25-30% crude protein, although 12-18% would be more common. When growing conditions deteriorate (onset

---

* USDA leaflet #561 Controlling Grass Tetany.

Figure 25-6 A&B. A cow herd can be utilized to clean up crop residues. Well meaning urban groups have often criticized beef consumption as an inefficient use of grain. The fact is that ruminants such as cattle can utilize roughages and/or crop residues that can otherwise not be used by man or monogastric animals. In actual practice, cattle are fed grain for only about 100-150 days during the finishing period. The entire rest of their lives they are maintained on grass, pasture and other roughages.

of winter or summer, heat, drought, etc.,) and the grass turns from green to yellow or brown, the nutritive value drops off very quickly. Protein contents can go as low as 4%, and TDN can go from 65-70% to 45 and 50%.

When conditions such as these occur, beef cows must be supplemented if they are to remain on the pasture. *Supplementation* is defined as providing a small amount of a concentrate in order to allow the cattle to more fully utilize the range forage that is available. This is opposed to *feeding*, which is defined as providing a large amount of the animal's total intake from a feed source other than range forage.

The distinction between supplementation and feeding may seem to be a matter of semantics . . . it is not! One of the most crucial questions to any cow-calf operation is, "Are we engaged in a feeding program, or a supplementation program?" The basic enterprise in cow-calf production is to utilize forage. Indeed, the forage is, or at least should be, the cheapest feed available. Under nearly all historic, economic conditions, feeding harvested feed to beef cows has been a losing proposition. In addition, whenever cattle must routinely be "fed", the pastures are obviously overstocked. At some point then, overgrazing will almost always occur.

The strategy should be to supplement as little concentrate feed as possible. To understand what and how to supplement, one must understand *why* supplementation is so important.

When the protein and energy values of grasses fall off, the fiber content goes up. Fiber is difficult to digest, which reduces consumption considerably.

Essentially, the problem is this: A dry, pregnant cow needs about 8.0 lbs. of TDN to maintain her body weight and produce a healthy calf. Dry, mature grass will have a TDN which may average about 47%. Without supplementation, a cow will eat about 1.75% her body weight in this type of material. A 900 lb. cow will therefore eat about 15.75 lbs. (900 x .0175) of that kind of grass. The reason for the decreased intake is the buildup of fiber in the rumen; i.e. the fiber is difficult to digest, and therefore, it builds up in the rumen and the cow feels "full". With an intake of 15.75 lbs. and a TDN of 47%, the cow's TDN intake is only 7.4 lbs. (.47 x 15.75). Since she needs 8.0 lbs. TDN, but only receives 7.4 lbs. TDN, the cow will lose weight.

More important than the weight loss, are the adverse effects upon reproductive efficiency. These are caused by what are known as priorities for nutrients. This means that certain physiological functions are given priorities for available nutrients. The priorities that pertain to reproductive functions are (in order of priority):

    I. Lactation
    II. Body Maintenance
    III. Fetal Growth, or
    IV. Estrus Cycle

Lactation is given priority over all other body functions. The cow will deplete her own body stores of energy, protein, and minerals in order to meet her genetic capability to give milk. This is why the cows that produce the heaviest calves at weaning will usually be in the poorest condition.

After lactation, body maintenance is given priority. Body maintenance is given preference over estrus and fetal growth, although the division is not as clear-cut as with lactation. In borderline cases of underfeeding, the fetus will continue to grow, and a small percentage of the cows in the herd will fail to come into heat. If malnutrition is substantial, the fetus will be born small and weak, or in extreme cases, may be reabsorbed. (Underfeeding is the most critical during the last three months of pregnancy.) Likewise, a disastrously large percentage of cows may fail to breed if serious malnutrition occurs during the breeding season.

In terms of survival of the species, these phenomena seem easy to rationalize. In the case of lactation being given priority over fetal growth and estrus, it seems logical to conclude that nature has, in effect, given priority to the newborn animals, saying "Look, we've got a calf on the ground, let's take care of it, and then if there is anything left over, we'll see about making another one". In the case of body maintenance given priority over fetal growth, it seems obvious that a weak or dying cow (from starvation), cannot give birth to a live calf.

What to Supplement

Since range cows cannot meet their TDN requirements from dry, mature, range grass, it should be apparent that some type of supplementation is required. The question is, "What kind of supplement?"

It is energy that controls the estrus, therefore, many stockmen have reasoned that it is energy (grain) that they should supplement with. While feeding grain will certainly help cows to cycle, it is relatively inefficient and usually uneconomical. In the example given, .6 additional pounds of TDN are needed (cow needs 8.0 lb. and is getting 7.4 lbs.). Theoretically, less than 1 lb. of grain would be needed to rectify the situation. In actual practice, it would take from 3 to 6 lbs. of grain to keep a cow in good shape.

There are two reasons for this. The primary reason is easily understood. Grain is much more palatable than dried up grass, and therefore cattle will reduce their consumption of grass. Moreover if they know that they get fed grain at a particular time, they will quit grazing early, and go to the area where they are usually fed. If they are fed with a self-feeder utilizing some sort of intake inhibitor (salt, gypsum, etc.), they will hang around the feeder at least 2 or 3 hours a day.

The second reason is more complex. Rumen microorganisms which do the actual digesting, develop specific populations for certain feeds. In general, there are two types, cellulose users (cellulolytic) and starch users (amylolytic). If cattle are consuming feeds high in fiber, a population of cellulose using bacteria will

Figure 25-7. The cows that are in the poorest condition at weaning time, are often the ones that produce the heaviest calves. This is because these usually are the heaviest milking cows.

develop. If cattle are fed both fiber and starch intermittently, the rumen population will be in constant turmoil, and a good population of neither kind of microorganism will develop. The end result will be inefficient utilization of both feeds.

The feeding of small amounts of protein, however, usually results in an increased forage intake. Low quality roughage is quite low in digestible protein ... so low, that the rumen microbial population often cannot be maintained on it. Supplementation with a small amount of protein allows the microorganisms to reproduce, and thereby increases the total number of microorganisms available for digestion. With more organisms working on the fiber present in the rumen, it is broken down and digested faster. This makes the cow feel less full, and therefore she is able to eat more forage.

With proper protein supplementation, intake of low quality roughage will increase from 10 to as much as 50%. Going back to the example of the cow that eats only 15.75 lbs. of dry grass, the advantage of only a 10% increase in intake can readily be seen; that is, if the cow now eats 17.3 lbs. of a 47% TDN grass, she will consume 8.1 lbs. TDN (instead of 7.4), and will therefore meet her requirement (8.0 lbs. of TDN).

Along with helping to meet energy requirements through increased forage intake, protein also quite obviously helps to eliminate the protein deficiency created by grazing low quality forage. *Protein deficiencies during gestation can result in decreased fetal growth.*

How to Supplement

Protein supplements are available in many forms. Cottonseed and soybean meal have been the old standbys, but "range" cubes containing non-protein nitrogen, protein blocks, and liquid feeds have also become very popular.

Cottonseed cake is one of the best, and is used by a large number of stockmen. Similar in form is the range cube, a product or products produced by innumerable feed companies and country mills. Range cubes are usually represented on a crude protein basis; e.g. 30% range cubes, 40% range cubes, etc. They almost always contain urea and are therefore usually cheaper on a protein basis than cottonseed cake. It is important to realize that range cattle cannot utilize as much urea as feedlot cattle. A small amount of urea can be used quite successfully, but no more than ¼ - ⅓ the crude protein equivalency in most cases. The stockman should be wary of products containing more than ⅓ the crude protein equivalency from urea. Likewise, many range cube products contain grain by-product fillers.

As explained earlier, starch is not of particular value in range supplementation. There are, however, a large number of quality products available which are primarily oilseed meals with some added urea. In addition, the better products will usually contain some extra phosphorous and trace minerals, as dry forage is almost always lacking in phosphorous, and usually low in trace minerals.

The only drawback to cottonseed cake or good quality range cubes is that they require hand feeding. As with feeding grain, this will cause cattle to quit grazing early and possibly eat less forage.

In order to reduce the amount of labor required to hand feed cake or range cubes, it is possible to feed twice as much every other day. Experiments done by extension services and several universities have compared every day with every other day feeding of cottonseed cake, and results have been comparable. Care should be exercised in the use of supplements containing urea, however, as every other day feeding will reduce the amount of urea that can be utilized.

The use of protein blocks is another popular way of supplementation. The big advantage to the blocks is that when properly used, they are self-feeding. Salt, other minerals, special binders, sometimes unpalatable ingredients such as ammonium sulfate, as well as the degree of hardness in the block itself are used to control consumption. Reputable companies will usually market several different blocks, which vary in hardness and palatability, so as to be able to more closely control consumption under different conditions. As with most manufactured goods, there are some inferior products on the market. Unlike range cubes, high urea contents are not usually the problem with cheap block products because if urea levels get too high, the block form will not hold together. Rather, poor quality block products will usually contain relatively large amounts of grain and grain by-products which make the blocks quite palatable, and thereby precludes self-feeding.

In the past, liquid feeds have been very popular. Unfortunately, there is an enormous amount of variation in the quality of these products. Some liquid proteins, such as corn steep liquor, or corn gluten, can be successfully used as range supplements. More commonly, however, molasses is used as the base, and urea or other forms of NPN as the "protein" source. Phosphoric acid is usually added to regulate viscosity, as well as to add phosphorous to the mix. In addition, phosphoric acid is also often used as a palatability regulator. Some products also contain trace minerals, although settling out is often a problem.

COW-CALF CONFINEMENT

In Japan, Korea and China, cow-calf confinement is a relatively common practice. In the countries usually associated with beef production (U.S., Canada, Australia, Argentina, etc.), it is not practiced at all.

The reason has to do with economics; that is, the cost of purchasing all the feed a cow consumes. In Japan and some other Asian countries, cattle prices are several times what they are in the U.S., Australia, etc., and therefore they can afford to do that. In the U.S., for example, the cost of the feed for just the cow, in most years, would be equal to the sale value of the calf that cow produces.

The actual physical act of producing calves from confined cows, however presents no particular problem. It requires more management than traditional systems, but it is feasible. Indeed, as mentioned, confinement rearing is the primary means of production in several Asian countries.

MANAGEMENT PROBLEMS ASSOCIATED WITH CONFINED COW-CALF PRODUCTION

Aside from economics, there are a number of problems associated with cow-calf confinement. The most commonly reported problem has been calf disease, bacterial scours in particular. Outbreaks of calf diseases would certainly be affected by cleanliness of the pens, weather, etc. However, something that can precipitate calf disease is the fact that during calving, calves in confinement will suckle cows other than their dams. Oftentimes these are cows that have not yet calved, but have begun secreting colostrum. *E. coli* are the organisms most often involved in bacterial scours, and it has been shown that colostral antibodies are required for the prevention of *E. coli* infections. To ensure that each calf gets its full portion of colostrum, pregnant cows should be kept separate from cows that have already calved.

The second biggest problem associated with cow-calf confinement is seeing to it that each animal gets its share of feed. In a feedlot this creates no problem since the cattle are fed ad libitum (all they will eat). This, of course, is because it is desired that the animals gain weight. However, with a brood cow the intent is only that she maintain her weight. This means that with most feeds, the cows must be limit fed; that is, given a predetermined amount rather than all they will eat. Cows will establish a pecking order rather quickly, which can result in the more timid cows receiving as little as 50 to 60% of the feed allotted them. In other words, if silage is being fed at 35 lbs./cow, a ration of 3,500 lb. would be fed to a pen of 100 cows. Under most conditions what would happen is that 15 or 20 of the most timid cows would only get about 20 lbs.

Bunk space per animal must therefore be increased considerably (over normal feedlot standards). In feedlots, 9 to 12 inches of bunk space per animal is usually considered adequate, but for cow-calf confinement feeding, the author would recommend an absolute minimum of three feet. Still, pecking order problems may persist as aggressive cows can intimidate more timid cows 10 to 15 feet away. The most timid cows may not even come to the feed bunk until the other cows have had their fill and moved away. (Note the sparse population of cows in comparison to the available bunk space in Fig. 25-8.)

In feedlots, a common practice is to feed twice a day to ensure fresh feed. For cow-calf confinement, a better practice is to feed every other day. By feeding twice as much every other day, the "boss" cows aren't able to monopolize the feed quite so much. They eat their fill, but it isn't as great a percentage of the total feed available, leaving more in the bunk for the timid cows.

GENERAL MANAGEMENT RECOMMENDATIONS FOR COW-CALF CONFINEMENT

Pen Requirements

Pens for cows and calves should provide no less than 300 square feet per cow-calf pair. If muddy conditions develop, that space should be doubled or even tripled.

Figure 25-8. A large cow-calf confinement operation (largest in the world), located in Japan. The breed utilized is the Wagyu, the breed used in producing Kobe beef. Only because the animals are valuable is it economically feasible to maintain them in confinement. (Asahi Farm, near Obihiro, Japan).

Figure 25-9 A&B. In confinement of beef cows a "pecking order" of dominant and more timid cows will develop. The dominant cows will thereby intimidate more timid cows (photo above), and eat a disproportionate share of feed. Photo below is of one of the more timid cows in a confinement feeding project - note the flesh of the other cow in the photo. (These cows were fed 100% the nutrient requirements as established by the National Research Council).

Also some type of shelter should be available. It is highly desirable that the flooring in the shelter be concrete, to facilitate cleaning. When the calves are young, clean bedding should be made available in the sheltered area.

As mentioned in the previous section, a minimum of 3 feet of bunk space per cow-calf pair should be provided. In addition, an area of bunk space inaccessible to the cows, should be provided for the calves. This will allow them access to some of the roughage, otherwise the cows will monopolize it. Likewise, this area can be used to feed a higher quality ration to the calves if creep feeding is desired.

Ordinary feedlot water troughs are suitable for the cows, but are quite often too tall for young calves. Waterers should therefore be set lower, or two waterers at different heights should be used.

BREEDING PROGRAM

Because the cows are confined, artificial insemination is very applicable. Natural service obviously requires bulls to be kept in the pens (unless hand mating is practiced). When kept in confinement, bulls inevitably do some type of damage to the facilities. Calves are usually from two to four months old during the breeding season, and bulls will occasionally cripple or kill calves in confinement with them. If natural service is used, a higher bull to cow ratio can be utilized than under pasture conditions, particularly range conditions.

Weaning Program

It would seem that early weaning would be advantageous under most confinement situations. The reason is that it is inherently more efficient for a calf to consume and convert feed into growth directly, rather than the cow consuming the feed, converting it into milk, and then the calf consuming the milk.

As mentioned earlier, the cow's nutritional requirements increase greatly with lactation. Early weaning obviously cuts down on the amount of cow feed required, by reducing the amount of time a lactation ration needs to be fed.

CONCLUSIONS CONCERNING CONFINED COW-CALF PRODUCTION

On a physical basis, cow-calf confinement is totally feasible. It does, however, require very tight management. The biggest problem being calf scours. Indeed, calf scours can result in extremely high calf death losses. But given proper sanitation and other management, calf losses can be minimized and good results obtained.

In traditional cattle producing countries, however, low cattle values makes cow-calf confinement economically infeasible. In many Asian countries, where cattle values are much higher, cow-calf confinement is practiced, and is a major form of cow-calf production.

STOCKER OPERATIONS

Stocker operations are usually defined as taking calves (250-550 lbs.) and putting them on some sort of pasture for sale later on as yearlings (600-850 lbs.). Most operations will consist of either: (1.) Spring and summer grazing of native or improved pastures, (2.) Winter grazing of cereal pastures (wheat, oats, rye or barley), or (3.) Fall and winter utilization of stalkfields.

STOCKER OPERATIONS ON RANGE PASTURES

This type of operation is normally limited to spring and summer grazing as that is the only time when range pastures are usually good enough to support weight gain. Indeed, there are some areas in the western U.S. that will not afford reasonable weight gains (at least 1.0 lb./day) even at these times, and therefore stocker cattle are not a viable option to some ranchers. Breeding cattle need only maintain their weight, and such areas are better suited for that use.

Knowing when a pasture or ranch is capable of being used in a stocker operation is relatively simple. In most cases all one need do is look to see if large quantities of green grass are available.

STOCKER OPERATIONS UTILIZING IMPROVED PASTURES

Improved pastures, both irrigated and dryland, are used all over the world for stocker cattle (small grain pastures are discussed in a later section). Their use usually stems from the fact that they are of marginal value for crop farming, but can support more productive cover crops than native grasses.

Forage Species Used

The first consideration in most improved pasture systems is to provide grazing for the longest time period possible. To accomplish this, two or more species are usually required. Oftentimes cool season grasses are mixed with legumes. These grasses, such as fescue or rye grow well during the cooler months of spring and fall, and legumes, such as clover or vetch, will grow during the hot summer months. In addition, the nitrogen fixing properties of legumes decreases fertilization requirements. The actual species chosen, of course, will depend upon a myriad of factors pertaining to local conditions.

STOCKER OPERATIONS ON SMALL GRAIN PASTURES

Winter small grain pastures (wheat, barley, oats, and rye) make exceptionally good grazing and are used extensively for stocker cattle in the south central plains of the U.S. Many small grain fields are used specifically for grazing, but by far the majority are used for both grazing and grain production.

With proper management, grazing does not necessarily reduce grain production. Actually, periodically cutting off part of the blade or leaf area of the plant (grass) tends to make the roots spread out and send up more shoots (blades of grass). The plant actually becomes hardier. An analogy would be mowing a lawn. Most lawn grasses respond to mowing with increased growth and density (assuming proper fertilization and irrigation).

Small grains are, of course, annual plants. Planting usually takes place in August or September and grazing will normally start in late October to November. If grain is to be harvested, the cattle must be removed before mid-spring (before the boot stage); grain harvest normally takes place in early summer.

Small grains (wheat in particular) have a tremendous re-growth ability, and can withstand extremely heavy grazing pressure. However, this great re-growth ability can cause problems in the form of bloat. Whenever weather conditions have stifled re-growth (extreme cold, cloud cover, etc.), and then very favorable weather conditions appear, bloat can be a substantial problem.

Supplementation on Small Grain Pasture

Since small grain grasses are quite high in protein, protein supplementation is not needed. Mineral supplements are needed since small grain pastures are seldom fertilized with anything except N, P, and K. In addition to trace elements, mineral mixes on small grain pastures should also be relatively high in magnesium, calcium, and phosphorous.

STOCKER OPERATIONS UTILIZING CROP RESIDUES

Running cattle on crop residue pastures is a paradox for many cattlemen. Some cattlemen get excellent gains (up to 2 lb./day) while other cattlemen experience very poor performance on the same type of residue pastures.

The difference lies in the understanding of a few basic facts concerning residue grazing. *The most important fact is that cattle cannot gain weight on crop residue (cornstalks, straw, etc.) . . . cattle gain weight by eating the wasted crop (grain, etc.) left in the field after harvest.*

Of the commonly grown crops, cornstalk fields usually support the greatest weight gain. Typically, 3 to 8% of the corn grain is lost during harvest, and it is the grain not the stalks, that the cattle gain weight on. Cattle are remarkably adept at rooting through stalks and leaves to sort out the ears lost during harvest. Likewise, cattle (calves in particular) are able to chew and utilize the corn quite well.

Knowledgeable cattlemen watch their cattle closely and when corn fails to appear in the manure, the cattle are moved. At this point the fields will not appear to have been well utilized as there will be a considerable amount of stalks, leaves, etc. left on the ground. This is because cattle will first seek out the leftover grain, and will eat the stalks only when forced to.

If cattle are allowed to remain in the field after the grain has been removed, gain will decrease rapidly. All too often the situation is that cattle will gain well for a period (while utilizing the available grain), and will later lose weight for a period (when forced to eat the stalks). The end result is often that the owner concludes that stalkfield grazing doesn't work, as overall gains are rather poor.

Other grain crops do not allow as much weight gain as corn. This is primarily due to the inability of cattle to chew whole wheat, milo, etc. (compared to corn). In addition, much of this grain is shattered and difficult for cattle to pick up off the ground (rather than being neatly contained on a cob). For the same amount of grain left in the field, one can only expect 40 to about 70% the gain of cattle on milo, wheat, or barley vs. corn.

Specialty crops, vegetables etc., are quite variable. Performance will vary with the individual crop and the maturity of that crop; i.e. often vegetable crops are pastured because adverse weather or other problems prevented harvest and the crop became too mature for sale as produce. Overly mature vegetable crops will not support the same performance that an equal amount of wasted vegetable will support at the correct stage of maturity. As with other forage, this is because of an increase in fiber and a decrease in soluble sugars and carbohydrates. Some specialty crops, such as onions, contain alkaloids which can create problems if they comprise a large part of a cattle diet.

Supplementation on Stalkfields

Cattle pastured on stalkfields should always be supplemented with protein, vitamins and minerals as grain and stalks are deficient.

Vitamin A supplementation is an absolute must for calves on stalkfield forage as there is no appreciable vit. A or carotene activity in either stalkfields or grain. Calcium, phosphorous, and trace mineral content of stalkfield forage is also very low. Mineral supplementation is therefore required. This may be accomplished via the inclusion of a mineral "pack" in the protein supplement, or the use of free choice mineral feeding.

COMMON PHYSIOLOGICAL PROBLEMS WITH STOCKER CATTLE

Pasture Bloat

By far the most common cause of death in stocker cattle on small grain pasture is bloat. It can occur at any time, but increased incidences can be expected whenever there is a period of rapid plant growth. The most dangerous time is when very favorable weather condi-

Figure 25-10. Holstein steer with bloat. Notice how the swelling will be most severe on the left side. In an emergency situation when the rumen must be punctured, always puncture the left side. The liver is in the way on the right side, and severe damage would obviously result.

tions appear after a period of unfavorable weather conditions; that is rapid plant growth after a period of dormancy. The longer the plant has been dormant and the more the subsequent conditions favor growth, the more dangerous the situation.

Once bloat occurs, the animal can die within a matter of minutes or hours. If found soon enough, simply forcing the animal to move around is often effective. If the bloating is severe, the preferred treatment is to drench with mineral or vegetable oil. This is accomplished by running a tube down the esophagus and pouring a pint to a quart of mineral or vegetable oil down the tube.

When the animal is in extreme distress and there is not time to drench with mineral oil, the rumen may be punctured to release the gas pressure. This should only be done as a last resort. The puncture should only be made on the left side, just behind the ribs. Indeed, this is where the swelling will be the most noticeable, since there are no organs between the hide and the rumen in this area. On the right side, the liver is in the way. The instrument used to make the puncture is known as a trocar. Essentially, it is a large, pointed and sharpened spike, contained within a barrel or cannula. The entire unit (spike and cannula) is driven into the rumen. When the spike is removed, the cannula remains in place and gas escapes through the opening. The cannula may be left in place as long as necessary to continue to release gas.

The trocar is used instead of a knife because it provides a tube for the gas to pass through. An ordinary knife wound often allows a small part of the gas and ingesta to gain entrance into the abdominal cavity; i.e. get between the skin and the rumen. Rumen ingesta is teeming with bacteria, and an infection usually occurs.

Regardless of whether the wound was made by a knife or a trocar, it will be very slow in healing. Gas will continue to push through and thereby greatly hinder healing. Peritonitis (infection within the gut cavity) is an ever-present danger. Obviously then, puncturing the rumen should only be done when there is no time for any other type of treatment.

Prevention of a malady is often said to be more important than the cure. In the case of pasture bloat, prevention can be quite difficult. There is a commercially available surfactant known as Poloxalene which has been shown to be effective in reducing bloat. The problem is getting pasture cattle to eat it when they need it most. Soybean meal, molasses, and other agents are normally added to enhance consumption. The problem is that bloat occurs most often during rapid growth, which is also when pasture grasses are the most palatable. In actual practice it has been extremely difficult to get pasture cattle to consume products containing Poloxalene when there is lush, green forage available. If the product is sprayed directly on the pasture, as is often done in Australia and New Zealand, it works quite effectively. Likewise, the direct spraying of mineral or vegetable oil has also been shown to be effective.

In the U.S., the direct spraying of a product on pastures is usually not feasible. Probably the only practical way to eliminate bloat is to keep plenty of dry hay on hand, and watch the weather and cattle very closely. Whenever good growing weather appears after an extended period of poor growing weather, pull the cattle off the fields and fill them with hay before returning them. The hay dilutes the effect of the lush grass and stimulates rumen motility (causes the rumen to "churn" the digesta more). The use of ionophores, such as Rumensin, has been shown to be of benefit. Therefore it is usually a good idea to supplement stocker cattle with ionophores.

Nitrate Toxicity

Nitrate toxicity is a relatively common problem with cattle grazed on fertilized pastures or stalkfields. Also there have been cases of cattle contracting nitrate toxicities from contaminated water supplies. Ground water can be contaminated by seepage from manure or sewage, and surface water can be contaminated by runoff from fertilized fields or manure (from corrals, etc.).

The classic symptom of nitrate toxicity is blood that appears a chocolate brown color. Other symptoms include poor performance and debilitation. If the concentration is high enough to cause death, it may come slowly or quickly.

The treatment for cattle with acute cases is intravenous injection of methylene blue, which acts as an oxygen carrier. Prevention of nitrate toxicity from harvested feeds and water is usually rather simple, since analyzing for nitrates is relatively inexpensive. If a feed is found to be relatively high in nitrates, it can be fed by diluting with other feeds known to be low in nitrates. (The rumen microorganisms can utilize a low level of nitrates as a source of non-protein nitrogen.)

Pasture situations are more difficult to prevent since the concentration of nitrates can fluctuate. Plants transport nitrogen in the form of nitrate, and when growth of the plant is slowed, excess nitrates may accumulate. Thus, whenever plant growth is retarded due to drouth, etc., and/or when heavy applications of N fertilizer are made, the cattleman should be wary. If pasture grasses have about 8,000 or more ppm nitrate (dry matter basis), the cattle should be moved. When normal growth of the plant is resumed, the nitrate level will decrease, and the cattle can be returned. Obviously this can create an extra inconvenience and expense to the cattleman, but death loss can be sudden and severe.

Nitrate toxicity occasionally occurs with stalkfield grazing. The greatest concentrations of nitrates in plants are at the base of the stalk, and since stalk stubble makes up the majority of available forage, nitrate toxicity problems can become serious. When toxicities do occur, it is usually after the cattle have been on the stalkfield for a period of time. This is because cattle will seek out and eat the grain and leaves, before they will eat the stalks.

A relatively common source of nitrates and subsequent toxicities is caused by allowing cattle access to irrigation tailwater. Rather than setting up water troughs and hauling water, some operators are tempted to allow cattle to drink out of irrigation canals or tailwater pits. Once water has run across land that has been fertilized, it is almost certain to contain high levels of nitrates. Even if the concentration is not high enough to create a visible toxicity, it can reduce performance. **Cattle should never be given access to tailwaters.** Even if tested and proven safe, a good rain can fill pits or canals with enough runoff to create nitrate problems.

Urinary Calculi (Waterbelly)

Urinary calculi are caused by the precipitation of dietary minerals in the urine to form a stone-like object. In steers, where the size of the urethra is reduced due to castration, the stones can become lodged and block the flow of urine. This, of course, causes the animal great pain, and can eventually cause the bladder to rupture and kill the animal. There is a surgical procedure for saving the animal in which the penis is surgically severed above the S curve (sigmoid flexure see figure 21-2). It is then rerouted, so that the animal urinates similar to a heifer. For urinary calculi occurring in pasture cattle, the mineral most often involved is silica. In areas where grasses contain high levels of silica, the problem can be acute. There is no generally recognized and accepted preventative treatment.

CATTLE FEEDING AND FEEDLOT MANAGEMENT

Cattle feeding and feedlot management is in itself an enormously broad subject, and therefore in the confines of this chapter only very general information can be presented. (For information on ration formulation and development, see Chap. 20. of the Feeds and Feeding section.)

GROWING PROGRAMS

A growing ration (as opposed to a finishing ration) is supposed to allow weight gains without actually fattening cattle. A true growing ration should allow weight gains of about 1.25-1.75 lbs./day. Finishing rations typically put on gains of 2.5 - 3.0 lbs./day.

Most steers are put on growing rations at weights of 300-500 lbs., and are put on finishing rations at weights of 600-800 lbs. (heifers 500-700 lbs.). That is, 200-400 lbs. are put on calves before they are ready to be put on finishing rations. If cattle are put on finishing rations at too light a weight, they become fat too quickly and their performance will be greatly reduced if they are fed to normal slaughter weights.

Growing rations are designed to contain just enough energy to allow the calf to put on muscle tissue (grow) without putting on much fat. Whatever amount of fat is put on, will reduce performance when the cattle are put on finishing rations. This is because of the phenomenon of *compensatory gain*, and because the fat increases the animal's maintenance requirement and reduces intake.

FINISHING OR FATTENING PROGRAMS

The idea behind finishing programs is to get the cattle to gain as rapidly as possible. To do this, rations as high in energy as possible are used, and they are presented in a manner to obtain maximum intake.

The animals are normally fed until they reach their maximum muscle growth plus a certain amount of fat. Over-finishing should be avoided as the laying down of fat is a very inefficient process which can greatly increase feed costs. This is easy to understand since muscle tissue contains 50-70% water, whereas fat contains very little water. Thus, when an animal is growing and laying down muscle tissue, 50-70% of the weight increase is due to water. For that reason, gain drops off sharply as cattle become fat. That is, fat is only about 4% water.

The Need for the Dual Program Feeding Method

A fairly common but misguided practice, especially among farmer-feeders, is to hold cattle from growing weights to slaughter weights on a ration containing approximately a 50/50 mix of roughage and grain.

A 50/50 grain/roughage ration is inefficient. The primary reason for inefficiency is that a 50/50 mix will not create the proper rumen environment for either amylolytic (starch using) or cellulolytic (fiber using) microorganisms. As a result, neither the roughage nor the grain is digested efficiently.

In addition to the inherent inefficiency of the 50/50 ration, holding cattle on that type of ration from grower weights further decreases total overall performance by making them too fleshy (fat) at too light a weight; i.e. if calves are put on that type ration at 400-500 lbs., they will be carrying considerable fat by the time they reach about 700 lbs. In order to be merchantable to most packers however, heifers must weigh 800-900 lbs. and steers 1,000-1,100 lbs. Carrying cattle up another 300-400 lbs. when they are already fleshy increases gain costs substantially.

For the sake of efficiency, cattle should go through what is known as a two stage or two phase program; (1.) a genuine low energy growing type ration from weaning to feeder weights (for gains of no more than about 1.75 lbs./hd./day), (2.) a high energy finishing ration from feeder to slaughter weights.

Acidosis and Founder

Whenever ruminants are to be fed high concentrate rations, they must be brought "up on feed" gradually. Otherwise, high grain rations can cause the rumen to turn acid which can result in what is known as *acidosis*.

Acidosis is an acute condition, in which lactic acid is absorbed from the rumen into the blood in large amounts. This can cause the blood pH to turn acid which can kill the animal. If the animal survives, a chronic condition known as "founder" may occur.

To avoid these problems, cattle (and other ruminants) are fed rations which gradually increase the grain or other concentrates over about a 21 to 30 day period. (For more information, see Chap. 20.)

Figure 25-11. A foundered feedlot steer.

The Feeding of Grain to Livestock

Urban societies have criticized beef production as an inefficient use of grain. The basis of this thought being that feedlot cattle convert grain into a live weight gain at a ratio of about 7.5 : 1, whereas swine convert at about 3.5 : 1 and poultry at about 2 : 1.

The fallacy in this idea is that only about 40% of the final slaughter weight of a steer is attributable to grain feeding. The typical feedlot steer doesn't come into the feedlot until it weighs about 650 lbs. If the amount of grain fed during that period is divided by the animal's final slaughter weight (about 1050 lbs.), a grain conversion of about 3 : 1 is obtained. In addition, breeding herds of cattle receive no grain (held on pasture), whereas swine and poultry breeding stock must be fed grain diets.

BEEF CATTLE BREEDS

Cattle breeds can be broken down into two basic groups; *Bos taurus* and *Bos indicus*. For scientific purposes it is important to realize that these are actually two different genera of animals. Bos indicus includes the Zebu breeds originating from India, and Bos taurus all the breeds originating from Europe.

For practical purposes and application, however, beef cattle breeds may be broken down into four basic groups: 1. the classic British breeds, 2. Continental (European) "beef only" breeds, 3. Continental dual purpose breeds, and 4. The (Bos indicus) Zebu type breeds. In addition, in most countries of the world one would also need to mention "dairy beef", as steers from dairy farms make up a large percentage of the beef produced.

The classic British Breeds. In the U.S. and Canada, when one thinks of beef cattle, the breeds that usually come to mind are Hereford and Angus. Indeed, Hereford and Angus were two of the first breeds ever brought to the North American continent. Along with the Shorthorn, they comprise what are known as the British breeds, since these breeds originally came from the British Isles.

Figure 25-12. A group of two year old Hereford bulls. As range cattle, Hereford cows are noted for their strong mothering instincts.

As a general rule it can be said that these breeds have medium type frames, are relatively early maturing, and have moderate, mature body, and slaughter weights. However, it should be pointed out that within these breeds there are enormous differences. Some lines have been subjected to inbreeding* and some have not. As a result, within the Hereford and Angus breeds especially, there are enormous differences. Although one must always consider the actual source of a breed rather than

---

* Back in the 1930's the ideal cattle type was arbitrarily deemed to be "small, blocky and compact". In addition cosmetic traits such as purity of coat color were heavily selected for. To achieve these goals some breeders resorted to inbreeding.

generalizing this is particularly true with Hereford and Angus.

Figure 25-13. A Shorthorn steer in a feedlot. The Shorthorn was one of the foundation breeds in the Santa Gertrudis breed (a Zebu cross breed).

One generalization that can be made about these breeds is that they make good range animals. They have good mothering instincts, and are protective of their calves. This is especially true for the Hereford.

Another aspect of the Hereford that lends itself well to range, especially western arid type range, is a relatively low output of milk. While accelerated milking ability is often reputed to be a positive attribute, under harsh range conditions it can be a detriment. As explained in Chap. 21, lactation takes precedence over estrous. That is, a cow's nutrient requirements for lactation must be met before the cow will ovulate. Thus, cows capable of higher milk yields may be at a disadvantage when forage is sparse or of poor quality. The end result is reduced conception rates.

One of the unique traits of the Angus is a propensity to marble. Due to the emphasis placed on marbling in the U.S. quality grading system, carcasses with Angus breeding usually score well. As a result, Angus and Angus crossbred cattle have particular favor with feedlot operators and packer buyers.

The Continental (European) Beef Breeds. Back in the 1970's the advent of artificial insemination brought about the availability of a great many new breeds. Prior to that time endemic hoof and mouth disease on the continent of Europe had prevented wholesale importation of cattle. That is, the extensive quarantine required excluded the importation of large numbers of cattle due to the cost.

However, with artificial insemination, only a few head of highly valuable bulls could be imported, and the semen used as foundation for building herds in the U.S. By continuous crossbreeding, within 3 generations (less than 10 yrs.) cattle of 7/8 purity could be raised. That is, the first breeding produced a ½ blood; the second breeding a ¾ blood, and a third breeding a 7/8 blood. As a result, within a few years a large number of new breeds appeared in the U.S., Canada and to a certain extent Mexico.

Probably the most popular of all these breeds was the Charolois (although the importation of Charolois actually preceeded the artificial insemination boom of the 70's). The Charolois had been developed in France to produce large, lean and meaty carcasses, and the breed in general is very good for that purpose. Another breed from France that became established, albeit much less popular, was the Limousine. The Limousine had also been bred to produce very lean and meaty carcasses, and was superb for that purpose. Indeed, many of the cattle winning carcass shows have been and continue to be Limousine crosses.

The Limousine is somewhat smaller than the Charolois, but both are very muscular animals. The rounds and loins of these breeds are very full and well-proportioned. This conformation contributes to high cutability. Charolois and Limousine have very good meat

Figure 25-14. The Mertolenga breed. Brought to the U.S. by Spanish conquistadors, the Mertolenga became the foundation of the Texas Longhorn breed. (Photo taken in Portugal)

Figure 25-15. Purebred Charolois bulls. (Photo taken in France)

to bone ratios, and their natural tendency toward leanness makes for maximum overall meat yield.

For use in crossbreeding, however, the muscularity can be a detriment. That is, when small cows are bred to Limousine or (especially) Charolois, calving problems can result. The muscularity of the calves tends to make them thicker through the hips and shoulders which can create problems at birth. Likewise, just the overall size of the Charolois animal tends to cause problems (birth weight is positively correlated with mature weight). Obviously then, in crossbreeding programs these breeds should not be used on small cows or first calf heifers.

Another breed that enjoyed a great deal of attention and speculation was the Chiania. What created all the interest was the fact that the Chiania is the largest cattle breed in the world; mature animals reaching weights nearly double those of British breeds.

The Chiania is an Italian breed, which had been developed over several centuries as a draft animal. The Chiania has very long legs and an overall general body composition that would suggest athletic capability. Early on, the Chiania became very popular in Mexico, but as Mexican cowboys discovered how athletic Chiania cattle can be, popularity waned. (Chiania cattle were reported to be difficult to catch or control on horseback.) However, the real detriment to Chiania cattle is late maturity. Chiania heifers take up to 6 months longer to reach puberty, which means they cannot be calved as two year olds. This undoubtedly has been responsible for the reduction in interest in this breed.

The continental (European) Dual Purpose Breeds. During the "boom" of the continental breeds during the 1970's there was intensive speculation in these newly acquired breeds. Three-quarter and even half-blood animals of breeding age brought extremely high prices as speculators sought to develop herds.

What fueled the fire with respect to speculation were the weaning weights many of these breeds were capable of producing. Many of these new breeds were reported to wean 7 month old calves of 600 and even 700 lbs. (50 to 75% larger than what the existing breeds were capable of). What many of these speculators were apparently not aware of is that many of these "new" breeds were actually dual purpose animals. In Europe they had been developed to produce milk as well as meat. Thus, many of these breeds were capable of producing nearly twice as much milk as many of the more common beef breeds (such as Hereford and Angus). As a result, they produced larger calves.

But as explained previously, increased milk production is not always an advantage. Indeed, under sparse western range conditions it can be a disadvantage.

As explained in Chapter 21, the nutrient requirements for lactation must be met before a cow will rebreed. If the quality of available forage is not capable of meeting the lactation requirements of a dual purpose animal, then reduced conception will result.

This is not to say that many of these dual purpose breeds have not been used, because they have. However, most of the use has been in the area of crossbreeding. Most commonly, a percentage of these breeds have been incorporated into existing breeds or herds.

Undoubtedly the most popular of the dual purpose breeds has been the Simmental. An exceptionally beautiful breed, the Simmental varies from a golden yellow to a golden brown with alternating patches of white. When crossbred with Hereford, the white face of the Hereford is retained. The body color, however, usually carries the golden color of the Simmental, which usually creates a strikingly beautiful animal. This, to be sure, has had a great deal to do with the popularity of the breed. But the Simmental breed also has excellent growth characteristics which have been used to improve the genetics of a wide array of commercial herds. The only drawback to the growth rates have been calving problems in small cows. As discussed in the previous section, this is something that must be considered in evaluating the suitability of a large growthy sire or otherwise small commercial cows or heifers.

The Gelbvieh is another dual purpose breed very similar to the Simmental. Indeed, the only real difference is coat color. Actually, the Gelbvieh is claimed to be the German relative of the Simmental. (The Simmental was developed in Switzerland.) Unlike the Simmental, the Gelvieh is usually a solid color, reddish-brown being the most common.

Another dual purpose breed is the Maine-Anjou. Traits appear similar to the Simmental and Gelbvieh, including good milking ability and excellent growth rates.

Zebu breeds. Many parts of the world where cattle are raised are hot and sometimes quite humid; environmental conditions that are not at all well suited for the traditional British breeds, nor the continental European breeds. The Zebu breeds, however, were developed in India which is a tropical area. The Zebu breeds were therefore much better adapted to the southern area of the U.S., Mexico, Central America and the tropical areas of South America. As a result, Zebu and Zebu crossed cattle have become the dominant breeds in most of these regions.

By itself, the straight bred Zebu breeds are not particularly productive. They have a slower rate of gain than British or Continental breeds, they are late maturing, and are known to have very mediocre conception rates. But when crossed with British or Continental breeds, the performance characteristics of the offspring are greatly enhanced. As a result, a great number of third breeds have developed from Zebu crosses.

The Santa Gertrudis, developed by the famed King Ranch of Texas, was the first Zebu cross to gain wide acceptance as a recognized breed. Originally the two parent breeds (Shorthorn and Brahma) were present on the ranch as purebreds. The development of the crossbreed came out of necessity as neither breed was particularly productive by itself. The Brahman breed has its inherent fertility problem, and the Shorthorn simply was not adapted to the intense heat and humidity of the South Texas Gulf Coast. By crossing the two breeds, both fertility and rate of gain were increased dramatically. The cross ratio of 5/8 Shorthorn and 3/8 Brahma was set as the standard for the breed. The resulting animal carries most of the Brahman characteristics, except for the red coat color, and an increased size and muscularity.

Next to the Santa Gertrudis, the Brangus breed is probably the most popular of the Zebu crossbreeds. The breed is standardized at a 5/8 Brahma and 3/8 Angus ratio, carries the black coat color of the Angus, and is somewhat smaller than the other Zebu crosses. Due to the Angus genes, Brangus cattle marble and finish quicker than other Brahman crosses in the feedlot.

Braford, a cross between Hereford and Brahma cattle, is an extremely popular cross, particularly in South and East Texas, and much of Mexico. There is a Braford Breed Registry, and the official percentages are 5/8 Hereford and 3/8 Brahma. In commercial cattle circles, however, most any kind of cross between Hereford and Brahma is usually called a Braford. First cross F1 heifers (½ Hereford - ½ Brahma) are very much in demand as herd replacements in the southern parts of Texas. The breed usually carries the red coat color and white face of the Hereford. At times, Braford cattle will display a brindle color.

Other Zebu crosses that have become relatively popular are Beefmaster (American Brahman, Hereford, and Shorthorn), Barzona (Brahman, Hereford, Shorthorn, Angus, Santa Gertrudis, and the South African breed, Africander), and Simbrah (Brahma and Simmental).

Dairy beef. In most countries of the world a large percentage of the beef produced comes from dairy steers. Indeed, some 20 to 40% of the beef produced in Europe comes from feedlot finished dairy steers, in Great Britian probably 50 to 60%, and in Japan nearly 70%. In the U.S. and Canada the percentages are much lower, but this is only because there is a much larger beef herd. The actual tonnage of meat produced from dairy steers is quite large.

In dairying, as well as in dairy beef production, the most popular breed is the Holstein. Indeed, Holstein steers make excellent feedlot animals. They have rates of gain that will equal or exceed any beef breed. The only drawbacks in the U.S. is that they cannot be fed to the choice grade economically. Also, they have a greater amount of cod fat (which is typical of dairy breeds). Usually Holsteins will have a carcass yield about 2 to 3 percentage points below what fed beef breeds yield; i.e. most beef breeds fed to 1050-1100 lbs. will dress from about 61 to 63%, whereas Holsteins fed to about 1100 lbs. will dress from about 58 to 60%. The net result is that packers will not pay as much for Holsteins as they

Figure 25-16. A first calf Santa Gertrudis heifer.

Figure 25-17. A Beefmaster bull, another very popular standardized Zebu cross breed.

Figure 25-18. Grade Braford steer with the characteristic brindle stripes. A very popular Zebu cross in Texas and much of Mexico.

will beef breeds.

The Brown Swiss, like the Holstein, is a large dairy breed capable of heavy milk production. In the feedlot, Brown Swiss steers are very similar to Holsteins; i.e. excellent growth rates and somewhat reduced carcass yields. As terminal crosses, they are, or could be, very useful for the production of lean meat.

The Jersey breed is used as a dairy breed in much of the tropics, as the Jersey has more heat tolerance than other dairy breeds. In temperate climates Jerseys are sometimes used as a novelty. Jersey (and Guernsey) milk is a golden color and occasionally brings a premium over Holstein milk. Jersey steers sometimes find their way into feedlots, but little can be said for them. Jersey steers have one of the slowest growth rates of all beef breeds, as well as the poorest carcass conformations.

SUMMARY

Beef production is an enormously broad and multi-faceted industry. This chapter has attempted to cover only broad generalizations concerning the various segments of the industry.

Traditional western style ranching worldwide represents the major source of cow-calf production. In the U.S., however, cow-calf farming in the South and Southeastern states has actually exceeded traditional ranching in terms of the number of calves produced. In several Asian countries, cow-calf confinement is a major source of calf production.

After weaning, most calves are "grown" for 6 to 10 months before being put on high concentrate finishing rations. "Growing" may occur on high quality pastures, or on forage type feedlot rations.

Ultimately, in the U.S., Canada, and most European nations, cattle receive some grain finishing before going to market. Normally this will consist of the last 300 to 400 lbs. of the animal's finish weight (about 1000 lbs.).

CHAPTER 26 GENERAL CHARACTERISTICS OF THE SWINE INDUSTRY

E.T. Kornegay, Ph.D.
Virgina Polytechnic University

Archaeological evidence indicates that swine were first domesticated in the East Indies and Southeastern Asia as early as 9000 B.C. Despite some ancient cultural and religious taboos forbidding the consumption of pork, the domestication of the pig as a source of human food has persisted throughout the ages.

Columbus first brought hogs to the West Indies on his second voyage in 1493; however, the arrival of pigs on American soil (Florida) is attributed to Hernando De Soto in 1539. It is recorded that De Soto brought 13 sows (probably included boars) which were released in the woods. On De Soto's death in 1542, the number had grown to about 700. Other early explorers also brought pigs to America.

In the early years of the United States the pig was a backyard inhabitant, and as late as the first half of the 20th century, almost every farm family had a few pigs. They were used primarily as scavengers to utilize human food waste. Because of changing cultural and supply needs, that situation has changed in the United States as well as in many other countries of the world. Now, a majority of the pigs in the United States are raised on medium and large pig farms.

DISTRIBUTION IN THE WORLD AND UNITED STATES

In general, swine production is highest in temperate zones of the world and areas where population is relatively dense, with the exception of areas where food habits or religious restriction prohibit the consumption of pork. Asia produces about half (48%) of the pigs produced in the world. Most of the pigs in Asia are produced in China, which has the largest hog population of any nation.

Europe is the second largest swine producing area: with most countries in both eastern and western Europe being large producers and consumers of pork, as well as exporters in many instances. Hog production in the USSR (includes both Asia and Europe) has risen in recent years and now ranks second in total numbers.

Per capita consumption of pork is largest in Denmark, East Germany, West Germany, Austria and Switzerland, with all countries consuming over 100 lbs. of pork per person per year. In both Asia and Europe, swine production makes up 50% or more of the red meat produced.

North America is the third largest swine producing area with the United States being the largest contributor. Hog numbers have advanced rapidly in South America since the late 1950's, with Brazil accounting for most of the production in that area.

Since about 1950, pork production in the United States has varied between 80 and 100 million head of pigs a year. In the past, pork production was a way to market home grown corn. However, this is beginning to change as large commercial farms are being established which purchase all of their feed, including grain. The trend is toward fewer, but larger operations.

IMPORTANCE, FUNCTIONS AND USES OF THE PIG

The pig is capable of converting inedible feeds and poor quality feeds, which would generally not be acceptable for human consumption, into pork - a valuable product well-known for its protein quality. Although not utilized in the United States to a major extent now, in the past pigs were an excellent way to convert waste feeds, such as garbage, bakery waste, garden waste, culled or damaged grain, root crops and fruit, to a valuable product.

Pigs also provide a large and flexible outlet for farmers producing grain. Moreover, pigs have played an important role in agriculture because they have been a profitable enterprise.

In addition to providing meat, many by-products of hog slaughter are used for the health, medical, and clothing industry. For example, pig heart valves are used as replacements for humans, pig bristles (hair) are used in paint brushes, adrenal glands provide en extract which is used to treat Addison's disease, pigskin is used for many items of clothing and as an aid in the treatment of severe burns, gelatin is derived from pigskin, and pepsin (an enzyme) is used in chewing gum.

Pigs also serve as a valuable research animal in biomedical research. The pig has numerous digestive and physiological similarities to man and serves as a valuable model for conducting medical research. For example, baby pigs can be used to test the nutritional adequacy of baby (infant) diets because of similar food requirements and the digestive system. The pig's heart and major blood vessels are much like man's, and thus, the pig is an excellent model for studying atherosclerosis and congestive heart failure. The pig has also been used as a research animal for a variety of other areas relating to man: dental, dermatology, nuclear radiation, immunology, arthritis, kidney function, obesity, toxicology, and alcoholism.

ORGANIZING AND ESTABLISHING A SWINE ENTERPRISE

COMMERCIAL AND BREEDING STOCK PRODUCERS

Swine producers may generally be grouped into those that produce commercial hogs (slaughtered and used as food) and those that produce breeding stock.

PRODUCTION SYSTEMS

Swine producers are also classified on the basis of purchases and/or marketing. There are three major systems: 1) farrow-to-finish producers, 2) feeder pig producers, and 3) pig finishers. A farrow to finish producer maintains a breeding herd, and thus farrows pigs and grows these pigs to market (slaughter) weight. A feeder pig producer maintains a breeding herd and farrows pigs which are sold after weaning at 40 to 80 lbs. A pig finisher purchases feeder pigs at 40 to 80 lbs. and grows them to market weight. While farrow-to-finish operations are found most often among all sizes of producers, the proportion of farrow-to-finish operations decline as the operation gets larger.

TYPE OF HOUSING

There is a great diversity of facilities used by swine producers. These facilities range from pasture or dirt lot operations to totally confined, environmentally controlled buildings. Over 90% of the large pig finishing operations use some type of confined facility with only a few of the large operations using dirt lots. The very large operations tend to have more complete environmentally controlled confinement facilities. The degree of total confinement decreases as the operation gets smaller. The advantage for, or priority for, confined facilities for farrow to finish operations is as follows: 1) farrowing, 2) nursery, 3) growing-finishing, and 4) breeding-gestation.

TYPES OF OWNERSHIP

The type of ownership is closely related to size of operation. Operations below 1000 head in size are generally individually owned, with some partnerships and a few family corporations. As the size of the operation gets larger, ownership by corporation increases. About one-half of the largest operations are owned by corporations, of which nearly one-half are non-family.

TYPES AND BREEDS OF SWINE

The type of swine produced is a result of three contributing factors: 1) demand of the consumer; 2) character of the available feed, and 3) breeding and show-ring fads of breeders. Historically, three distinct types of hogs have been recognized: (1) lard type, (2) bacon type, and (3) meat type. At the present, the goal for all swine breeds is for a meat type hog, which has a maximum amount of muscle with a minimum amount of fat.

Worldwide, there are probably over 400 breeds of swine; however, presently in the United States there are only about 8 major breeds. With the exception of American Landrace, Berkshire, and Yorkshire, the breeds common to the United States are strictly American creations.

In addition to purebred producers, which are normally privately owned by individuals, a number of commercial companies have developed and sell breeding stock. Often these are called hybrids or synthetics though some may be referred to as crossbreds. Hybrid boars usually contain genetic material from more than two breeds in various percentages. Some companies have several lines of both boars and gilts and they are recommended in planned crossbreeding programs. Some of the companies operating in the U.S. are as follows:

Farmers Hybrid Company, Des Moines, Iowa
Babcock Swine, Inc., Rochester, Minnesota
DeKalb Swine Breeders, Inc., DeKalb, Illinois
Pig Improvement Co., Franklin, Kentucky
Lieske Genetics, Henderson, Minnesota
Kleen Leen, Cedar Rapids, Iowa

COMMON BREEDS

The **American Landrace** is white in color, although black skin spots or freckles are acceptable. The breed is characterized by its very long side, level top, well defined underline, deep flanks, trim line, straight snout, and medium-lop ears. The sows are noted for their prolificacy (litter size), milking, and mothering ability. American Landrace probably originated in Denmark.

Figure 26-1. American Landrace gilt. Courtesy of American Landrace Association.

The **Berkshire**, although one of the oldest of the imported breeds of swine (south central England), is relatively unimportant in terms of total numbers. The distinctive peculiarity of the Berkshire breed is the short and sometimes upturned nose. The face is somewhat dished and the ears are erect but incline slightly forward. The color is black with six white points - four white feet, some white in the face and a white switch on the tail. Splashes of white may also be located on any part of the body.

Chester Whites are increasing in popularity although the numbers are still much less than some of the other breeds. As the name indicates, it is a white breed with ears that are medium sized and tipped forward, although not drooped. Chester Whites are usually prolific, good milkers, and make excellent mothers. They originated in southeastern Pennsylvania.

The **Duroc** is one of the leading breeds of swine in America, originating in the northeastern United States. The color of the Duroc is solid red with the shade ranging from golden to deep brick. Their ears are medium sized and tipped forward. The popularity of the Duroc may be attributed to the valuable combination of size, feeding capacity, prolificacy, and heartiness. They have been very popular in crossbreeding systems.

The **Hampshire** is probably one of the youngest breeds of swine, but increasing rapidly in popularity. Although the ancestors of the American Hampshire undoubtedly came from southern England, where black hogs with white belts were known in several localities, the American Hampshire breed was developed in Kentucky. The most striking characteristic of the Hampshire is the white belt around the shoulder and body, including the front legs. Hampshire breeders have always stressed great quality and smoothness. The Hampshire breed has been known for meatiness and carcass quality.

The **Poland China** is the oldest breed of swine to have originated in the United States. Modern Poland Chinas are black in color with six white points - the feet, nose and tip of tail. Prior to 1872 they were generally mixed black and white spotted. Poland China yields a generally high quality carcass and have been used in crossbreeding programs. No breed has been subject to any more radical shifts in type as have Poland Chinas.

The **Spotted breed**, once called Spotted Polands, was developed in the north central part of the United States. The popularity of the Spots is chiefly attributed to the success of breeders in preserving the utility value of the old Spotted Polands, while making certain improvements in the breed. The color requirements of the breed specify that, to be eligible for registration, an animal must be spotted black and white, with either the black or white dominating - never less than 20% of either color (exclusive of head and legs).

The **Yorkshire** is one of the more popular breeds, originating in England. Yorkshires should be entirely white in color, although black pigment spots called freckles do not constitute a defect; they are however, frowned upon by breeders. The face is slightly dished and the ears erect. Yorkshire sows are noted for their prolificacy, milking ability, and are good mothers.

Figure 26-3a & b. Hereford gilt (above) and boar (below). Courtesy of National Hereford Hog Record Association.

Figure 26-2. Poland China boar. Courtesy of Poland China Record Association.

Yorkshires are a very popular breed and are used a great deal in crossbreeding programs. The Yorkshire is a large, smooth, long-bodied and active hog.

Two other pure breeds that are often listed when breeds of swine are discussed are **Herefords** and **Tamworths**. However, they are of minor importance. There have also been a number of inbred breeds of swine such as Beltsville #1 and #2, Maryland #1, Minnesota #1, #2, and #3 and Montana #1. However, there presently are only a few, if any, of these animals being bred in the United States.

Several miniature breeds have been developed for use in bio-medical research: Hormel, Nebraska, Hanford, Gottingen (German) French, Greer, Yucatan and Vietnamese.

CROSSBREEDING SYSTEM

Crossbreeding systems are widely accepted and used for commercial swine production - more than 95% of slaughter hogs are produced from crossbreeding systems. Crossbreeding is used to combine desirable characteristics of different breeds and to capitalize on hybrid vigor (heterosis). As discussed in Chap. 22, heterosis is defined as the average superiority of the crossbred progeny over the average of their parents. Heterosis occurs when genetically different lines or breeds are crossed, and it is greatest for traits with low heritability. Traits such as litter size, litter weaning weight and survival rate respond best to crossbreeding. Carcass traits and performance values are generally highly heritable and are, therefore, not improved by crossbreeding per se, but can be improved by selecting the proper type of animal to use in the crossbreeding program. Unplanned crossbreeding programs may not be successful. A crossbreeding system must be selected that will capitalize on heterosis, take advantage of breed strengths, and fit the management program of the producer (Table 26-1).

Table 26-1. Percentage of the maximum heterosis obtained from various crossbreeding systems. (Pork Industry Handbook #106)

System	% Heterosis Offspring	Maternal
F_1 (inital cross, A x B)	100	0
Backcross (A x A-B)	50	100
2-breed rotation	67	67
3-breed rotation	86	86
4-breed rotation	93	93
Terminal cross using F_1 sows	100	100
Rotaterminal using a 2-breed rotation	100	67
Rotaterminal using a 3-breed rotation	100	86

Two basic systems of crossbreeding may be considered - rotational cross or terminal cross. The rotational cross system combines two or more breeds where a different breed of boar is mated to the replacement crossbred females produced the previous generation. In a terminal cross system, slaughter hogs are sired by the same breed boar with all offspring marketed. The female stock is usually selected primarily for reproductive performance.

The three breed rotational cross has been the most popular crossbreeding system. Sires from the three breeds are systematically rotated each generation, and replacement crossbred females are selected each generation. These females are mated to the sire breed furthest removed in the pedigree.

Figure 26-4. Three-breed rotational cross system. (Pork Industry Handbook #39)

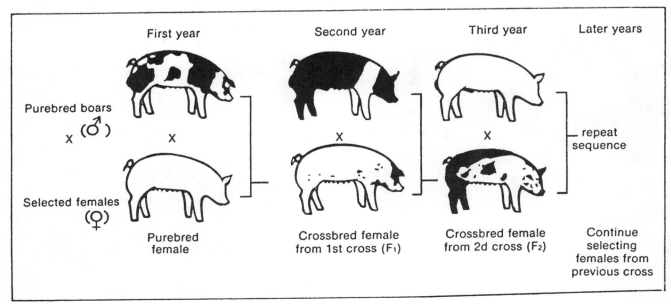

A four breed terminal cross system would provide for maximum offspring and maternal heterosis; but is more difficult to operate on a single farm because two breeding herds (each including two different breeds) must be maintained to produce the two-way cross females and the two-way cross boars that will be used in the terminal cross.

A rotaterminal system using a 3-breed rotation for gilt production and a terminal cross with a fourth breed for market hog production is increasing in popularity. The prolific white breeds are normally used in the 3-breed rotation program for replacements and to produce crossbred females for market hog production; and one of the colored breeds is usually used as the sire for terminal cross. This system gives 100% heterosis for the offspring and 86% maternal heterosis.

LOCATION OF ENTERPRISE

The availability and cost of feed, are certainly the most important factors in the suitability of an area for swine production. However, access to markets and prevailing climate are also major factors to be considered. Within a given area, a new producer must also consider: 1. distance from other swine units and neighbors and, 2. the contour of the land (with respect to drainage and waste disposal). If possible, the facility should be located so that prevailing winds are blowing away from neighbors as odors and dust might create problems. Screening of the area and the use of natural barriers can be helpful in preventing nuisance complaints.

It is preferable that the facility be located on a dead end road to prevent through traffic. Paved roads are highly desirable, as access is necessary in all kinds of weather (snow, ice, rain) for movement of feed to the farm and for transportation of market hogs. Electricity is needed for ventilation, waste movement, lighting and many other purposes. Water is needed, not only for drinking, but for cleaning operation of liquid waste management systems. Although expansion is not planned, producers often expand later. Therefore, future expansion should be considered in the initial planning of an operation.

The type of operation and the production system will effect the type of housing and general layout of the facility. Points to consider in the general layout of a facility are: fence or barrier needs to be considered around the entire unit to prevent and control movement of animals and people, (primarily for disease control). Delivery of supplies (feed, gasoline, etc.) and the introduction of new breeding stock needs to be planned so that they can be made without coming directly into the facility. It is advisable to have the loading ramp for market hogs and sows located on the perimeter of the operation so that trucks do not have to come into the

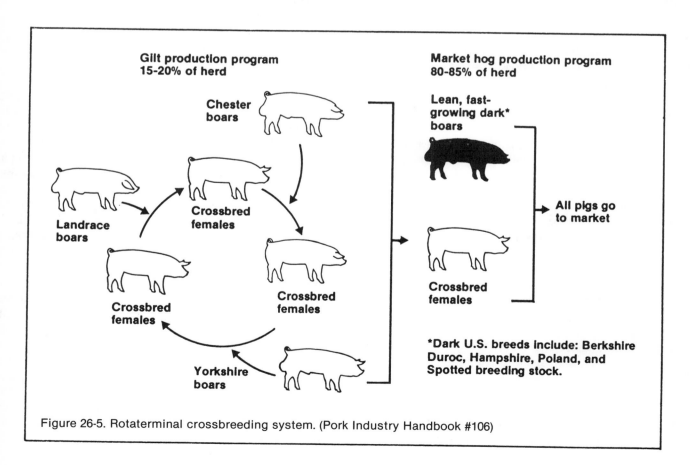

Figure 26-5. Rotaterminal crossbreeding system. (Pork Industry Handbook #106)

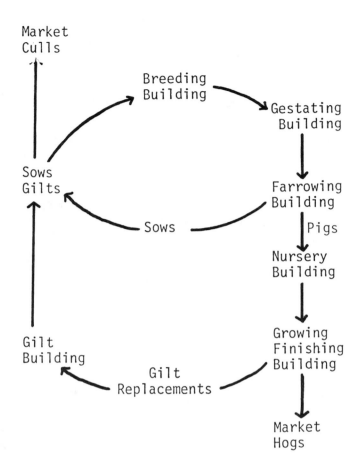

Figure 26-7. A Swine building flow diagram which provides for the orderly movement of animals from one set of buildings (or rooms) to the next.

facility. In general, the facility should be laid out in such a manner as to facilitate the flow of animals, feed, and waste. Further, they should be laid out in such a way that they are easy to clean, disinfect, and to provide good disease control. There should be plans to receive visitors so that they do not walk or drive into the facility. An office area can also provide a central point for visitors as well as providing a change area for the employees.

Components of a farrow-to-finish operation would normally include: 1) breeding and gestation unit - boar housing, breeding area, sow housing and developing gilt pens; 2) farrowing and nursery unit - sow preparation room, farrowing house and nursery; 3) growing-finishing unit - growing and finishing pens; 4) feed preparation and handling unit - grinder, mixer, storage and feeding system; 5) waste disposal system. The feed preparation and handling unit and the waste disposal system would be superimposed on the other units of the facility.

SELECTION BASIS FOR GILTS AND BOARS

The factors or traits that should be considered when selecting breeding stock may be broadly grouped into six categories: 1) feedlot performance, 2) carcass merit, 3) sow productivity, 4) behavioral traits, 5) soundness and conformation, and 6) health.

Feedlot performance includes growth rate and feed efficiency. These traits have one of the highest economic values. When selecting for these traits, more emphasis should be placed on the individual animal's record than on the record of relatives. Both growth rate and feed efficiency respond well to selection because they have medium to high heritabilities (20-50%), but they show only an average to poor heterosis response (5-15%) with crossbreeding.

Table 26-2. Heritability estimated on some economically important traits (see also pg.162) (Source: Pork Industry Handbook #106)

Trait	Heritability %
Litter survival to weaning	0
Number farrowed	10
Number weaned	10
Birth weight	15
Weaning weight	15
Feed efficiency	30
Growth rate	30
Carcass quality	30
Loin-eye area	50
Backfat thickness	45
Age at puberty	35

Body composition or carcass merit can be evaluated by taking measurements of back fat thickness and loin eye area, and by estimating percent muscle. Of these measurements, back fat thickness is the single most important trait. These traits have high heritability values (30 to 70%) and show low heterosis response with crossbreeding. They are medium to average in overall economic importance, but due to high heritability, respond extremely well to selection.

Production costs are influenced greatly by the number of animals produced per sow per year. Sow productivity traits include reproductive ability, prolificacy (litter size), mothering ability, and milking ability. Breeding and farrowing rate, and number and size of pigs farrowed and weaned are the most common measurements. Litter weight at 21 days expressed per breeding female (sows and gilts exposed to the boar) is probably the best single measure of sow productivity, as this takes into account all sow productivity traits. The heritability of these traits is generally considered to be low. Therefore they respond best to crossbreeding. It should always be remembered that these traits are of immense economic importance.

Behavioral traits include docility, temperament, and the complex traits associated with reproductive potential (sexual-development, maturity, and aggressiveness). The behavior of the sow must be such that she will adjust to confinement housing. She must be docile so as to allow the herdsman to move her and to work with her litter.

Soundness in boars and gilts means being free from flaws or defects which would interfere with normal reproduction and maternal functions. Areas of particular concern are: 1) reproductive; 2) mammary; 3) skeletal.

Replacement gilts should exhibit normal reproductive development, both anatomically and behaviorally. They should be free from genetic abnormalities. Unfortunately, most anatomical defects of the reproductive system are internal and not visible. However, gilts should not be selected that have small vulva which are indicative of infantile reproductive tracts. Gilts should begin to show signs of puberty at an early age, at least a month prior to anticipated breeding.

Current industry standards stipulate at least six well spaced functional teats on each side. Gilts with inverted or scarred nipples should not be saved. New concrete, or otherwise rough floors and corrosive chemical compounds on the floor of farrowing houses, can cause abrasions to the gilts' underlines which can result in nonfunctional teats. Sows which have difficulty farrowing, are extremely slow farrowing, or have damaged reproductive tracts (uterine prolapses or severe uterine infections) should be culled and replacements should not be kept from these animals.

Boars and gilts with feet and leg problems which interfere with normal breeding, farrowing, and nursing functions should not be saved. Breeders commonly consider skeletal unsoundness as one of the results of confinement, but in truth, confinement rearing only makes this trait more noticeable. A sound pig reared on pasture or dirt lots may be an unsound pig in confinement. **Commercial producers should buy their pigs from seed stock producers who raise their pigs in confinement housing similar to their own.**

The conformational-type traits used in visually evaluating boars and gilts generally include body length, depth, height and skeletal size (ruggedness, frame and body capacity); muscle size and shape; boar masculinity characteristics and testicular development; gilt fertility. Traits such as length, height, and underline have high heritability values and give low heterosis (crossbreeding) responses. The physical soundness traits (structural soundness, bone size and strength) have average heritability values, give average heterosis responses with crossbreeding, and have medium to high economic values. Thus, select a boar or gilt on the basis of their own record, with some attention to litter mates and other close relatives. The genetic abnormalities and mating ability traits have high economic importance; therefore, insist that relatives be free of these defects.

PERFORMANCE GOALS AND TESTING PROGRAMS

Every producer should have selection standards or goals for both boars and replacement gilts. These would normally include litter size, underline, feet and legs, age at 230 lb, feed per gain, daily gain, and back fat thickness. As pointed out earlier, visual appraisal is important in evaluating structural soundness, length of body, underlines, and general conformation and physical defects. However, performance records, which can only be obtained through some type of testing program, are **absolutely essential** for most of the important economic traits such as growth rate, feed efficiency, and carcass quality. Production records may be collected on the farm by the producer or may result from independent testing at central facilities (such as boar test stations). On the farm testing is probably most valuable to the producer in identifing outstanding female replacements. Most states, as well as a number of private organizations, operate boar test stations for testing of boars from a variety of farms in a common environment.

NUTRITION AND FEEDING

IMPORTANCE OF FEED COSTS

Feed costs make up the single largest cost for a market hog. Currently, feed costs represents 60 to 70% of the total cost in a farrow-to-finish operation, 50 to 60% in a feeder pig operation and 65 to 71% in a finishing operation.

NUTRIENT CLASSIFICATION, FUNCTIONS AND DEFICIENCY SYMPTOMS

Pigs do not utilize feed as such, but rather nutrients which are released by digestion and then absorbed into the body fluids and tissues. Nutrients are generally classified into carbohydrates, fats, proteins, minerals, vitamins and water. For a detailed discussion of nutrients, turn to Chap. 17.

Pigs are monogastric animals, and therefore protein quality (amino-acid balance) as well as quantity is important. Also, because pigs are monogastric, roughages cannot be utilized to much extent and therefore grains are the most abundant energy source in swine diets. Fats and oils are often included in swine rations, but for a number of reasons must be limited to a small percentage of the overall energy provided.

NUTRIENT REQUIREMENTS AND FEEDING STANDARDS

Nutrient requirements that should be considered when feeding cereal grain-plant protein based rations to swine are shown in Table 26-3. For information on how to apply nutrient requirements to practical ration formulation, turn to Chap. 20.

PROCESSING OF FEEDS

Many potentially useful feeds for swine would be unusable if appropriate processing were not possible. Grinding, one of the most common methods of processing, is often necessary to reduce particle size which may result in more efficient utilization. Usually grinding is more beneficial for the very young and the very old pig. The young pig will perform best on a finely ground mixture (1/8 to 3/16 in screen); whereas, the finishing pig will perform as well or better with more coarsely ground grains (1/4 to 3/8 in screen). The problem with fine grinding for the finishing pig is that stomach ulcers often develop.

Pelleting is the processing technique that has attracted more interest over the years. Pelleting requires that the feedstuff be finely ground, and sometimes is responsible for stomach ulcers in finishing pigs. Nevertheless, an improvement in feed efficiency of 5 to 10% is sometimes experienced with pelleting. Usually the greatest improvement in pelleting results with high fiber feedstuffs. The cost of pelleting can partially offset the

Figure 26-8. Pelleted feeds are commonly used in swine feeding, but can cause incidences of stomach ulcers. Pelleting is best reserved for feeds with high fiber contents.

TABLE 26-3. NUTRIENT REQUIREMENTS OF SWINE WHEN FEEDING A CEREAL GRAIN-PLANT PROTEIN BASED RATION. ADAPTED OR MODIFIED FROM NRC (1988)

	Class of Swine					
Nutrient	Early Wean	Starting 20-50 lb	Grower 50-125 lb	Finishing 125-240 lb	Bred gilts, sows & boars[a]	Lactating sows & gilts[a]
Dig. energy, kcal/lb	1550	1555	1560	1565	1520	1520
Protein, %	20	18	16	14	14	14
Lysine, %	1.15	.95	.75	.60	.43	.60
Calcium, %	.80	.70	.60	.50	.75	.75
Phosphorus, %	.65	.60	.50	.40	.60	.60
NaCl, %	.25	.25	.25	.25	.35	.35
Iron, ppm	100	80	60	50	80	80
Zinc, ppm	100	80	60	50	100	100
Copper, ppm	6.0	5.0	4.0	3.0	5	5
Manganese, ppm	4.0	3.0	2.0	2.0	10	10
Iodine, ppm	.14	.14	.14	.14	.14	.14
Selenium, ppm	.3	.3	.3	.3	.3	.3
Vitamin A, IU/lb	1000	800	600	600	2000	1000
Vitamin D, IU/lb	100	100	70	70	100	100
Vitamin E, IU/lb	7	5	5	5	10	10
Vitamin K, mg/lb	1	1	1	1	1	1
Riboflavin, mg/lb	1.6	1.4	1.2	1.0	1.7	1.7
Pantothenic acid, mg/lb	10	8.2	6.5	5.5	5.5	5.5
Niacin, mg/lb	7	6	5	4	5	5
Choline, mg/lb	250	200	150	150	600	500
Vitamin B_{12}, ug/lb	8	7	5	3	7	7
Biotin, ug/lb	100	50	50	50	100	100
Folic acid, ug/lb	150	150	150	150	500	300

[a] May wish to increase energy level (added fat) two weeks before farrowing and during lactation.

improvement in efficiency.

Heating swine feeds can be like a double-edged sword. It can improve the utilization of some feed ingredients, if properly applied. But **over**-heating can **reduce** the efficiency of utilization, degrade protein quality, and destroy certain vitamins.

Heating is required for soybeans in order to destroy toxins and antidigestive factors. Also, heating is important in processing animal by-products, such as fishmeal and meat and bone meal, in order to sterilize the product. Heating may be necessary with certain ingredients, such as potatoes, to rupture the starch granules so that the starch is more digestible to the animal. However, most studies indicate that heat does not improve the nutritive value of properly supplemented feed grains.

Other processing methods, such as roasting, steam-flaking, micronizing and extruding which involve heat, (and commonly used in ruminant rations) generally have no effect on feed intake or feed efficiency for swine.

FEED ADDITIVES

Feed additives have been used in swine rations for more than 25 years. Broadly speaking, feed additives refer to non-nutritive compounds that may be classified into two categories: 1) those intended to stimulate growth and improve feed efficiency, and 2) those which are used to improve or control some other function such as parasites and/or diseases.

A number of chemical compounds including sugar, saccharine and many flavoring agents have also been used with the idea that they can enhance feed consumption. Research has generally shown that the pig may have a preference for diets containing certain flavoring agents; however, if not given a choice the pig will eat as much of the control diet without the flavoring agent. It has been suggested, although not proven, that flavoring agents are more beneficial when the quality of the ingredients in the diet are poor. Whey and milk products have been used in swine starter diets with varying degrees of success. Although they may act as flavoring agents, their main effect is through improved protein quality.

A number of hormones used in other species, for example, thyroxine or thyroprotein (iodinated casein), thiouracil (antigrowth hormone), testosterone and estrogens have been used in swine with no practical improvement. The hormone that appears to have the greatest potential is the growth hormone or somatotrophin; however, problems of cost and method of administration must be solved before this can be of importance for practical swine production.

Feed is used as a carrier for a number of anthelmintics (wormers). These are primarily hygromycin B, dichlorvas, and levamisole. The feed may also be a carrier for drugs used in estrus synchronization.

Some feeds have been reported to contain unidentified growth factors. However, in well balanced diets these feeds appear to give no response or a questionable response in swine. Bacterial and yeast fermentation products have met with limited success and acceptance. The idea is to seed the intestinal tract with good microorganisms, and thus reduce harmful microorganisms and their effect. In general, the yeast products, primarily *Sarccharomyces cervisae*, have not shown any benefit when included in swine diets. *Strepococcus faecium* and *Lactobacillus acidophilus*, both gram positive lactic acid producing organisms, are the most commonly used bacteria in fermentation products. They generally decrease pH in the intestinal tract, thereby giving a bacteriostatic effect. It is believed that they produce antibiotic-like substances that destroy or decrease the production of certain coliforms; thus, allowing for a high lactobacillus:coliform ratio which may be beneficial in reducing scouring. However, research results have been very mixed with only slight or no improvement observed.

Antibiotics have been shown to stimulate feed intake, rate of gain and improve feed efficiency. The improvement is much greater with young vs older animals and also tends to be greater in situations where the level of performance is poor compared with situations where the level of performance is very high. Also, it has been suggested that the response is greater in old vs new facilities.

The mode of action of antibiotics is not clearly understood, but may be involved with decreasing subclinical infection, increased nutrient absorption and/or decreased production of toxin by microorganisms. It also has been suggested that antibiotics may spare certain nutrients perhaps, by decreasing the number of microorganisms using nutrients in the digestive tract. Another theory has been that the rate of passage is decreased when antibiotics are fed thereby enhancing digestibility.

Antibacterial or chemotherapeutic agents are chemically produced compounds that have antibiotic-like properties. These include arsenicals, carbodox, sulfonamfides, furazolidone, nitrofurazone and copper sulfate. Citric acid and fumaric acid may also be included in this group. It is believed that these antibacterial agents produce antibiotic-like effects with mode of action similar, although not identical, to that of the antibiotics. Starter diets commonly have a combination of antibiotic and antibacterial agents.

Enzymes are products that are frequently promoted by commercial feed companies. However, enzymes have generally not been successful in improving the feed intake and digestibility of most feedstuffs. Certain fungal enzymes have been reported to improve western barley and rye but have had no effect on eastern barley. As a general statement, enzymes have had no effect in swine diets.

Antioxidants, such as ethoxyquin, BHT and BHA, may be included in diets that are high in fat to prevent oxidation. Organic acids are used in some cases as preserving agents, especially when high moisture grain is to be stored.

The public continues to be concerned about the use of feed additives in livestock feed. The antibiotics or antibacterial agents used in human medicines are of the most concern. Sulfa drugs have especially been of concern. Withdrawal times are necessary for some additives to prevent tissue residues. The producer should realize this and always observe withdrawal requirements of the additives being used. Failure to observe withdrawal times has resulted in meat residues of additives, especially with sulphas. Failure to observe withdrawal times is irresponsible and jeopardizes the trust of the public, as well as the continued availability and legal use of additives.

Table 26-4. APPROVED LEVELS AND WITHDRAWAL PERIODS FOR COMMONLY USED FEED ADDITIVES (Pork Industry Handbook #31).

Feed additive	Approved growth promotion level, grams/ton	Pre-slaughter withdrawal period*
Antibiotics		
Bacitracin	10-50	None
Bacitracin, M.D.	10-50	None
Bacitracin, Zinc	10-50	None
Bambermycins	2	None
Chlortetracycline	10-50	None
Erythromycin	10-50	None
Oleandomycin	5-11.25	None
Oxytetracycline	7.5-50	None
Penicillin	10-50	None
Tylosin	10-100	None
Virgininiamycin	10	120 lb
Chemotherapeutics		
Arsanilic acid	45-90	5 days
Sodium arsanilate	45-90	5 days
Carbadox	10-25	10 wk (75 lb)
Furazolidone	---	5 days
Nitrofurazone	---	5 days
Roxarsone	22.7-68.1	5 days
Combinations**		
Chlortetracycle +sulfamethazine (or sulfathiazole)	100 100	15 days 7 days
+penicillin	50	
Penicillin + streptomycin	1.5-8.5 7.5-41.5	None
Tylosin + sulfamethazine	100 100	15 days

*Period of time the drug must be removed from the diet before slaughter.
**Not a complete list of approved combinations; for further information consult the Feed Additive Compendium, Miller Publishing Co., Minneapolis, MN.

FORMULATION OF RATIONS

Balanced rations contain the necessary nutrients in the correct proportion to properly meet the pig's requirements throughout its life cycle. Ration formulation is the process of combining the available sources of energy, supplemental protein, minerals, vitamins, antibiotics, and other feed additives to meet these needs. For detailed instruction see chapter 20.

Traditionally, medium to small farmers who produced their own grain or buy their grain locally, use a commercial protein/vitamin/mineral supplement to provide a balanced ration. The supplement is typically used to carry all the additional nutrients needed, including additives. Larger operations often formulate their own supplements for blending into the ration via an on-farm feed mill.

METHODS OF FEEDING

There are several methods of feeding and associated terms: 1) ad libitum vs. restricted feeding, 2) self-feeding vs. hand feeding, 3) individual vs. group feeding 4) wet vs. dry, 5) frequency of feeding, 6) alternate feeding. The method of feeding will depend upon the number and type of pigs to be fed, kind and amount of ration to be fed, type of facilities and the amount and quality of labor.

Ad libitum, or full-feeding as it may be called, refers to feed being kept before the pigs at all times. Self-feeders are usually used, and care must be exercised to keep the feeders adjusted properly to prevent feed wastage. Generally, growing and finishing animals are fed ad libitum because we want to maximize daily feed intake.

Restricted feeding, or limit feeding as it may be called, refers to a situation where the animals are fed once or several times a day, by hand or automatically, but the amount fed is less than full-feeding. This method is used widely with gestating sows where it is desirable to limit daily intake. Restricted feeding has also been a common practice in finishing hogs in many areas of the world where producers wish to slow down the finishing rate to improve carcass quality.

Self-choice, or free-choice feeding, is a system where the grain portion of the ration is fed in one compartment of a self-feeder and a complete (protein) supplement is fed in a separate compartment of the feeder. The pigs choose the feed and the amount they eat. It is important that the palatability of both grain and supplement be nearly equal so that pigs will eat the proper proportion of each. Free choice feeding has been widely used for growing-finishing hogs on farms which had home grown grains.

Hand feeding is a system in which the daily amount of feed is placed before the animal once or several times per day. A hand feeding system lends itself very well to restricted feeding. A complete ration or a combination of grain and supplement may be fed. Gestating and lactating sows would normally be hand-fed.

In group feeding systems, it is important to provide adequate feeder space and to size animals accordingly. That is, so larger, more aggressive animals do not get a

Figures 26-9a & b. The inside and outside of a large finishing operation utilizing automated feeding equipment. A complete balanced ration is delivered to the bulk tanks outside of each building. From there the ration is augered into the building and deposited into self feeders.

disproportionate share.

In an individual feeding system, the animals (usually gestating-lactating sows or boars) are usually housed in individual crates or at least in individual feeding stalls. The advantage of an individual feeding system is that the correct amount of feed for each animals' condition can be fed to that animal. That is, in a group feeding system, there is no assurance that an individual animal will actually get its share of the feed.

In a wet feeding system the feed is fed as a liquid or

paste type feed. Water is mixed with the feed prior to, or as the feed is being fed. Proponents of the liquid or wet feeding suggest that dustiness and thus, feed wastage is reduced with an overall improvement in feed efficiency. However, wet and moldy feed can be a serious problem in wet feeding systems.

Some have suggested that feed intake is stimulated when feed is fed in a moist condition. However, research results have not shown any real advantages for wet feeding. If the moisture level is too high in a liquid feeding system, dry matter intake can actually be depressed because of the diluting effect of the water and the limited capacity of the digestive tract.

Frequency of feeding refers to the number of times per day that the animal is fed and refers to a system of restricted feeding. For example in **alternating type** feeding systems (also called skip feeding), the animal is fed every other day, every third day, three times per week, etc. Alternating feeding has been used for feeding gestating sows when restriction of intake is desired.

INFLUENCE OF NUTRITION ON CARCASS QUALITY

Although nutrition may not have a major effect on carcass quality, it is known that a deficiency of protein will result in a fatter carcass and that increasing the level of protein (up to the protein requirement) can increase muscling. Also, increasing the fat concentration in the ration can increase fat deposition in the carcass and result in fatter carcasses. Usually, if recommended levels of protein are fed, additional protein will only make minor improvements in carcass leanness and the economics of this is questionable. Normally, fat or oil levels in the ration of 7 to 8% will not have a major effect on the amount of fat in the carcass. However, higher levels of fat may increase carcass fat. The type of fat will influence the type of fat that is laid down in the carcass. That is, unsaturated oils should be avoided as they tend to cause soft, oily carcasses.

REPRODUCTION AND BREEDING HERD MANAGEMENT

A high reproductive rate is essential for the successful operation of a purebred or commercial swine farm. (For detailed information on reproduction, see Chap. 21.)

Puberty (onset of the estrous cycle) in gilts occurs at 5 to 7 months of age. The number of ova released per estrus increases gradually over the first few estrous cycles. Unless mating and conception occurs, the estrous cycle is repeated about every 21 days (16 to 25 day range). The heat period, or estrus, is the time during the estrous cycle in which the sow will accept the boar. It lasts from 1 to 5 days with an average of 2 to 3 days.

Older sows generally remain in heat longer than gilts. The average gestation period of the sow is 114 days, although extremes of 98 to 124 days have been reported.

Sexual maturity in the boar is a gradual process in which sperm production and sex desire begin concurrently in increasing intensity at about four months of age. However, it is generally considered that a boar should not be used in service before eight months of age, and even then with very limited service until about one year of age.

Mating (copulation) is a prolonged process in swine, varying from 3 to 25 minutes in which waves of high and low sperm concentration exist in the flow of the ejaculate. Thus, it is important that copulation take place without disturbance.

SELECTION AND MANAGEMENT OF GILTS

Gilts should be selected that reach puberty at an early age and have intense estrus periods. Also, gilts should meet all of the anatomical, production, carcass, and soundness characteristics previously discussed.

Crossbred gilts usually reach puberty one to four weeks sooner than purebreds, but there are some breed differences. Confinement rearing has been shown to delay the onset of puberty in developing gilts and thus, special attention needs to be given to their housing and management. Total darkness has been shown to delay puberty; about eight hours of light appears adequate.

High temperatures should be avoided whenever possible for both boars and gilts. In outside lots shade should be provided. In confinement facilities, sprinklers or evaporative cooling may be useful in some situation to reduce the detrimental effects of high temperatures. Gilts should be housed in groups of about six to ten. Gilts should be exposed to a mature boar at 150 to 165 days of age to enhance puberty, and checked for heat to identify early maturing gilts. Gilts should be fed a properly formulated ration at a level that will cause a modest growth rate (1 lb/head/day), **but at a level which prevents the animal from getting fat**. Overfeeding gilts is not only wasteful from a standpoint of using excessive feed, but it causes poor breeding and farrowing performance. Fat gilts do not cycle, breed or farrow as well as gilts in modest flesh. Breed gilts on 2nd or 3rd heat after puberty.

HEAT DETECTION AND TIME OF MATING

Heat detection is critical in the reproductive process and can be the cause of many breeding problems. To determine heat, apply back pressure to each female in the presence of the boar. Females that are in heat will allow a man to sit on their back and will respond by standing solidly and attempting to stiffen their ears. Particularly in gilts, the vulva may be swollen and/or nervousness may be noticed before and after standing heat. Females identified in heat can later be mated to a boar or they can be mated artificially.

Live sperm must be in the reproductive tract two to three hours before ovulation occurs, otherwise litter size will be reduced. This is referred to as capacitation; a yet to be explained phenomenon. Ovulation usually occurs 38 to 42 hours after the start of heat. Sperm can live in the reproductive tract 24 to 48 hours. However, the eggs are viable for a shorter period of time (12 to 24 hours) after ovulation. Therefore, for conception to occur, breeding must take place at the right time.

If heat is checked daily, females can be bred each day they will accept the boar. If heat is checked twice each day, females can be bred at 12 and 24 hours after they are first detected in heat. Gilts will sometimes have heat periods less than two days and may have to be bred as soon as they are detected, and then each succeeding 12 hours that they will stand for the boar.

ESTRUS SYNCHRONIZATION

For many producers controlling the time of breeding and farrowing is highly desireable. Synchronization of estrus allows the control of breeding and thus farrowing, which allows for more efficient use of labor.

The lactating sow does not normally ovulate (come into estrus) during lactation, but will show estrus during the first three to eight days following weaning. Weaning, therefore, is an effective way of synchronizing estrus in reproducing sows.

Naturally occurring hormones (PMSG - follicle stimulating hormone and HCG - ovulating hormone), when used in combination, are effective in regulating estrus in gilts. However, the use of these hormones is not practical because they have to be injected on a time schedule. Research is underway to obtain clearance from the Food and Drug Administration to use a new synthetic progesterone-like compound (Altrenogest) which has been found very effective in synchronizing estrus in cycling gilts. Estrus is blocked during the time that the drug is being fed (14 days), and upon drug withdrawal, gilts will come into estrus within a few days (usually four to six days). Thus, by timing the feeding of Altrenogest, gilts can be brought into heat at a time that fits into a sow group. (A means of synchronizing estrus in sows is to wean all sows at the same time.)

MATING AND BREEDING PROCEDURES

Gilts and sows may be hand mated or pen mated depending on the production system. In hand mating, the female in heat is placed with the boar and their activity is observed. This system allows the producer to control the number of inseminations thus maximizing the use of the boar. Also, the producer can record the breeding dates of all females mated. In pen mating, a boar is placed in a pen of females. The boar breeds females as they come into heat and for as many times as he likes. Less labor, but more boars, are required and accurate breeding records are not feasible. When using pen mating, divide sows or gilts into groups of six to ten and put one boar with each group. Rotate boars among pens every 12 to 24 hours.

Young boars from 8.5 to 12 months of age should have no more than one service per day with a maximum of five services per week. Mature boars (over 12 months of age) may be used twice per day for a period of seven days.

ARTIFICIAL INSEMINATION

Artificial insemination (AI) techniques are available for swine. Fresh semen is usually preferred because of the lower conception rates obtained with frozen semen. AI allows for the more extensive use of older proven boars on lighter weight females, and AI decreases the number of boars required. The average ejaculate contains enough sperm to inseminate 6 to 8 females.

One drawback to the use of AI is the higher level of management required. Heat detection, as discussed earlier, is a critical factor in achieving maximum conception with AI because of the need to inseminate females at exactly the right time. AI in swine is not used widely in the United States; but its use should increase in the future with the development of better techniques to freeze semen, as well as the availability of drugs to synchronize estrus.

PREGNANCY DIAGNOSIS

Pregnancy diagnosis is a management tool that can increase the overall reproductive efficiency of the breeding herd by allowing producers to cull unproductive females early after breeding. Thus, feed can be saved and better use of facilities can be made. Several methods are currently being marketed; but, ultrasonic detectors used between 30 and 45 days after mating are one of the more popular pregnancy diagnosis techniques. Checking return to estrus after mating is still the most accurate means of predicting if a sow will farrow. Blood testing for certain hormones, progesterone which peaks 27 to 30 days after mating, has been found to be about 90% accurate (equal to ultrasound). Assay procedures that can be performed on the farm are being developed.

SOW MANAGEMENT DURING BREEDING AND GESTATION

A dry, comfortable, well-ventilated environment should be provided for gilts and sows before, during, and after breeding. Avoid any stress-producing situations, especially during the first three weeks after breeding, as stress at this time increases embryonic mortality.

Temperature is important. Avoid exposing gilts or sows to high temperatures. Exposure to high temperatures (above 85 F) before breeding delays or prevents the occurrence of estrus and reduces ovulation. Exposure during and after breeding increases early embryonic deaths by as much as 35 to 40%. Sows housed in groups can easily tolerate a temperature as low as 50 F.

After breeding, group sows according to their stage of pregnancy and body condition to facilitate feeding management. **Do not mix gilts with mature sows and keep group size small.**

Weaned sows should be fed a well-balanced ration at a level commensurate with their body condition, season of the year, and type of facility. Feed withdrawal and high levels of feed following weaning have both been shown to be detrimental to breeding and embryonic survival, as well as being costly. As a rule of thumb, 4 to 5 lbs. of a well-balanced ration is usually adequate.

Thin sows, especially during the last half of gestation, should be fed a higher level of feed. On the average, gilts should gain about 90 lbs. during gestation, while 60 lbs. is adequate for sows. Don't overfeed sows; it is costly and fat sows do not conceive as well, and have more difficulty during farrowing.

BOAR MANAGEMENT

Proper management of the boar is required to ensure high levels of libido (sex drive) and high quality sperm. New boars should be selected carefully (performance tested if possible) and purchased early, so as to allow 30, and preferably 60, days of isolation. The boar should be transported, handled carefully, and placed in clean (disinfected), comfortable housing (at least 100 yds from the herd). Isolation is necessary to identify any disease problems, to build immunity, and to make adjustments to changes in climate, farm conditions, etc. Boars should be blood tested during this period for pseudorabies, brucellosis, and leptospirosis. The producer may also want to vaccinate for erysipelas, leptospirosis and atrophic rhinitis as well as treat for internal and external parasites. It is also a good practice to test mate (using market gilts as sentinel animals) as well as evaluate the sperm of new boars. New boars could be co-mingled with a few culled gilts or market hogs to develop immunity to specific microorganisms on the new farm.

The producer will also want to test for aggressiveness and mating habits during this time. Remember, be gentle and helpful with young new boars. Frequency of use is important in that young boars (8.5 to 12 months of age) should not be overused (see previous section). A boar showing signs of overwork should be rested for seven to 10 days. Boars should be kept cool in the summer to prevent reduced fertility which usually results three to six

Figure 26-10. Recommended sow body weight change (Swine Management, AGDEX 440-20, Manitoba Agriculture)

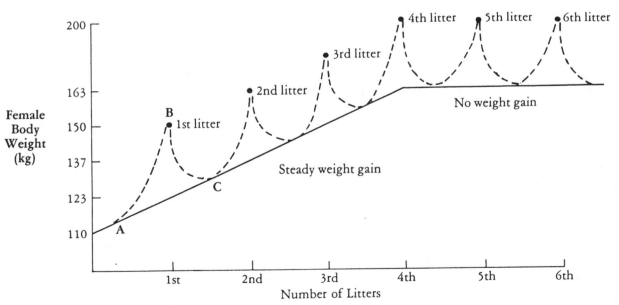

A — Weight of gilt at breeding
B — Weight of gilt before farrowing first litter
C — Weight of gilt after weaning litter

weeks after exposure to high temperatures occurs. Hot temperatures can also reduce the aggressiveness of a boar. Provide an adequate breeding area (good footing for the boars) and always use a properly sized boar in relationship to the female. Boars should be fed a properly balanced ration at a level to maintain good body condition; but, do not overfeed so that boars become fat and lazy.

SOW AND PIG MANAGEMENT

The feeding and care of the sow during gestation and farrowing will greatly influence the number of pigs that a sow will farrow and wean. On the average, producers only wean 70 to 80% of the live pigs farrowed. Thus, there is great potential for improving sow productivity.

PREFARROWING MANAGEMENT OF SOWS

The farrowing facility should be properly washed and sanitized. Also, the pregnant gilt or sow should be carefully scrubbed and washed, and placed in the cleaned farrowing house at least three, and preferably five days before her due date. This gives the sow a few days to become acclimated to the facility, and also allows adequate safeguard for sows that may farrow early. A farrowing crate or pen should provide for the comfort of the sow and her litter. It should be designed so as to prevent smothering of the piglets (laying on the litter by the

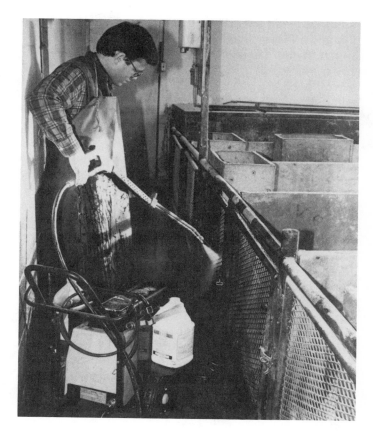

Figure 26-11. Sanitation is an absolute must in any successful farrowing program. The best type of sanitation program is what is known as the all-in all-out. In the all-in, all-out system, the confinement building is totally vacated and sanitized before new pigs are brought in. In this illustration a steam cleaner is being used to clean the pens.

Figure 26-12. The purpose of a farrowing crate is to allow baby pigs access to the sow's nipples, but prevent the sow from rolling on top of the pigs and smothering them.

sow). It should also provide easy access during farrowing and after the litter is born.

The ration fed during gestation can be continued when the sow is moved to the farrowing house. However, the feeding of a laxative ration during late gestation and early lactation is believed by some to be helpful in preventing constipation. Ground oats or wheat bran have been used at a level of 20 to 25% to create such a ration. However, the inclusion of these high fiber ingredients can reduce the energy density which can be of concern during lactation. Therefore, after farrowing fat usually needs to be added to offset the reduction in energy. Water should be freely available at all times but spillage should be avoided to prevent wet pens.

Temperatures in the farrowing house and farrowing pen are critical. The sow is most comfortable at a temperature of 60 to 70 degrees F, but baby pigs should have access to a much warmer temperature (90 to 95 degrees) for the first week or two. Heat lamps, heating pads or some other kind of spot heating is usually required to provide the higher temperature for the baby pigs. A sow at a temperature of 90 to 95 degrees F would be extremely uncomfortable and, as a result, might have difficulty during farrowing and would not eat well during lactation. The overall temperature in the farrowing house can be raised to (70 to 75 degrees F) during farrowing and for a few days following, to prevent chilling of the newborn pigs.

Sows should be treated (sprayed or dusted) for external parasites and should be treated for internal parasites prior to being brought into the farrowing house. The newborn pig is very susceptible to most parasites. Therefore, the herd health program should be designed to prevent exposure of the baby pig to internal and external parasites.

MANAGEMENT OF THE SOW DURING AND AFTER FARROWING

As the farrowing date approaches, sows should be observed carefully. Usually, within a few hours of farrowing, the sow's milk will be let down and she will often show signs of restlessness by constantly getting up and down and moving around. Duration of labor may range from 30 minutes to more than five hours. Prolonged labor is often associated with difficult births. Therefore, a sow that has been in labor over five hours may require assistance.

It is important that the newborn pig nurse quickly to get needed energy for body functions. (Energy stores of the newborn pig are limited.) Also, colostrum supplies necessary antibodies for both immediate and temporary protection against bacterial infection. As pointed out earlier, supplemental heat should be provided for the newborn pig during cold weather to prevent chilling.

Sows normally do not eat much for the first 12 to 24 hours after farrowing, but water should be continually available. After farrowing, the sow can be fed according to her appetite. It has been shown that sows can be successfully full-fed following farrowing. Inadequate feed consumption is usually a much greater problem than over-consumption.

A majority of deaths of baby pigs occur during the first 3 days after farrowing. Therefore the sow and litter should be observed carefully during the first few days after farrowing. Lack of appetite, listlessness and failure of the sow to positively respond to the nursing activity of the pigs indicate a need for corrective treatment. If problems are observed early they can often be corrected; otherwise if the problem persists for very long, the sow may actually dry up with the resultant starvation of the pigs.

BABY PIG MANAGEMENT

A number of baby pig management practices need to be considered and performed during the first few days after farrowing. Navel cords should be dipped with tincture of iodine as soon as possible after birth. It may be desirable to cut the cord 3 to 4 inches from the body. An undipped navel cord provides easy entrance of bacteria into the body which can result in infection and possible abscesses. If excessive bleeding occurs when the navel cord is cut, it can be tied off with a piece of string (about an inch from the body).

The baby pig has needle teeth which should be clipped the first day to prevent injury to the sow's udder and to other pigs. Care must be taken in clipping the teeth so that the pig's mouth and gums are not injured. Clean and disinfected clippers should be used to prevent spreading of bacteria. Tails are usually docked during the first day; this prevents tail biting later in life. (Most feeder pig sales require that tails be docked.) The tail can be clipped one-half to one inch from the body. Care should be taken not to clip the tail too short which could cause serious injury to the pig.

At three to five days of age, pigs are given supplemental iron to prevent baby pig anemia. An intramuscular injection (ham or neck muscle) of an iron solution is the preferred method of treatment. Some producers routinely given an antibiotic injection to aid in scour prevention. Pigs may also be vaccinated for atrophic rhinitis at seven and 21 to 28 days of age. Male pigs need to be castrated prior to two weeks of age. For castration, use a clean, sharp instrument and make the incision low to promote good drainage. Use antiseptic procedures as far as possible.

In operations where several sows are farrowing near the same time, it is often a good practice to equalize litter sizes according to the number of functional teats and estimated milking ability of the sows. Pigs can be moved from one sow to another before three days of age without problems. However, be sure that the pigs have received colostrum before they are transferred.

IDENTIFICATION

Ear notching is the most common method of identifying swine. It is permanent, grows with the pig and is easy to read. Purebred breeders are required to identify each individual for purposes of registration and herd records. Even in commercial herds, a system of identification is necessary for selection of replacements and for culling. Branding and tattooing are not particularly useful in swine identification. Plastic ear tags are useful for older animals such as gilts and sows.

REBREEDING THE SOW

In most commercial operations, the production system is designed so that sows will be rebred as soon as they come into heat after weaning. As discussed earlier, weaning is an excellent means of synchronizing estrus. Thus, pigs will normally be weaned in batches rather than when they reach a specific age.

The success of rebreeding the sow is dependent, to some degree, on her condition at the time of weaning. A very thin sow will often fail to come into heat at the proper time. Thus, feed intake (and energy content of the feed) during lactation is important from the standpoint of rebreeding as well as milking ability. First litter gilts are most prone to excessive weight loss, and thus poor rebreeding. Special attention must be given to them.

CULLING

The maintenance of unproductive females in the breeding herd results in reduced sow productivity and thus profit. A culling policy must therefore be developed and followed. Individual identification of each female and adequate records are required. Poor breeding performance (failure to come in heat or return to estrus, delayed estrus or late puberty in gilts, poor conception rate and/or repeat breeding) is the major cause of unproductive females and the reason for culling. Lameness, farrowing difficulties and milking problems are other clear-cut reasons for culling. Culling sows, and especially gilts, just for small litters is questionable because of the very low heritability of litter size. Culling rates of 30 to 40% annually are common. Guidelines for an effective culling program follow:
1. Identify and quickly eliminate freeloaders (sows that are not bred, not in heat or not nursing a litter).
2. Sell sows that have farrowing difficulties, defective udders or teats, are poor milkers, have poor temperament or structural unsoundness which interferes with breeding.
3. Give a sound, good milking gilt or second litter sow another chance if her litter size is the only problem. Older sows usually produce larger litters.

WEANLING PIG MANAGEMENT

Following farrowing, weaning is the second most critical period in the life of the pig. Successful weaning programs must minimize stress and provide for the major needs of the young pig: comfort, sanitation, and nutrition. In most commercial operations, an environmentally controlled nursery is necessary to provide comfort and sanitation. A nursery, in addition to providing for the environmental needs of the pig, should also facilitate the manager's job of observing, handling, feeding, and removing waste.

As pigs are weaned earlier to increase the overall productivity of the sow, the potential for death loss increases. Therefore, the younger the pigs are weaned, the greater the care that is required to meet their needs. Comfort includes several environmental and social factors: temperature, ventilation, flooring, group size, uniformity, and stocking density.

An effective temperature should be provided at the pig's level and will vary depending upon the age at weaning. Effective temperature is a function of air temperature, radiant environment (walls and ceiling), air speed and floor characteristics. Critical temperature is the effective temperature below which pigs must increase body heat production to keep warm. A suggested temperature schedule for pigs weaned at three to four weeks of age is shown in Table 26-5.

Table 26-5. Temperature Optimums and Ranges for Housed Swine. (Pork Industry Handbook #84)

Animal	Temperature (°F) at animal level	
	Optimum	Desirable Limits
Lactating sow	60	50-70
Litter-newborn	95	90-100
Litter 3 weeks old	80	75-85
Pre-nursery (12-30 lb)	80	75-85
Nursery (30-50 lb)	75	70-80
Nursery (50-75 lb)	65	60-70
Growing-finishing	60	50-70
Gestating sows	60	50-70
Boars	60	50-70

It is extremely important to provide a warm environment for pigs during the first and second weeks after weaning. This is because pigs are usually just learning to eat solid food during this time, and their feed intake is often less than desirable. Fluctuating temperatures should be avoided. A ventilation system should be designed to remove gases, odors, and moisture, and to distribute fresh air. In addition, ventilation rates are influenced by the size of the pig and the season of the year. A suggested ventilation schedule is shown in Table 26-6.

Table 26-6. Recommended Ventilation Rates (CFM, at 1/8 in static pressure) Per Pig or Sow and Litter. (Pork Industry Handbook #41 and 60)

Hog unit	Minimum (winter) rate	Normal rate	Hot weather rate
Sow & litter (400 lb)	20	100	450
Pigs			
20-40 lb	2	15	36
40-100	5	20	48
100-150	7	25	72
150-210	01	35	100
Sow or boar, limit fed			
200-250 lb	10	35	120
250-300	12	40	180
300-500	15	45	375

Flooring materials should be nonabrasive and comfortable, self-cleaning, easy to sanitize, and durable. Plastic coated, welded wire makes a good floor, but is expensive. Galvanized flat expanded metal and woven wires are cheaper, and are used widely.

Small pen sizes, and thus a small to moderate number of pigs per pen, provides more flexibility to ensure that pigs are of a uniform size and that each pen is kept fully stocked. Pigs in a fully stocked pen usually develop better feeding, resting and dunging patterns. As mentioned previously, six to 10 pigs per pen is desirable, and gives the producer the opportunity to keep litters as pen groups after weaning. However, mixing of uniform pigs from different litters is not detrimental to their performance.

Table 26-7 displays space allowances suggested for weanling pigs. Remember, the stocking density you start with is dependent upon the length of time the pigs will stay in the nursery.

SANITATION AND DISEASE CONTROL INTERRELATE

It is vitally important to realize that the ability of the

Table 26-7. Space Recommendations For Nursery-Growing-Finishing Pigs Using Partial or Total Slats. (Pork Industry Handbook #55)

Pig weight or class	Space for partial or total slats
lb	sq ft
15 - 30	1.7-2.5
30 - 60	3-4
60-100	5
100 - 150	6
150 - market	8*

*Adjusting pig numbers per pen seasonally may result in improved performance. For example, increasing the number per pen by 1 or 2 pigs during the winter or decreasing the number in summer may be desireable.

pig to respond immunologically to disease causing organisms is at its lowest point when pigs are two to five weeks of age. Also, newly weaned pigs are no longer protected by antibodies that are normally present in sows' milk.

An all-in, all-out concept of sanitation is necessary to minimize disease risk to which the young pig is exposed. All facilities and equipment must be thoroughly cleaned and disinfected between groups of pigs, and then kept as clean as possible during the time pigs are in the facility.

Nutrition and feeding are also important. A hungry pig is more likely to be stressed by cold and other conditions. Newly weaned pigs should be offered **fresh, highly palatable feed daily** in small amounts to encourage them to eat. Pigs should be observed carefully to be sure that all pigs are eating. Post weaning digestive disorders, are believed to be caused by overeating of newly weaned pigs. Therefore to minimize problems, restrict the daily amount of starter diet given to newly weaned pigs for a few days after weaning.

Remember, in confinement nurseries, you control the pig's environment. The earlier pigs are weaned, the greater the attention that must be given to providing for comfort, sanitation and nutritional needs.

FEEDING AND MANAGEMENT OF GROWING-FINISHING SWINE

Pigs at 40 to 80 lbs. are not as susceptible to stresses

caused by the environment as the young pig, but they should not be exposed to severe temperatures, improper ventilation, improper sanitation, and/or unnecessarily exposed to potential pathogens. Feed efficiency is extremely important during this phase of production as about 75% of feed costs are involved in the growing-finishing phase. An environmental temperature should be provided that produces maximum efficiency. When environmental temperatures are above or below the critical temperature of the pigs, energy (feed) is used to cool or heat the body.

Although an all-in, all-out sanitation program may not be necessary with growing-finishing hogs, it is important that facilities be cleaned and disinfected before new groups of pigs are moved into the facility. If disease problems are encountered, more complete sanitation may be necessary to break the chain. External and internal parasites should be monitored and treated as necessary.

Most growing-finishing pigs in the U.S. are self-fed a complete diet. It is extremely important to have self-feeders that can be properly adjusted to prevent feed wastage. A properly formulated diet will result in maximizing the growth rate and feed conversion potential of the pig. Usually 14 to 16% protein diets with properly balanced amino acid levels (usually soybean meal diets) are fed to growing-finishing pigs. As a general rule, a higher level of protein is fed to the younger pig, with the level decreasing as the animal gets older.

It is advisable that a producer determine the growth rate and feed efficiency of his pigs during the growing-finishing phase. Therefore, the days to reach market weight can be calculated. An estimate of feed efficiency can be obtained by keeping a record of the amount of feed eaten by a certain number of pigs during a given period of time, and compared to their weight gains.

HERD HEALTH MANAGEMENT

HERD HEALTH PROGRAM

A successful swine operation must have a well-planned and managed herd health program. The basis of a herd health program involves: the 1) prevention of exposure to disease-producing organisms, 2) the maintenance of a high level of resistance, and 3) the early, accurate diagnosis and treatment of all sick animals. A herd health program must be a team approach involving the owner/manager, the veterinarian, the nutritionist, and other specialists (such as agricultural engineers).

PREVENTION OF EXPOSURE

The prevention of exposure to disease-producing organisms can be accomplished by a proper isolation program for newly purchased animals, coupled with a proper sanitation and disinfection program. Only healthy breeding stock should be purchased from producers known to have disease-free herds. These purchased animals should be isolated for a minimum of 30 days, preferably 60 days. During this period of time (or prior to purchase), they should be tested for brucellosis, leptospirsis and pseuorabies as well as treated for internal and external parasites. Also, newly purchased animals should be vaccinated for leptospirsis, erysipelas and atrophic rhinitis during the isolation period. Care should be taken that disease organisms are not brought onto the farm through the feed, water supply, boots and shoes of visitors, and vehicles that travel from farm to farm. Rodents, dogs, cats, birds, and insects are possible carriers of disease organisms and their contact should be restricted as much as possible.

Cleaning and disinfecting is very important in controlling the accumulation and spread of disease-causing microorganisms and is the basis of a sanitation program. This is especially true in modern swine buildings where continuous use and high concentrations of animals accentuate the problem.

An all-in, all-out concept is most desirable for farrowing houses, nurseries, and situations where disease has been a problem. As mentioned in the caption of Figure 26-11, the all-in, all-out concept means the facility is completely emptied of animals, thoroughly cleaned, and disinfected before the next group is brought in.

IMMUNITY

The maintenance of a high level of resistance to disease is an equally important part of a swine herd health program. Proper nutrition is important because nutritionally deficient animals are more susceptible to disease. Severe stress should be minimized at all ages to maximize the animal's ability to respond to challenges by pathogens. It is important that newborn animals get sufficient colostrum to provide them with the passive immunity that is necessary to protect them until vaccinations and their active immune system can provide protection.

DIAGNOSIS AND TREATMENT OF DISEASE

The early and accurate diagnosis and treatment of sick animals is a very important aspect of a herd health program. The manager should continually observe the herd and seek veterinary assistance if problems are observed.

Slaughter checks can also be an important aid to the herd health program. The presence of abnormal tissues may indicate potential disease problems.

PARASITE CONTROL

With the exception of the very young animal, internal and external parasites rarely kill the animal. However parasites usually cause reduced performance, with the effect being more severe for the younger pig, especially under 100 lbs. Losses may be due to retarded growth, reduced feed efficiency, or increased susceptibility of the animal to other diseases. Condemnation of organs and tissues from the carcass can also be a result of parasites.

In general, animals cannot be immunized for internal and external parasites, but must depend upon a program of diagnosis and treatment, as well as management, to minimize exposure to parasites. Climate is no barrier to parasites and confinement facilities are not free of parasites, although well-planned and managed facilities may lend themselves to the control of parasites. Sanitation is a great help, but will not necessarily prevent parasitism.

The major species of internal parasites are large roundworm, nodularworm, whipworm, threadworm, lungworm, and kidneyworm. The large roundworm is perhaps the most common internal parasite. A number of commercial wormers (anthelmintics) are available with varying degrees of effectiveness for the various types of worms. Overuse of wormers can be expensive; and therefore one should thus get a diagnosis of the kind of parasites present, and then select the wormer that will be most effective.

Lice and mange are the two major external parasites. Lice is the most frequent and mange the more serious. It should be pointed out that there are several commercially available compounds which are effective for lice, but not mange. Mange mites bore into the skin and are difficult to reach. Lice are found on the skin and are easier to reach, therefore more compounds are effective.

DISEASES

There are many infectious, metabolic, and nutritional diseases that affect swine. Diseases caused by bacteria include: brucellosis, atrophic rhinitis, erysipelas, arthritis, pleuropneumonia, leptosporosis, pasteurellosis, salmonellosis, streptococcus and swine dysentery. Among viral diseases that are of concern are: swine influenza, pseudorabies, rotavirus, parvovirus, and transmissible gastroenteritis (TGE). Hog cholera was important in the past, but an eradication program has effectively eliminated or reduced its importance (in the U.S.). Colibacillosis, TGE, rotavirus and edema disease are major concerns in baby and weanling pigs. African swine fever and foot and mouth diseases are continual threats, but are not currently found in the U.S.

Parvovirus, leptospirosis, brucellosis and eperythrozoonosis are the major infectious diseases capable of reducing reproductive efficiency in swine.

Of the viruses (entroviruses, parvovirus, adenovirus and reovirus) that are commonly isolated from aborted swine fetuses, parvovirus is of the greatest concern. Herd outbreaks of parvovirus infection are characterized by small litters and mummified pigs. Repeat breeding is a common sign of infection, but without mummified fetuses to sample, the disease is difficult to diagnose. Leptospirosis, the most important bacterial disease, usually causes abortions, stillbirths and reduced baby pig survival. Effective vaccines for the different serotypes are available.

Eperythrozoonosis is a small parasite of red blood cells that can cause repeat breedings, occasional abortions, prolonged farrowing, weak, anemic pigs at birth, hemorrhagic death of newborn pigs, slow growth of pigs and/or acute death in growing pigs. The frequency of eperythrozoonosis is not as great as parvovirus and/or leprospirosis.

MYCOTOXIN

The problem of mycotoxin in corn and other feedstuff is a serious and chronic problem for the swine producer. Molds are found widely in nature, but only a few are problems for swine. Moldy grains have been associated with poor gain, reproductive problems, and vomiting.

The molds that cause problems are those that produce what are known as mycotoxins. Visibly moldy feed may or may not contain mycotoxins. Pregnant animals and younger animals are more susceptible. The mycotoxin of most importance is produced by the mold *Aspergillus flavus*. This mold produces a mycotoxin known as aflatoxin. It is incredibly powerful and can cause death. Other molds of concern are fusarium and tricothecenes. These molds produce a variety of mycotoxins which can cause vomiting, general ill health, and loss of performance.

The infection of grain by field fungi may be beyond the producers control. Continuous wet weather during harvest and severe drought during growth may increase mold growth. However, proper cleaning and disinfection of feed bins and equipment, regulation of temperature and moisture during storage, prevention of insect infestation, and the use of mold inhibitors can minimize mold growth during storage.

METABOLIC DISORDERS

There are a few metabolic disorders that may be important in commercial swine production, but generally they are not major problems. The occurrence of stomach ulcers in growing-finishing pigs has caused some problems in the past. As discussed previously, ulcers have been associated with pelleted and otherwise finely ground feeds.

A condition known as osteoporosis often results in lameness and spontaneous bone fractures in older sows. (For a discussion of osteoporosis, see pg. 211). Ketosis and milk fever occasionally occur, but are more of a problem in cattle than in swine (see pg. 242). Porcine stress syndrome (PSS) is a condition apparently of genetic origin, which essentially causes market-ready hogs to go into shock prior to slaughter. The result can be death loss or pale, soft, watery carcasses. Selection of breeding stock with no history of PSS is the only means of control.

NUTRITIONAL DISEASES

There are, of course, many nutritional diseases which can be caused by marginal levels of one or more nutrients. However, only those which have been of practical importance will be discussed here.

A zinc deficiency causes a condition known as parakeratosis which is typified by skin lesions. Swine are more susceptible to Zn deficiency than most other domesticated animals. Iron deficiency anemia is of considerable importance in suckling pigs, and is the reason for providing iron injections to baby pigs. In recent years, a selenium and vitamin E deficiency has been recognized as the cause of a number of sudden deaths, especially in young weanling pigs. Supplementation can either be dietary or by injection.

There are a few nutritional toxicity diseases that should be mentioned. Massive doses of vitamins A and D can cause toxicity. Also, selenium which is needed in small amounts, can be very toxic if fed at excessive levels (above 3 ppm). Copper is required by the animal, and is often used as a growth stimulant. However, Cu is toxic if fed at levels above 250 ppm for extended periods of time. Copper sulfate, the form often used, is a very toxic form of copper, and should be used with great care. (It should also be pointed out that other species of livestock, especially sheep, are much more susceptible to copper toxicity. Feeds containing what would be a moderate level of copper for swine, would be toxic for sheep.)

MARKETING AND MARKETING SYSTEMS

There are three basic types of animals that are sold: 1) breeding stock, 2) feeder pigs and 3) slaughter hogs, which would also include cull boars, sows, and stags.

BREEDING STOCK

Selling breeding stock (traditionally purebreds) can be a highly specialized business. They may be sold by private treaty and moved from farm to farm. Some purebred producers have private auctions either on their farm or at a designated location. There are also a number of public auctions at which one or more purebred producers will consign animals. There are also public auctions held to sell boars that come from the centralized performance testing stations. Most breeding companies sell their stock by private treaty, and move it directly from farm to farm. This is the best way to minimize the spread of disease.

FEEDER PIG

Feeder pigs normally range in weight from 40 to 100 pounds. Feeder pig grades normally follow the USDA standards of US 1, 2, 3, 4, utility and cull. Culls or rejects are not normally sold through these sales but must be disposed of separately or returned to the farm.

Most feeder pigs are sold by private treaty, and moved directly from farm to farm. For those that are sold through organized feeder pig sales, they are consigned, graded, and sold to the highest bidder. Electronic marketing is involved in a number of these sales. Pigs consigned to these sales must meet certain health requirements; their tails must have been docked, they must have been treated for worms, and males must have been castrated. All animals must be free of external parasites. In many states, feeder pigs cannot be sold through the normal livestock auction markets, but must be sold on a special day, with special sanitary requirements being met, to minimize the spread of disease.

SLAUGHTER HOGS

Market classes and grades of slaughter hogs influence the price paid for these animals. Market class is determined chiefly by age and sex. Market grade normally relates to factors of conformation, quality, and finish.

The market recognizes four classes of slaughter hogs: barrows and gilts, sows, stags, and boars. Standard grades of barrows and gilts consist of US 1, 2, 3, 4, and utility, and are based upon backfat thickness and muscling. Although it would be desirable to sell on a carcass grade basis, most of the hogs in the US are still sold on a liveweight basis.

Transportation of Slaughter Animals

Care should be taken in the handling and transportation of all classes of pigs. Losses due to shrinkage, death, and carcass damage such as bruises, broken bones, etc. can be minimized if comfort of the animal is kept in mind. It should be recognized that confinement hogs must be handled with even greater care than hogs raised in dirt and wood lots. **Also, producers should be aware of and adhere to, drug withdrawal times.**

When moving and loading hogs, chutes and alleys should be properly arranged to avoid over-excitement and stress. The animals should be moved quietly and slowly. Canvas slappers and panels should be used in lieu of chains, whips and electric shockers. Don't kick animals as they will bruise easily and bruise(s) may result in damage to a part of the carcass which must be cut out. The truck should be prepared for winter or summer depending upon the season (absence or presence of vents, etc.).

WASTE MANAGEMENT

Due to the advent of intensive production systems, and more stringent environmental regulations, waste management has become more and more important. In general, waste management is essentially automatic on pasture or dirt lots when hog numbers (density) are very small. However, as hogs are concentrated in dirt lots, the need for waste management increases.

In a confinement operation, there are five general components of a waste management system; collection, storage, treatment, transportation, and utilization or disposal. Swine waste can be handled as a solid or semi-solid, a slurry (feces, urine and waste water), or as a liquid such as a lagoon effluent or runoff from open lots.

A popular waste management system in recent years has been a liquid system in which the manure passes through a slatted floor and is collected into either a deep or shallow pit containing water. The pit is emptied on a regular basis into a lagoon, which serves as a storage and treatment facility. Excess effluent (water) that accumulates in the lagoon is normally used for irrigation.

A modification of that system is called a flush-recycle system. The pit remains empty except when being flushed. Lagoon water can be used to flush, thus minimizing the amount of liquid that must be disposed of. Pits normally are flushed once or twice per day. It is believed that a shallow pit-flush-recycle system minimizes odors that arise from fermentation in the pits. Because of a number of nuisance lawsuits in recent years, the location of the swine farm, and the design, location, and operation of the waste management system is of major importance. Most localities have zoning regulations at the local government level and guidelines for waste disposal at the state level. Producers can minimize problems if they follow established regulations and ensure that their property is zoned correctly. They should further locate their facility and waste management disposal system in an area that will minimize objectionable odors.

Disposal of the liquid and solid materials is a major problem where land application is limited and proximity of people is near. Methods of re-feeding all or part of the waste (solid material) back to some species of livestock is continually being researched. Also, the use of manure in production of methane continues to be evaluated as a method of waste disposal and utilization. However, re-feeding and methane production are currently not practical in most situations. Manure is a valuable source of nutrients when used as a fertilizer, and land application remains a viable method of disposal where land is available.

RECORD KEEPING

Sound business management becomes more important as swine enterprises become larger, more intensive, and highly capitalized. The conduct and use of a good record keeping system is fundamental to sound business management.

Financial statements, profit-loss statements, cash flow statements and budgets are necessary records for financial analysis which are required for obtaining credit. Information necessary for developing these statements is provided in Chap. 5. Indeed, financial and other business practices common to most agricultural enterprises apply to swine management.

The only thing that is really unique to swine management is the type of production records that must be kept. A nominal list of data to be tabulated would include:

1. Number of females in herd, including gilts eligible for breeding.
2. Conception and farrowing rates for gilts and sows in each breeding group and in total herd.
3. Number of sows and boars that died or were culled and the reason.
4. Number and weight of pigs farrowed and weaned per sow, and per breeding female.
5. Number and cause of abortions.
6. Pre-weaning, nursery, grower and finisher mortality.
7. Number of market hogs sold and their days to market.
8. Feed conversion ratios of growing-finishing hogs and of the total herd.

Records should be verified at regular intervals with actual physical inventories. The frequency will depend on the individual farm, but should occur at least annually.

Computers are finding greater use in swine enterprises as the proper software (programs) become available. However, hand-entry records are certainly useable. Indeed, there are many manners in which records may be kept. The most important thing is simply that they actually are kept.

The reason records are kept, of course, is for analysis. By analyzing and comparing with other farms or standards for the industry, problems and areas for improvement can be identified. Table 26-8 represents some production goals.

Table 26-8. PRODUCTION GOALS FOR THE YEAR 2000.

Litters per sow per year	2.2
Litters per breeding female per year	2.0
Pigs born per sow	11.4
Pigs born alive per sow	11.0
Pigs weaned per sow	10.5
Pigs marketed per sow	10.0
Days to market	130
Feed conversion from 15 to 240 lb	2.50
Feed conversion including breeding herd	2.75
Inches of backfat at market	.8
Pigs born alive per sow per year	24
Pigs born alive per breeding female per year	22
Pigs marketed per sow per year	22
Pigs marketed per breeding female per year	20
Farrowing rate for breeding herd, %	83
Mortality	
Birth (stillborn), %	4.
Birth to weaning, %	6.0
Nursery, %	2.0
Growing-finishing, %	1.0
Breeding herd, %	2.0

SUMMARY

In the past, swine production was often a secondary farm enterprise. However, swine production is becoming more of a primary, high intensity enterprise. Confinement, and large numbers of pigs per farm is becoming a standard practice.

Whenever large numbers of animals are housed together in confinement facilities, disease and potential disease problems increase at a rapid rate. Therefore, sanitation and health aspects have become major issues or requirements of modern swine production.

Although certainly not complete, this chapter has outlined most of the principles of swine husbandry. Principles of business management are discussed in Chapters 1 through 5. Persons wishing to engage and investment in swine production would be well advised to, of course, seek further information, as well as employ the services of professional consultants; e.g. nutritionists, veterinarians, and waste management engineers.

CHAPTER 27 SHEEP PRODUCTION AND MANAGEMENT

Verl Thomas, Ph.D.
Dept. of Animal Science
Montana State University

ORIGIN

It is believed that sheep originated in the dry mountainous regions of central Asia 10,000 years ago. These sheep were probably horned, colored, and possessed a hairy coat - an outer coat of coarse hair and an undercoat of short, fine wool. Sheep provided early man a soft, warm covering and food.

Sheep and wool spread to Europe between 3,000 and 1,000 B.C. by way of ancient Greece. Domesticated haired-sheep were gradually replaced by the more recently developed wooled-sheep, except in some tropical regions of the world. During the next thousand years, the Greeks, Romans and Persians contributed to improvements in sheep breeds. The Romans established a woolen manufactory in Winchester, England as early as 50 A.D.

PRODUCTS PRODUCED

Products produced from sheep are meat, milk, skins and wool. Milk production is confined primarily to the Near East countries (Turkey and Iran) and to southern and central Europe. Skins are by-products from the meat produced. Wool production in countries such as Australia and New Zealand is one of the primary functions of sheep. Wool may contribute up to 75 to 90% of the income in some sheep operations.

CHARACTERISTICS OF SHEEP

Sheep and goats are closely related. Sheep do not have a beard like billy goats, and lack the divided upper lip that goats possess. Sheep have mobile lips and distinct face and foot glands that secrete an oily substance.

Figure 27-1a & b. Extensive sheep production consists of grazing large flocks of sheep on lands too arid or otherwise unsuitable for crop production.

Most sheep are gregarious (flocking instinct) and this characteristic allows one herder to care for a large number of sheep. However some breeds, and most notably many of the British breeds, are not gregarious.

Sheep are a ruminant animal and therefore can utilize fibrous feeds. Mature sheep have 32 teeth (8 incisors, 12 pre-molars and 12 molars). They have no upper incisors; instead, they have a hard, bony dental pad.

Reproductive rate is variable and influenced by breed, age, nutrition, and the environment. Sheep have the ability to produce multiple births. Sheep are seasonably polyestrus and short day breeders. This means that the breeding season is initiated as the ratio of daylight to darkness decreases (as the days become shorter). Thus, fall is the usual breeding season. Average gestation length is 147 days. Sheep reach sexual maturity from 5-12 months of age depending upon the breed. Most sheep are fully grown by 2 to 3 years of age.

Sheep graze lower to the ground than cattle and are better adapted to steep topography than cattle. In general, goats are more selective than sheep, while sheep are more selective than cattle. Grass is preferred by sheep, whereas goats tend to eat more browse.

SYSTEMS OF PRODUCTION

It is impossible to describe in detail all systems of sheep production. Systems developed have been influenced by the social and economic environment of specific areas. In general, however, most sheep production systems are classified according to intensity of production. A broad classification would be: 1.) extensive and, 2.) intensive.

EXTENSIVE PRODUCTION is based on grazing large areas of relatively unproductive land because it is arid or otherwise unsuitable for crop production. Stocking rates are usually low and flocks are large. Pasture growth is limited by low rainfall or cold, hilly country. Normally minimal handling and housing facilities are available.

INTENSIVE SYSTEMS may include areas where high levels of pasture production can be achieved. In some cases, sheep are used to graze crop aftermath and stubble as a secondary source of income. Where a local market exists, the use of improved pastures are sometimes used specifically for sheep grazing. In that case, the use of high fertility breeds of sheep that breed more than once per year are often utilized. In comparison with the extensive production systems, flock size is lower but stocking rates are much higher.

REPRODUCTIVE CYCLE

Successful sheep production depends on proper management of the reproductive cycle of the ewe to attain maximum production efficiency. Therefore, one must understand and appreciate the ewe's needs as she progresses through the reproductive cycle.

GESTATION

Average gestation length of the ewe is 147 days. For management purposes, gestation can be divided into the first 15 weeks and the last 6 weeks.

FIRST 15 WEEKS OF GESTATION

It is very important to maintain an adequate level of nutrition the first 21 days after the end of the breeding season to prevent embryonic mortality. Ewes should generally be gaining slightly (1 to 2 ounces per day) and do not need to be fat. In cold climates, during the first 15 weeks of gestation, ewes can be maintained on winter range plus ⅓ to ½ pound of a 16% protein grain supplement; 2 to 3 pounds of a good quality hay; 1.5-2.0 pounds of a good quality hay plus 1.0 pound of grain; or 6-7 pounds of corn silage. An appropriate calcium and phosphorous supplement should be provided depending upon the combination of feed ingredients used.

Figure 27-2. Sheep are much more selective grazers than cattle. While sheep are often said to consume much browse, they do very much like grass. As selective grazers they will seek out the newest, and most tender of grasses (regrowth). Because of that, sheep must be tended to and moved often to prevent the over grazing of regrowth and/or the more palatable species of grass.

Fig. 27-3. Sheep are much more adept at grazing hilly rocky areas, than cattle, and are therefore capable of utilizing terrain that would be too steep or rugged for cattle.

TABLE 27-1. BIOLOGICAL CYCLE OF THE EWE.

Period	Duration In Days
First 15-weeks gestation	105
Last 6-weeks gestation	42
Lactation	112*
Post-weaning period	106*
	365

* Variable

LAST SIX WEEKS OF GESTATION
Nutrition

A ewe of average size and condition should gain 10 to 30 pounds by the time she lambs. Most of this gain should occur during the last 6 weeks of gestation. The major factor in the nutrition of the pregnant ewe is that of the unborn lamb. Fetal demand for nutrients is greatest the last 6 weeks of gestation when approximately 70-80% of the growth of the fetus occurs. Inadequate nutrition during the last 6 weeks of gestation may result in lighter, weaker lambs at birth and more death loss. **Ewes in late pregnancy (vs. early pregnancy) require 50% more feed if bearing a single lamb, and 75% more if bearing twin lambs.** If the ewe is fed a high-roughage diet, she will usually not be able to consume enough feed to satisfy her nutrient requirements for increased energy. When feeding a high-roughage ration it is advisable to supplement with ½ to 1 pound of grain during the last 3-4 weeks of pregnancy. In situations where a large number of twins and triplets are expected, it is often desirable to begin graining ewes at the beginning of the last 6 weeks of gestation.

Flushing

Another period in which nutrition is highly important is just prior to breeding. Known as flushing, feeding extra, high quality feed, or putting ewes on the best pasture available can substantially increase reproductive rate. Not only can it increase the total number of lambs for the flock, but it can also increase the number of lambs per ewe. That is, "flushing" can increase the number of twin and triplet lambs.

BODY CONDITION SCORING

The productive ewe or ram in any flock should be neither too fat nor too thin. Body weight alone is an inadequate measure because of differences in mature size between individuals and breeds. Recording both body weight and condition scores is the best system for nutritional decisions. Condition scoring is a system of classifying breeding animals on the basis of differences in body fat. It allows a producer to identify, record and adjust the feed intake of ewes determined to be too thin or fat.

A simple scoring system containing only 5 scores is represented in Figure 27-4. One represents the thinnest and 5 represents the fattest.

There are two methods to evaluate the condition of ewes: (1) hands-on and, (2) visual appraisal. In the hands-on method, the fingers (held together) and thumb are used to determine the sharpness of the spine behind the last rib and in front of the hip bone, and the sharpness of transverse processes. In addition, it may be helpful to determine the extent of fat covering over the foreribs because in many instances the handler may find sharpness over the spine (2.0), but will find considerable fat over the ribs (3.0). Then, one must arrive at some average for the condition score.

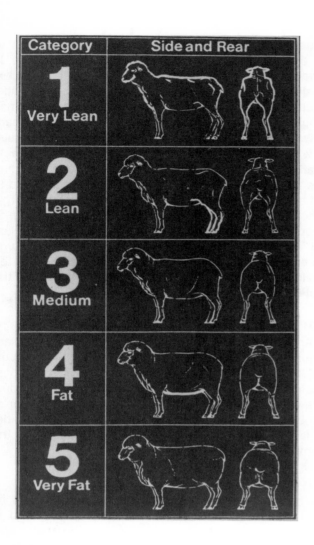

Figure 27-4. A simple system of scoring body condition.

Significance of Body Score

Continual observation is necessary during the dry period. Ewes tend to become excessively fat on little feed during this period. Those that condition score 1.0 to 2.0 at weaning can increase to 4.0+ during this period. When this happens, flushing will be difficult and reduced conception may occur. Therefore, a continual evaluation of ewe condition during the dry period is essential if economic efficiency is to be maintained. Do not let ewes attain condition scores above 3.0 during the dry period.

If the ewe is excessively fat at lambing, dystocia (difficult birth) may occur. If she is excessively thin, low birth weights, low milk production, and reduced survival is likely to occur. An optimum condition score at lambing would be 2.0 to 3.0.

PREGNANCY TOXEMIA

Pregnancy toxemia is also called Twin Lamb Disease and Ketosis. It occurs in ewes improperly fed during late gestation. Affected ewes are almost always carrying twin or triplet lambs. It is similar to ketosis in dairy cows in that the animal develops hypoglycemia (low blood sugar) due to losing more glucose (energy) than she can consume.

In dairy cows the situation usually occurs right after calving, when milk flow is the highest. The cow loses more energy in the milk than she can physically consume in feed.

In ewes, ketosis occurs before parturition. In ewes, ketosis is caused by the extreme demands created by the very rapid fetal growth toward the end of pregnancy. When the ewe is carrying twin or triplet lambs, the nutrient demand for fetal growth is more than the ewe can consume. This causes the ewe to burn off body fat, which can result in the accumulation of metabolites known as ketone bodies. When the concentration of ketones becomes too high, death can result.

TABLE 27-2. A SYSTEM OF CONDITION SCORING EWES.

Point of Evaluation	Condition Score				
	1	2	3	4	5
Spine	Prominent and sharp	Prominent and smooth	Smooth rounded	Detected only as line	Not detectable
Fat Cover	None	Thin	Moderate	Thick	Dense
Transverse processes	Prominent and sharp	Prominent and rounded	Smooth and rounded	Not detectable	Not detectable
Fore Ribs	Prominent	Prominent with slight covering	Smooth indentation of ribs	Slight detection of ribs	Smooth no rib detection

The first symptoms of ketosis are weakness and/or listlessness. The ewe will lag behind the flock, have difficulty getting up, and may have shortness of breath. Upon close examination, she may grind her teeth and the ketone bodies will impart a sweet odor to the breath.

Ketosis, or pregnancy toxemia can usually be prevented by feeding grain or other concentrates during late gestation. (Care must be exercised in supplementing concentrates gradually, over a period of time [pg. 179].)

Treatment is effective, if the condition is caught in its early stages. Treatment can consist of drenching with propylene glycol (a rapidly available form of energy), I.V. feeding with glucose (preferably also with amino acids), and ceasarean delivery of the lambs.

PREGNANCY TESTING

Pregnancy testing is a highly recommended practice in temperate and cold climates. It allows producers to cull non-pregnant ewes, before beginning the winter feeding program. That is, open ewes can be sold before the expense of winter feed is put into them.

A routine practice in cattle herds, pregnancy testing has not been universally accepted in sheep production. The reason has to do with ease of detection. With cattle, the technique is quite simple and inexpensive. Cattle are large enough that a man's arm can be inserted into the cow's rectum and the calf palpated in the uterus (which lies just below the rectum). Because sheep are so much smaller, that technique will not work. As a result, less direct and/or more technologically oriented methods are required.

Marking Harness

This technique involves fitting the ram with a marker around the brisket. Upon mating, crayon marks are left on the ewe's back. Not really a pregnancy determination procedure, this technique simply indicates which ewes have come into heat. However, if a ewe comes back into heat several times (and shows breeding marks), this is an indication that she is not conceiving.

Rectal-abdominal Palpation

This is a positive, direct examination method. With this method, a rod is inserted in the rectum and a sweep motion is made from side to side. Simultaneously, the abdomen is palpated with the hand. Experienced technicians can determine 90-95% of the pregnancies (at about 60 days) with this method.

The disadvantages to this method are that ewes must be fasted overnight, and they must be laid on their back during the examination. Also, with inexperienced operators, there is the potential of puncturing the rectal wall with the rod and creating peritonitis.

Ultrasound and X-ray Techniques

These are high-tech methods that should catch on rapidly as the cost of the equipment begins to decline. Ultrasound, a very common practice in humans, probably holds the greatest potential (due to safety).

Udder Examination

A very old-fashioned technique, this method can be used to detect approaching parturition within 20-30 days. Obviously not effective for reducing winter feed costs, the primary use for this technique is to separate ewes in preparation for lambing.

Twenty to 30 days before lambing, the udder will begin to fill and the teats will feel firm. In addition, a wax plug will begin to fill in the teat opening.

PREPARATIONS FOR LAMBING

As discussed in the previous section, ewes should be "bagged", and those nearest lambing separated, in order that they may be observed more carefully. Likewise, ewes lambing for the first time should be separated, since they are more likely to encounter problems.

Ideally, some sort of facility should be provided for actual parturition. Known as a jug, individual stalls with clean bedding (inside a barn) can help to reduce lamb mortality. (Figure 27-5 exemplifies such a facility.)

Shearing or Crutching

Ewes should be shorn or crutched prior to lambing to provide a cleaner udder for the nursing lamb. Crutching is also called tagging and involves the shearing of wool from the tailhead, vulva area, behind the hind legs, around the udder, and into the flanks. Crutching or shearing reduces soiled wool, improves sanitation, and allows newborn lambs access to the udder and teats.

A number of supplies should be on hand to cope with the various problems and contingencies that occur at this time. A rudimentary list would consist of:

Medication

Propylene glycol (or molasses) for treatment of pregnancy toxemia
K-Y jelly or other lubricant
7% iodine (to disinfect the navel)
Combiotic, 5 or 10cc hypodermic syringe, and no. 18 gauge needles for treatment of pneumonia, navel ill in lambs, or after difficult delivery (other antibiotics may also be recommended)
Uterine boluses (for retained afterbirth)
Vitamin E and selenium injectable for white muscle disease (in areas where selenium deficiency is a problem)
Colostrum from sheep or cow (kept frozen) (for weak

Figure 27-5. Ideally, at lambing time each ewe should have her own jug (individual stall). At the very least, ewes close to lambing should be kept inside a barn with clean, dry bedding (as depicted in Fig. 27-6).

or orphan lambs)
Mastitis ointment or bag balm
Combination of glucose, electrolytes, and amino acids for very weak, dehydrated lambs. Alcohol, tame iodine, or other similar disinfectants

Equipment

Lambing pens with water buckets (approx. 4 ft by 4 ft.)
Bedding materials
Heat lamps
Bottles and nipples (for weak or orphan lambs)
Thermometer (normal sheep temperature is 102.3 degrees F)
Old towels or burlap sacks (to wipe off and dry newborn lambs)
Ear tags and paint brands (for identification)
Lambing record book
Twine for use in difficult delivery and/or tying off umbilical cord. Lamb puller or snare (to correct malpresentations of the head)
Marking crayon (for short term identification)
Suturing material
Docking irons (for docking tails)
Plastic disposable gloves - for assisting ewes with difficult births
Shears (for tagging, if not previously crutched or tagged)
Bearing retainers (for treating prolapses)

LAMBING

Lambing is a labor intensive time. Proper lambing management is necessary to maximize the numbers of lambs produced per ewe. A good understanding of the birth process is a necessary prerequisite for a successful lambing season.

Figure 27-6. A group of Suffolk ewes brought into a barn prior to lambing. Note the fresh straw placed as bedding.

Birth process

Ewes that are close to lambing will show the following visible signs:
1. Restlessness
2. Frequent urination
3. Inflamed udder, vulva or extended genitalia
4. Isolation from the rest of the flock
5. Lying down with head extended and obvious straining

A normal presentation of the fetus is with the front legs forward and the head resting on or between the legs. The birth process normally takes 4 to 5 hours with first lambers requiring more time than ewes that have previously lambed. Three to four hours are required for dilation of the cervix and approximately one hour for delivery of the lamb. After the ewe is in hard labor, the nose and front feet of the lamb should appear (Figure 27-7). Birth should occur within one half hour to forty-five minutes after rupture of the water sac. If the nose and front legs of the lamb do not appear after 15-20 minutes of hard labor, a careful examination should be made to determine the position of the lamb.

If a ewe has difficulty, find out whether the lamb is being delivered in a normal presentation. Difficult lambing may be due to a large lamb, a ewe with a small pelvic area, lack of cervical dilation, or an abnormal presentation.

POST-LAMBING MANAGEMENT

Newborn lambs should have the mucous and membranes removed from around the nose and face (otherwise breathing may be hampered and the lamb can suffocate). The ewe's teats should be checked to see if the teat canals are open. If they are closed, remove the wax plug that commonly obstructs the teat canal. If the lambs are weak or fail to nurse, assist them in suckling or use a drenching tube to insure colostrum is consumed (colostrum is the first milk a ewe produces after lambing). Survival of the newborn is dependent on adequate colostrum consumption. Colostrum is rich in protein, energy and provides antibodies that protect the lamb against disease. Lambs should be given 1-2 ounces of colostrum at the first feeding and 4-6 ounces in subsequent feedings.

It is a recommended practice to clip the navel cord about one and one-half inches from the navel and apply tincture of iodine (Figure 27-8). It is best to do this by putting the iodine in a wide mouth jar about the size of a vaseline jar. Holding the lamb by its front legs, place the jar over the navel cord and hold it tightly against the navel. A few good shakes of the bottle will cover the navel and cord sufficiently to prevent most cases of navel ill.

Ewes with healthy single lambs normally only need to remain in a jug pen for 24 hours. Ewes with healthy twins or triplets can be removed from the jugs after 48 to 72 hours.

There is an occasional newborn lamb that is not brought into the barn before it becomes chilled, or that may have become wet and chilled during the first few days of life. A vigorous rub with a towel or gunny sack often helps to increase the blood circulation and to revive the lamb. However, it is best to get some other form of moderate heat as soon as possible.

Take lambs from ewes that will not accept them or are unable to provide enough milk. Dispose of these ewes at the earliest opportunity. Lambs that are orphans should be grafted to a ewe that has lost her lamb or to a heavy milking ewe. There are many lamb grafting methods but the most common is to place the pelt of a dead lamb

Figure 27-7. A normal presentation. Normal position for the lamb is head first, between the legs in a "diving" position. This is the position of least resistance. If the lamb is in another position delivery will be difficult, and assistance may be required.

Figure 27-8. Dipping the naval cord in iodine. At birth, the naval cord should be cut about 1½ inches from the naval and dipped in tincture of iodine.

over the lamb to be grafted for a few days, until the dead lamb's mother "takes" the orphan lamb. Household deodorizers and many other methods are used to varying degrees of success.

After lambing, the ewe should be given all the clean water and good quality hay or hay pellets she desires for the first few days. However, grain should be fed sparingly to avoid swollen udders or tender teats caused by stimulating too much milk (mastitis is a major problem in sheep). After the lambs are 24 to 48 hours old, the ewe can receive some grain. Recommendations are 1.0 pound of grain per ewe with a single lamb and 1.5 to 2.0 pounds of grain per ewe with twins (Table 27-3). If possible, pens should be arranged so that ewes nursing single lambs can be separated from those with multiple lambs.

Milk production during the last 8 weeks is much lower than the first 8 weeks of lactation. Consequently, the ewe's nutrient requirements are also lower during this phase of lactation, versus the first 8 weeks.

DOCKING AND CASTRATION

Dock and castrate all lambs by the time they are ten days to two weeks of age. Both operations can be performed at the same time. With an elastrator and rubber bands, this can be done as early as two or three days of age. Place the elastrator band on the tail, one to one and one-half inches from the body. For castration, place the rubber ring over the scrotum, making sure both testicles are down and outside of the elastrator ring. The ring cuts off the supply of blood and causes the tail and testicle to fall off. If a knife, or other means of docking is used, the lambs should be a few days older. Cut off the tail in the same place you would place an elastrator ring. Watch for excessive bleeding and stop it by using astringent powders or tie the tail stump with a string for an hour or so. Be sure to remove the strings within an hour. Hot irons are another method that can be used (by cauterizing the wound). A rapid heating soldering iron works very nicely for this job.

To castrate a lamb with a knife, pull the hind legs up toward the front shoulders. Place the rump of the lamb on a table or platform and cut off the lower one-third to one-half of the scrotum. Squeeze the testicles out one at a time by catching them in the crook of the forefinger,

TABLE 27-3. RATIONS FOR A 154-POUND EWE USING ALFALFA HAY AS THE FORAGE.[1]

Ingredient	1st 15-wk gestation	last 6-wk gestation	1st 8-wk lactation	last 8-wk lactation[2]	post-weaning
	(lb)	(lb)	(lb)	(lb)	(lb)
Alfalfa hay	3.3	4.1	5.0	5.5	2.8
Grain	---	.75	2.0	1.0	---
Total	3.3	4.85	7.0	6.5	2.8

[1] Quantity of feed expressed on an as-fed basis.

[2] Some producers may discontinue grain feeding during the last 8 weeks of lactation.

using the thumb to keep them from slipping. Then cut the cord by using the knife in a scraping motion.

Be certain all equipment used in these operations is clean and put the lambs into a clean, well-bedded pen immediately. Use good disinfectants on all wounds and repellant if there are flies.

CREEP FEEDING

Creep feeding is a common practice with intensive systems of production. If lambs are to be early weaned, creep feeding is almost essential.

Lambs will begin to nibble at grain and hay when they are about a week old, so access to creep feeders can begin 7 to 10 days after birth. Although lambs will eat only small amounts the first 3 to 4 weeks, early creep feeding will establish both rumen function and the habit of eating. In most situations, lambs will eat about 0.5 pound of a creep ration per day at 20 to 30 days of age, and from 1.0 to 1.5 pounds per day at 40 to 50 days.

TABLE 27-4. INGREDIENT AND CHEMICAL COMPOSITION OF CREEP RATIONS FED UNDER VARIED CONDITIONS.[a]

Item	Ration 1[b]	2[c]	3[c]	4[c]
Ingredients %				
Corn	67.5	--	--	58.5
Oats	--	37.0	30.0	30.0
Barley	--	38.0	70.0	--
Beet Pulp	--	25.0	--	--
Soybean Meal	15.0	--	--	10.0
Alfalfa Hay	15.0	--	--	--
TM Salt	.5	--	--	.5
Limestone	2.0	--	--	1.0
Chemical Composition %				
Crude Protein	15.0	10.1	10.1	13.2
TDN	72.0	68.2	69.4	74.9
Calcium	.9	.2	.04	.4
Phosphorus	.3	.2	.3	.3

[a] As-fed basis.
[b] Ration fed as a meal after being ground through a ½" screen in a grinder mixer.
[c] Feed the highest quality, alfalfa hay in a separate rack. Offer trace mineralized salt free choice, plus dicalcium phosphate (2:1 ration of TM salt to dicalcium phosphate).

The feed composition of a creep ration will vary depending upon the availability and relative cost of feed ingredients and the anticipated market date for the lambs. A creep ration does not need to be complicated. Many producers successfully use rolled or whole oats as their grain creep, plus good quality alfalfa hay. However, more cost effective rations can be formulated.

Wean lambs on the same ration that is fed during the pre-weaning period. The best way to combat "weaning shrink" is to be sure the lambs have adjusted to weaning. Therefore, leave them on the same ration 14 to 21 days after weaning.

WEANING MANAGEMENT

The proper age to wean lambs depends primarily on the system of management. Typically, it is more efficient to wean at an early age. The primary reason for this is that milk production usually reaches a peak approximately four weeks after lambing and decreases thereafter. Three or four months after lambing, most ewes will be producing very little milk and it would be more economical to wean the lambs and turn the ewes out to pasture or at least to decrease their daily ration. Many flock owners now wean lambs at 60 days of age with good results, and some are weaning lambs successfully at younger ages. Successful early weaning depends upon how well the lambs are eating supplemental feed (creep) at the time they are weaned.

DRYING UP EWES

Special care must be taken in drying up ewes that have had their lambs weaned. To minimize udder problems, milk flow must be reduced by immediate and drastic limitation of both feed and water for a brief period. Suggested steps for drying up ewes are:

1. Take ewes completely off feed and water 12 hours before weaning the lambs, but allow lambs continued access to creep feed. (Creep feeding lambs on the diet they will be on immediately after weaning is important for good lamb performance).
2. Wean the lambs early in the morning. Move ewes completely away from the lambs to a dry corral of poor quality pasture or range without water.
3. After 24 hours without water or feed, give the ewes water once in the morning. Give them a light feed of poor quality roughage such as straw or low quality, grass hay, or continue them on poor quality pasture or range.
4. Continue the ewes on once-per day watering and a restricted quantity of low quality feed or pasture for about 1 week.

POST-WEANING NUTRITION FOR THE EWE

This is a time of rest for the ewes. In order to realize the effect of flushing, ewes must not be allowed to become fat.

Most ewes are overfed during this period. Mature ewes should be fed so they only gain back the weight they lost during lactation. A ewe can be maintained on about 2-1/2 to 3 pounds of alfalfa hay per day. If on pasture, it is sometimes beneficial to limit the time the ewe can graze (i.e., two hours in the morning and two hours in the evening). Also, it is desirable to utilize the poorer quality pasture or feed during this stage, to save the better quality forage for the periods of the production cycle which are nutritionally more critical.

POST-WEANING MANAGEMENT OF LAMBS

Management of lambs after weaning will depend upon the age at which the lambs are weaned and the quality of pastures available.

In extensive operations utilizing semi-arid, marginal rangelands, lambs must be either shipped to commercial feedlots or fed special rations at the ranch (or station). This is because high quality pastures are required for weanling lambs. Marginal lands, such as in much of the western U.S. and Australia, can support bands of ewes but are totally unsuited for growing young lambs.

If the lambs are to be fed, several options are available. If the lambs have been creeped, the creep ration plus free choice alfalfa hay is a simple, yet safe option. Lambs are adjusted and adapted to the ration; therefore, the chance of digestive disturbances is minimized. At the University of Idaho, post-weaning lambs fed free choice hay plus grain will consume, on the average, a ration that is 60 to 70 percent grain and 30 to 40 percent forage.

Table 27-5 is a guide for hand-feeding lambs a simple diet of shelled corn or rolled barley, supplement, and hay. It is based on the feeding of alfalfa hay as 20 percent of the daily ration, feeding a 34 to 36 percent protein supplement at a constant half-pound per lamb per day, and increasing the corn or barley as the lambs get heavier. This program will meet the major nutrient requirements of growing-finishing lambs. Protein levels are reduced consistently with the lower requirements of the lambs as they get older and heavier. Lambs should be fed twice daily with the grain and protein supplement fed first, followed by the hay.

Feeding a complete ration in a pelleted form is another option. Pelleting, although costly, improves palatability, rate of passage and feed intake. Pelleting also insures that lambs will eat grain and roughage in the desired proportions. Table 27-6 demonstrates a simple pellet formulation developed at the Univ. of Idaho.

BREEDING MANAGEMENT - EWES

Ewe Lambs

The post-weaning nutritional needs of replacement ewe lambs vary with the lambs age at first breeding.

TABLE 27-6. INGREDIENT COMPOSITION OF UNIVERSITY OF IDAHO SHEEP PELLET (DM BASIS).

Ingredient	%
Alfalfa hay	58.5
Barley	20.0
Oats	20.0
TM Salt	1.0
Ammonium Chloride*	.5
	100.0

*The ammonium chloride is used as a prophylaxis for urinary calculi (kidneystones).

TABLE 27-5. FEEDING GUIDE - SIMPLIFIED RATION.

Daily Consumption (lb)	Grain (lb)	Calculated CP* (%)	Supplement (lb)	Calculated TDN (%)	Alfalfa hay (lb)	Calculated Ca (%)	Calculated P (%)	Approx. Lamb Wt. (lb)
2.0	1.1	18.4	0.5	73.4	.4	.85	.46	50
2.5	1.5	16.7	0.5	75.4	.5	.73	.43	70
3.0	1.9	15.6	0.5	76.1	.6	.65	.40	80
3.5	2.3	14.8	0.5	76.6	.7	.60	.39	90
4.0	2.7	14.2	0.5	77.0	.8	.55	.37	100
4.5	3.1	13.8	0.5	77.3	.9	.52	.36	105

*CP = crude protein

Under range conditions, producers generally do not attempt to breed ewe lambs. With this management schedule, nutrition is not nearly so critical. The ewes generally go through a green feed season before breeding, and will usually gain enough weight during this period to breed as yearlings.

In intensive systems ewes are usually bred at 12 to 14 months of age. If this is to be done successfully, nutrition is critical. (Finn ewes, in intensive production can be bred as early as 7 months.) As a general rule, ewe lamb fertility will increase as their weight increases. **In most situations ewe lambs must weigh at least 90 pounds and be gaining ½ pound daily to breed successfully.** (If the ewe lambs are part Finn breeding, they will usually breed at slightly lower weights.)

Mature Ewes

The practice of "flushing" can influence the number of eggs a ewe ovulates. Flushing is accomplished by placing ewes that are in fair to poor condition on an increased plane of nutrition for 2 to 3 weeks before the start of the breeding season. Some sheep producers feel they get a better flushing response by increasing the plane of nutrition the day the rams are turned into the breeding pasture. Giving extra feed to ewes already in good condition does not usually result in the desired response. Thus, fat or fleshy ewes at breeding time will not respond to flushing as well as ewes in average or thin condition.

Methods of flushing vary between flocks. If ewes are on a pasture that is not green and lush before the breeding season, changing them to an excellent or good pasture 2 to 3 weeks prior to breeding will usually provide the desired response. If better pasture is not available, the flushing response can be obtained from a wide variety of supplemental feeds - corn, oats, barley or sorghum fed at the rate of ½ to ¾ pound per day.

Ewes should be tagged prior to the start of the breeding season. Tagging means trimming the wool around the dock area so the genital area will be exposed for the ram.

RAM MANAGEMENT

AFTER BREEDING

Cull healthy rams that no longer fit into the breeding program. Mature rams should be put on a maintenance ration, while ram lambs and yearlings should be fed better to promote proper growth and development. Good quality hay (or pasture) plus 1 to 2 pounds of a grain supplement should be adequate for ram lambs and yearlings. Rams should be treated for internal and external parasites.

Non-vaccinated rams should be vaccinated for epididymitis at least two months before breeding and a second shot be given 4 - 6 weeks later. Previously vaccinated rams should be given an annual booster 4 - 6 weeks prior to breeding. Epididymitis is a bacterial infection that causes the epididymis to swell and the testes to shrink, resulting in infertility. Palpation of the testis by an experienced shepherd can detect rams that have had epididymitis.

LAST MONTH BEFORE BREEDING

Evaluate the condition of all rams and adjust the feeding program based on body condition. If rams are in poor body condition, it is advisable to supplement the ram's forage based diet with some grain. Each ram's reproductive system should be evaluated and the semen tested. All rams should be shorn if breeding in warm weather. All feet should be trimmed.

Number of Ewes Serviced by a Ram

A mature ram can service 25 to 30 ewes. In intensive systems, it is not uncommon to see a ram to ewe ratio of 1 to 50 and occaisonally as much as 1 to 75.

BREEDING SYSTEMS AND SELECTION

Sheep producers should select for a minimum number of traits if they wish to make progress (see also pg. 213). Traits selected should be of economic importance. The selection of the proper breed of sheep for existing environmental conditions is very important.

Traits that should receive emphasis in nearly any selection program are:

1. Prolificacy
2. Growth potential
3. Carcass merit
4. Fleece traits

A practical means of selection for improving the above traits is outlined below:

PROLIFICACY

1. Identify twin or triplet lambs and select replacement ewe lambs from this group. Twin lambs from young ewes would have a greater potential for twinning than twins from older ewes.
2. If additional replacements are required, select single-ewe lambs from young ewes.
3. Use twin-born rams or rams from ewes with a high level of twinning throughout their lifetime.

GROWTH

1. Weigh all lambs at weaning. If records are available, correct weights for age, type of birth (single or twin), sex, and age of dam. **In most selection programs, lambs from twin births will all be held for breeding purposes.**

Among single births, the heaviest (after age adjustment) will be held. Under range conditions, it many not be necessary to adjust for age (biggest lambs came from the ewes that conceived the earliest).
2. Obtain daily weight gain during a post-weaning period during which lambs receive uniform treatment. At least a 60 day feeding period is required.

CARCASS MERIT

Carcass merit selection is similar to selection for growth. That is, select primarily for rate of gain, as rate of gain will favor protein deposition rather than fat deposition.

FLEECE TRAITS

Reports indicate there is a negative correlation between fleece traits and carcass traits. Fleece weight and staple length are negatively related to meatiness but positively related to fatness. Therefore selection for fleece will retard carcass traits. Therefore the trait with the greatest market value would be given the most emphasis. As a general rule, in the U.S. carcass traits will be of more economic value and in Australia and New Zealand fleece traits will be more valuable.

BREEDING PROGRAMS

Breeds are used in three ways to influence the genetic makeup of the breeding population.
1. Corrective breeding
2. Grading up
3. Crossbreeding
4. New breed formation

Corrective breeding can be used to introduce a trait that the flock is deficient in. In other words, if a flock is deficient in some characteristic, rams from a breed that has excelled in that particular trait may be introduced. This same corrective breeding technique can be used within a breed and is termed "outcrossing"; that is, utilizing specially selected lines within a breed.

Grading Up

This is the practice of mating purebred sires of one breed to females of another breed. It is a method of replacing one breed by another which is regarded as being superior. After a number of generations, the percentage of the original breed in the replacement ewes will be quite low, and the new breed quite high.

Crossbreeding

Systematic crossbreeding as a mating system for sheep production will be the best mating system in most commercial situations. This is due to the phenomenon of heterosis or hybrid vigor (see pg. 211).

When several breeds are combined in a systematic crossbreeding program, the manner in which the breeds are used becomes of considerable importance. The adapted crossbred ewe is more fertile than a straightbred ewe. Therefore, the use of as many crossbred ewes as possible in commercial flocks is usually a sound decision. If crossbred ewes are going to be used and mated to rams of an unrelated breed, a minimum of three breeds must be involved. It would seem wise then to develop in certain ewe breeds those characteristics which contribute to making a highly productive female. The actual breeds selected would depend upon the

Figure 27-9. For use in tropical and subtropical areas, special haired and even hairless breeds of sheep have been developed to avoid the heat stress endured by wooled breeds.

climatic conditions of the area as well as the local market conditions.

Traits to be emphasized (in a crossbreeding system)

The traits to be emphasized in a crossbreeding system are basically the same as in a pure-breeding system, except that separate traits for the ewe and ram breed may be specialized. For example, in a terminal cross situation (where all the lambs are sold and none are kept for breeding), there is no need to be concerned with such traits as maternal instinct or milking ability.

Another example might be in cold climates where heavy wool is required for winter ewe survival, but wool growth is not as economically important as carcass quality in the lambs (remember, selection for wool quality retards progress in carcass characteristics). In that case, if the lambs are born in the spring and sold in the summer, rams with exceptional carcass characteristics could be used on ewes with heavy wool characteristics. Therefore, by being able to select the ram and ewe breeds separately more specialized flocks can be developed.

TYPES OF BREEDS

Breeds of sheep are normally classified according to their commercial use.

1. Ewe Breeds: Recognized as the white-faced sheep producing fine, medium, and long wool or crosses among these types. Ewe breeds are selected for adaptability to environmental conditions, reproductive efficiency, wool production, size, milking ability, and longevity. Replacement ewe lambs should be raised from these breed types or from crosses among these breeds.
2. Ram breeds: Recognized as meat-type breeds, or crosses of two of these breeds. Ram breeds are selected for growth rate and carcass qualities.
3. Dual Purpose Breeds: Recognized as breeds that may be used as ewe breeds or ram breeds, depending upon the production situation.

The following table (Table 27-7) attempts to classify the more common breeds of sheep in the United States based on expected performance levels assuming optimum environmental conditions. Photos, and a brief description of some of the more important breeds of sheep, are found at the end of the chapter.

COMMON SHEEP DISEASES AND DISORDERS

ENTEROTOXEMIA

Enterotoxemia, or "overeating disease" is a very common and costly disease of sheep. It is caused by a toxin produced by the bacteria Clostridium perfringens type C and D. The disease is most prevalent in lambs nursing heavy milking ewes or lambs fed high grain diets. Incidence can be as high as 40 to 50% in non-vaccinated lambs. Prevention by vaccination is the best means of control. Ewes not previously vaccinated should be vaccinated 4 to 5 weeks prior to lambing and a booster should be given to ewes previously vaccinated. Lambs will be protected by colostral antibodies for 3 to 5 weeks. Lambs born to ewes that have not been vaccinated should be vaccinated during the first week of life with a

TABLE 27-7. CLASSIFICATION OF BREEDS.

FUNCTION	BREEDS
Foundation Ewe Breeds	Rambouillet, Merino, Columbia, Corriedale, Targhee
Breeds used largely in crossbreeding programs in which the production of F_1 ewes is an important goal.	This breed should be one that has good growth rate and sires the best F_1 ewes. Large Dorsets would generally fit into this classification. Hampshire and Suffolk rams are probably more commonly used for this purpose, but ideally they should be used for the terminal cross. On a world basis, the Leicester is widely used for this purpose.
Sires of Market Lambs Only	Suffolk, Hampshire, Shropshire, Southdown, etc.

Source: Montana Sheep Producers Handbook.

½ dose followed by a full dose of the vaccine at 5-6 weeks of age.

INFECTIOUS SCOURS

Infectious scours is a very common problem in sheep raised under intensive conditions. Lambs that have scours will be weak, feverish, dehydrated, will tend to stand humped up, and will have a posting of feces around the tail. Proper sanitation is important in preventing an outbreak of scours since the more severe cases of scours are caused by a bacterial contamination (usually *E. coli*). Treatment for scours is a combination of replacing fluids lost with electrolytes and the use of antibiotics.

VAGINAL AND UTERINE PROLAPSES

Prolapses tend to be of genetic origin. They can involve only the vagina being forced out (usually before lambing) or the vagina and the uterus (usually after lambing). If the vagina moves in and out, it may only be necessary to treat with a topical antibiotic to prevent infection. However, if the vagina comes out and stays out, it may need to be stitched. A bearing (vaginal) retainer can be used as an alternative to stitching. A bearing retainer is a plastic tool that is inserted into the vagina and holds in the vagina by applying inward pressure. It is held in place by being tied to the wool near the forward part of the hind leg. The advantage of a bearing retainer is that no surgery is required. Lambs are either born over the retainer or the retainer is forced out during labor.

As mentioned, uterine prolapses normally occur after lambing. Correction of a uterine prolapse is similar to that described for a vaginal prolapse. All ewes that prolapse should be culled.

ABORTIONS

The two major causes of abortion in sheep are Vibriosis and enzootic abortion (EAE). Both normally occur in late pregnancy and can be prevented by vaccination and good sanitation.

MASTITIS

Mastitis is a severe inflamation of the mammary gland caused by bacteria, trauma, wounds, and soremouth lesions. The first sign may be lameness or refusal to allow lambs to nurse because of udder soreness. Ewes should be immediately isolated and lambs removed and fed artificially.

Systemic antibiotics are necessary and intramammary medication may be helpful. Pre-lambing shearing allows better observation of udders and earlier detection of problems, as well as reducing bacterial contamination. Other prevention includes removing obstacles (in pens, corrals, etc.) that could cause physical injury to the udder.

WHITE MUSCLE DISEASE

Nutritional muscular dystrophy or white muscle disease is a common problem in the northwestern part of the U.S., as well as parts of Australia and New Zealand. A combined deficiency of selenium and Vitamin E, the condition causes degeneration of the skeletal and cardiac (heart) muscles. Lambs exhibit stiffness and rapid breathing; sudden death may sometimes occur from heart failure. Symptoms can resemble those of several other lamb diseases. Consequently, an autopsy is needed to determine precisely which condition is present. Upon removing the skin, the muscles will be white or streaked with white, and it is very distinctive.

Prevention is the most effective route to follow. Feed supplementation of selenium at 0.1 ppm is now allowed by the Food and Drug Administration. Salt and mineral supplementation up to 3 ppm is allowed as an alternative. Another common method of prevention is to inject lambs subcutaneously with selenium-vitamin E at birth and again at 30 days of age.

Ewes may be injected with Vitamin E and selenium three to six weeks before lambing. Satisfactory amounts will be transferred to the unborn offspring and into the colostrum.

INTERNAL PARASITE CONTROL

Internal parasite control is a constant battle. The most common way to control internal parasites is to interrupt their life cycle and thereby prevent infective larvae from gaining access to the sheep. In general, the more concentrated the sheep and the higher the rainfall, the greater the accumulation of internal parasites.

Flock owners need to develop a specific program for their flocks. Total numbers, facilities, management practices, goals, and available acreage all must be taken into consideration when developing a worming program. In some parts of the world it is necessary to worm as often as 4 or 5 times per year while some flocks may only need worming once per year. Veterinarians can do periodic fecal examinations to determine the necessity of worming.

BASICS OF WOOL GRADING

Certain qualities such as fineness, length, color and appearance, etc., determine the end use and value of wool. Fineness, the fiber diameter and its distribution, is one of the most important of these quality factors. It largely determines whether the wool is used in a suit, sweater, blanket, or pair of socks.

GRADE

In general, grade refers to the average diameter or thickness of the fibers. Three systems of wool grading are commonly used in the United States: 1) American or

Blood system; 2) the English or Spinning Count system; and 3) the Micron system. All three systems are measures of average fiber diameter and can be related to each other (Table 27-8).

The American or Blood System. The American system of grading wool was developed in the early 1800s when the native coarse-wooled sheep were being bred to fine-wooled Merino rams imported from Spain. It assumes that the offspring of the cross would have fleece that was intermediate in fineness between the two parents. The wool grade is defined as the percentage of Merino blood carried by the sheep that typically would produce a particular fineness of wool. The grade or fiber diameter came to be expressed as fine, 1/2 blood, 3/8 blood, 1/4 blood, low 1/4 blood, common, and braid. Today these terms are not as exact as the trade would prefer, and the spread within a grade is too wide for the purposes of most wool manufacturers.

The English or Spinning Count System. The English system of grading wool provides narrower ranges and a more exact nomenclature than the American system. It uses a measurement called the "Spinning Count" and is based on the number of "hanks" of yarn which could be spun from one pound of clean wool on the equipment available at the time the system was developed. As wool becomes finer, more hanks or yards of yarn can be spun from a pound of clean wool, and the spinning count becomes larger. In theory, one pound of clean 62's spinning count wool, could produce 62 hanks or 104,160 feet of yarn. (A hank of yarn is 560 yards in length.) Although wool is seldom spun to its maximum count, there is a limit to the number of fibers which will hold together in yarn. The English or Spinning Count system of grading wool provides a numerical designation of fineness. English or Spinning count grades of wool commonly used in the United States today are: 80's, 70's, 64's, 62's, 60's, 58's, 56's, 54's, 48's, 46's, 44's, 40's, and 36's (figure 27-8).

The Micron System. Increased emphasis on an exact and highly descriptive method of describing wool grade has produced a measuring system in which individual fibers are accurately measured. The unit of measure is the micron, which is one millionth of a meter or 1/25,000 of an inch. Fineness is expressed as the mean fiber diameter. Eventually, this system probably will become the standard for describing wool in the United States.

DISTRIBUTION OF GRADE OR FIBER DIAMETER

The use and value of a fleece or lot of wool is affected almost as much by the distribution of the individual fiber diameters as the average fiber diameter or grade. The more uniform the individual fibers are in diameter, the more valuable. An average fiber diameter or grade im-

TABLE 27-8. RELATIONSHIP BETWEEN THE AMERICAN, ENGLISH AND MICRON SYSTEMS OF GRADING WOOL.

Type of Wool	American or Blood Grade	English or spinning Count Grade	Micron (Range in Avg. Fiber Diameter)	Variability Limit for Standard Deviation Maximum (microns)
Fine	Fine	Finer than 80s	Under 17.70	3.59
Fine	Fine	80s	17.70-19.14	4.09
Fine	Fine	70s	19.15-20.59	4.59
Fine	Fine	64s	20.60-22.04	5.19
Medium	1/2 Blood	62s	22.05-23.49	5.89
Medium	1/2 Blood	60s	23.50-24.94	6.49
Medium	3/8 Blood	58s	24.95-26.39	7.09
Medium	3/8 Blood	56s	26.40-27.84	7.59
Medium	1/4 Blood	54s	27.85-29.29	8.19
Medium	1/4 Blood	50s	29.30-30.99	8.69
Coarse	Low 1/4 Blood	48s	31.00-32.69	9.09
Coarse	Low 1/4 Blood	46s	32.70-34.39	9.59
Coarse	Common	44s	34.40-36.19	10.09
Very Coarse	Braid	40s	36.20-38.09	10.69
Very Coarse	Braid	36s	38.10-40.20	11.19
Very Coarse	Braid	Coarser than 36s	Over 40.20	

plies nothing about its distribution within a given quantity of wool. For instance, a lot of wool may contain 50 percent of its fibers at 35 microns and the other 50 percent at 15 microns and still have the same average fiber diameter, 25 microns. Thus, it would be the same average as a wool with only 1 percent of its fibers at these extremes. In recent years, a measure of the amount of variation in fiber diameter within a given lot or fleece has been added as a quality factor.

CLEAN WOOL YIELD

The pounds of clean wool produced is one of the most important factors in evaluating a fleece. Clean wool is the amount of wool remaining after the dirt and non-wool material (shrinkage) has been removed. Expressed as a percentage it is 100% - Shrinkage = Yield. Expressed as a scoured wool, yield is the pounds of wool remaining after washing or scouring.

STAPLE LENGTH

Another important factor used in judging the value of wool is staple length. As a general rule, the longer the staple length the higher the value. Exceptionally long staple length, however, is of no advantage to the manufacturer.

Wool has to be classified into various lengths. This is because varied lengths of wool fiber are used in different types of fabrics or products, and usually are processed on different types of machinery.

WASTINESS

There are several things to look for in determining the wastiness or "noilage" of a fleece. Noils are the short fibers that are combed out in processing, and are a low value product with limited use.

Low tensile strength is the greatest contributing factor in increasing wastiness. Some fleeces have locks of wool fibers that pull apart very easily and in a definite area or section of the lock. These are called "breaks". In judging or evaluating fleeces, particular attention is paid to the area near the base of the fibers for breaks that have occurred during late winter or the lambing period.

PURITY

Purity is the degree of freedom of a fleece from fibers other than true white wool fibers. Black fibers are penalized very heavily. They are almost as long as the white fibers, but somewhat smaller in diameter. Look for black fibers in blackfaced or blackfaced crossbred sheep. Black and brown leg clippings are not desirable in a fleece, but they are more easily removed in processing than black fibers.

Also penalized are very coarse and hairy fibers, because they are medullated or hollow in part or all of the fiber. Medullated fibers do not have the same dyeing characteristics that true wool fibers have.

CHARACTER AND COLOR

Character is the general appearance and "handle" or "feel" of a fleece. Crimp, or the waviness found in wool fibers, is one of the more important factors indicating character. Distinct and uniform crimp is more valuable to the wool processor because it handles and spins more readily with less yarn or fiber breakage.

Fine wool has more crimp per inch than does medium wool, and medium wool has more crimp per inch than coarse wool. Fleeces should be soft to handle and free from harsh and frowsy ends.

The most desirable color of scoured wool is white. However, many fleeces have a light yellow or cream color that will scour or wash out and leave the wool white. Shades of gray, darker yellow or dark stains are discounted heavily.

BREEDS OF SHEEP

EWE BREEDS*

Border Leicester: The Border Leicester originated in England from Leicester X Cheviot crosses. Found in the U.S. and Canada, the breed is particularly adapted to high rainfall, good pasture areas. Border Leicesters are prolific, good milk producers, and useful in crossbred lamb production with good size and body length. This white-faced breed, with bare head and legs, yields a long, heavy wool fleece that spins well.

Figure 27-10. A Border Leicester ewe. Courtesy American Sheep Producers Council.

* Those breeds normally considered best for use as ewes in crossbreeding programs.

Corriedale: Corriedale sheep originated in New Zealand using Lincoln and Leicester X Merino crosses. Imported into the U.S. in 1914, Corriedales are well-adapted to farm flocks where feed is abundant, but may be used in range production. Corriedales are intermediate to large sized, white-faced with wool on the legs, produce desirable crossbred market lambs, are gregarious and early maturing, and yield heavy, high quality, medium wool fleeces with exceptional length, brightness, softness and crimp.

Figure 27-12. Delaine-Merino ewe. Courtesy American Sheep Producers Council.

Figure 27-11. A Corriedale ewe. Courtesy American Sheep Producers Council.

Debouillet: The Debouillet breed was developed in New Mexico in 1920 from Delaine-Merino and Rambouillet crosses. Well-adapted for Southwest range sheep production, the Debouillet is a medium sized, white-faced sheep with wool on the legs. They are hardy under arid conditions, gregarious, adaptable to unassisted pasture lambing, and produce a high quality fine-wool fleece with a deep, close crimp.

Delaine-Merino: The closely related Delaine-Merino and Texas Delaine were developed from the Spanish Merino having an unbroken line of breeding 1200 years old. The modern Delaine-Merinos are relatively smooth-bodied, intermediate sized, white-faced with wool on the legs. They are hardy, long-lived, and gregarious. Adapted for unassisted lambing, they produce well in extremely warm climates under relatively poor feed conditions. Capable of breeding year round, they produce a high quality fine-wool fleece.

Finnsheep: A native of Finland, the breed thrives in

Figure 27-13. Finnsheep ewe. Finnsheep are noted for high prolificy, and are the foundation breed in many crosses. (Finnsheep are capable of "littering" 3 and even 4 lambs.) Courtesy American Sheep Producers Council.

rugged climates on high roughage feeds. Imported into Canada in 1966, the Finnsheep later came to the U.S. where an association was formed in 1971. Finnsheep are adaptable to intensive management, are highly prolific, excellent milkers, and easy lambers. They have excellent maternal instincts and can be bred at six or seven months, are often used in crossbred replacement ewe production, and yield a medium grade wool.

Rambouillet: Developed from the Spanish Merino in France, the Rambouillet is the foundation of most western range flocks. The Rambouillet is a large, white-faced sheep with wool on the legs, rugged, fast-growing, long-lived, gregarious, and adaptable to various climatic and forage conditions. They are considered one of the best sheep for breeding year-round, and produce a high quality, fine-wool fleece.

Targhee: Developed in 1926 by the U.S. Sheep Experiment Station, Dubois, Idaho, the Targhee has ¾ fine-wool and ¼ long-wool breeding from Rambouillet X Lincoln, Rambouillet X Corriedale, and Columbia crosses. The Targhee is intermediate to large sized, white-faced with wool on the legs, durable, and adaptable to varied climate and forage conditions. The breed herds well, produces a high percentage of twins under range conditions, and crosses well to produce desirable market lambs. They yield a heavy medium fleece excellent for spinning.

Figure 27-14. A Rambouillet ewe. Courtesy American Sheep Producers Council.

Figure 27-16. Targhee ewe developed at U.S. Sheep Experiment Station, Dubois, Idaho in 1926. Courtesy American Sheep Producers Council.

Figure 27-15. A group of exceptionally well bred Rambouillet rams. Note the exceptional size of the rams. Cunningham Sheep Co., Pendleton, Oregon.

RAM BREEDS

Cheviot: One of the oldest breeds of sheep, the Cheviot originated in the hill country of Scotland where they were developed for hardiness and vitality. Imported into the U.S. in 1838, Cheviots are small sized, white-faced with bare head and legs. They are prolific, easy lambers, and good milkers with excellent maternal instincts. They are well adapted to rugged, harsh climates and grazing on hilly pastures. The breed produces meaty carcasses that may be marketed at light weights, and a medium, easy-to-spin wool.

Hampshire: The Hampshire was developed in England from Southdown X Wiltshire Horn and Berkshire Know

Figure 27-17. One of the hardiest, and oldest breeds of sheep, the Border Cheviot. Courtesy American Sheep Producers Council.

growing breed has excellent carcass merit and a medium, easy-to-spin wool.

Oxford: The Oxford originated in England from Hampshire X Cotswold crosses and was imported into the U.S. in 1846. Selected for size and productivity, the Oxford is medium large-sized with a dark brown to gray face and wool on the legs. The breed is prolific, heavy milking, with good maternal instincts, lambing ability, and is useful in farm flock production. Oxfords have good growth rates and carcass merit and produce a low ¼ to medium ¼ blood grade wool suitable for spinning.

Shropshire: The Shropshire was developed in England from native stock and Southdown, Leicester and Cotswold crosses. Imported into the U.S. in 1855, Shropshires became popular for rams in commercial farm flocks. Shropshires are medium to large with a dark face and wool on the legs, hardy, early maturing, and long-lived. They are prolific and heavy milking with good maternal instincts, easy lambing, and have good crossing ability. Shropshires produce meaty lamb carcasses at light weights and yield a medium wool with a distinct crimp.

Southdown: One of the oldest breeds of sheep, the Southdown originated in England where it contributed to the development of other breeds. Imported into the U.S. in 1803, the Southdown is best suited to farm flock production. It is medium to small-sized with gray to mouse-brown face and wool on the legs. This early-maturing breed has good lambing ability, average milk-

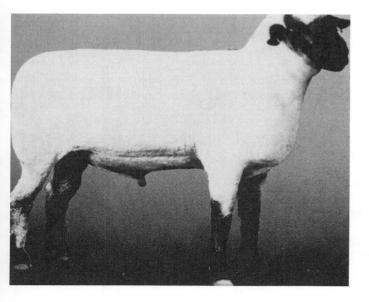

Figure 27-18. The Hampshire, a popular ram breed. Courtesy American Sheep Producers Council.

Figure 27-19. A Southdown ram, a breed known for early maturity. Courtesy American Sheep Producers Council.

crosses and imported into the U.S. in the 1880's. A popular meat breed, Hampshires are crossbred with white-faced ewes for market lamb production. Hampshires are large-sized with black faces and wool on the legs; adaptable to varied and wet climates. They are often used in farm flock production, and are prolific with good maternal instincts and milking ability. The fast-ing ability, and excellent crossing ability to produce meaty lamb carcasses at light weights. The Southdown is adaptable to varied and wet climates and yields a medium easy-to-spin wool.

Suffolk: The Suffolk originated in England from Southdown X Norfolk crosses and was imported into the U.S. in 1888. The breed is highly adapted to farm flock production and often crossbred with commercial white-faced ewes for market production. The Suffolk is large sized with bare head, black face, and bare, black legs. Hardy, highly prolific Suffolks have superior growth rates, milking and lambing abilities; they adapt well to heat and cold, and produce high quality meat carcasses and a medium, easy-to-spin wool. (For a photo, see Fig. 27-6.)

DUAL PURPOSE BREEDS

Columbia: Columbia, the first breed originated in the U.S., was developed in 1912 from Lincoln X Rambouillet crosses. Columbias are prolific, hardy, gregarious and good mothers with good milking ability. They are large with white faces and wool on the legs, useful in crossbred market lamb production, and yield heavy, medium wool fleeces with good length, free of black fiber.

Figure 27-21. A Dorset ewe. Courtesy American Sheep Producers Council.

milkers, will breed most any time of the year, and can produce more than one lamb crop each year. This hardy, long-lived breed yields medium wool fleeces free of black fiber and good for hand spinning.

Lincoln: The Lincoln originated in England and was imported into the U.S. in 1825. The breed was developed from Leicester X Old Lincoln crosses making important

Figure 27-20. The Columbia is a dual purpose breed. Courtesy American Sheep Producers Council.

Dorset: The Dorset Horn originated in Southern England and was imported into the U.S. in 1885. In 1948, a mutation occurred resulting in the Polled Dorsets, now a popular commercial breed. Dorsets are medium-sized, white-faced, with wool on the legs, they are heavy

Figure 27-22. The Lincoln breed is one of the largest breeds of sheep in the world. Courtesy American Sheep Producers Council.

contributions to other breeds. Lincolns are large, hardy, and white-faced with almost bare legs. They are adaptable to wet areas, show good maternal instincts, and wean heavy lambs from multiple births. They yield very long, coarse, strong, lustrous, common, and braid fleeces.

Montadale: The Montadale was developed in the U.S. from Cheviot X Columbia crosses to produce a multi-purpose sheep. The breed is medium sized, white-faced with bare head and legs, prolific, and an easy lamber with excellent maternal instincts. Montadales are hardy, have good carcass merit, can adapt to various climates, are selected for total pounds of lamb and wool, and yield an easy-to-spin wool.

Romney: Developed in England's Romney Marsh region to withstand cold, wet conditions, Romney sheep were imported into the U.S. in 1904 and are well-adapted in northwestern coastal areas. Romneys are medium-sized, white faced with wool on the legs, have sound legs and feet requiring little trimming, have good milking ability, maternal instincts and crossing ability. They produce a low ¼ to braid fleece well-suited for spinning.

Figure 27-23. The Romney breed was developed to withstand cold, wet conditions. Courtesy American Sheep Producers Council.

SUMMARY

There are many systems of sheep production, but basically they can be divided into extensive and intensive systems. Extensive would be synonymous with range sheep production as practiced in the western U.S., Canada and Australia. That is, large areas of marginal rangeland used to support sheep at relatively low stocking rates. Intensive would relate to what is also known as farm flock production; improved, high grade pastures used to support sheep in much greater concentrations.

Rate of growth on farm pastures is usually much more rapid. As a result, ewe-lambs are often used for breeding whereas, in extensive systems, ewes are usually first bred only as yearlings. Likewise, intensive systems often attempt to obtain two lamb crops per year versus one for most extensive systems (avg. gestation of sheep is 147 days).

In developing and managing ewes for breeding, nutrition is of primary importance. Ewes that are too thin will not conceive well and/or have increased lamb mortality as well as lighter birth weights. Ewes that are too fat will suffer more dystocia (difficult birth) and will typically have lower overall conception rates compared to ewes in more moderate condition.

"Flushing" is a term that refers to putting ewes on a higher plane of nutrition just prior and during the breeding season. If ewes are in thin or thrifty condition, "flushing" will not only increase the percentage of ewes conceiving, but the number of lambs born per ewe. That is, flushing also increases the number of twin lambs born.

Differences between breeds of sheep can be quite marked. One of the major differences relates to wool production versus carcass quality and growth rate. As a general rule, the carcass is of more value in the U.S. and Canada, whereas wool production is of more value in Australia and New Zealand. Therefore, the breeds of sheep selected for the various areas will be determined by marketing conditions as well as environmental conditions.

CHAPTER 28 AQUACULTURE AND MARICULTURE
D. Porter Price
Technical Consultant: David Mayer

INTRODUCTION

Aquaculture normally refers to the commercial raising of fish in fresh water, and mariculture refers to the raising of marine species in salt water. Together they comprise the newest, most challenging and potentially rewarding form of all animal agriculture.

Challenging in that when compared to conventional animal agriculture, the amount of information available is exceedingly small. Whereas conventional animal agriculture has been intensively researched for the last 50 years, aquatic animal research is only just now beginning. In every respect, aquaculture/mariculture is in its infancy.

Indeed, with the exception of only a few species, the elementary basics of nutrition, metabolism and reproduction are not known with any certainty. This leads not only to the challenge, but the potential rewards as well. For example, most current diets for fish are what are known as empirically derived, meaning non-scientific. That is, most fish diets in current use today are simply combinations of feeds that have been shown to be effective. As far as the basic nutrient requirements are concerned, in most cases they are not definitively known.

As a result, most fish formulas used today rely to a great extent upon fish meal and fish byproducts, (very high quality and expensive ingredients). If the true

Figure 28-1 A&B. The Yellowtail culture in Japan. Juvenile Yellowtail are caught in the wild and then fed in bamboo pens, supported by floats. Photo above shows a series of "pens" anchored in the coastal bays and lagoons of Japan. Photo below is of the individual cages.

nutrient requirements were known, then computer formulated least cost feeds could be used just as they are used in conventional livestock rations. In many cases, this could reduce feed costs by as much as 50% to 75%.

Understanding the reproduction of many fish, particularly marine species, could also be enormously rewarding. While fish and game departments have been able to reproduce a number of common game fishes; the reproduction of most fishes, particularly marine species, is poorly understood. As a result, many current marine fish farms rely upon juvenile fish caught in the wild. Being able to commercially reproduce these fish could substantially lower their costs. Also, it could ensure availability and in some cases allow the industry to expand.

Rewards. When we talk about the rewards of fish farming, we are talking about two different types of rewards: 1. increasing high quality protein foods for the ever expanding human population: and 2. monetary profits for the individual producer. Actually, of course it is reward no. 2, profit, that will allow us to achieve reward no. 1.

Indeed, the potential financial rewards are staggering. In many species of fish, we have animals that can convert feed even more efficiently than poultry, yet have a commercial value of several times the price of poultry. That is, there are many species of fish (which would include crustaceans) that can convert feed to live weight gain at ratios of less than 1.5:1, yet have a market value of $2, $3, and even as much as $5 per pound. This most certainly does not mean that all fish farmers are experiencing high profits, because they are not. But it does mean that the potential is there, once the technology is fully developed.

History of fish farming. While the introduction to this chapter stated that aquaculture/mariculture is very new, those knowledgeable on the subject may dispute that, since fish farming itself is very old. Many ancient cultures such as the Romans, the ancient Chinese, and a number of other Asian peoples have cultivated fish for centuries.

What is new is the technology. Much of the fish farming that exists today remains little changed from what has been practiced for centuries. In many cases, particularly with marine species, juvenile fish are caught in the wild and then grown out in cages or other enclosures. Most often the diets are nonscientific formulations of fish meal, and in some cases, simply chopped up "trash" fish obtained from the commercial fishing industry (see Fig. 28-2, and 28-4 a&b).

What is new, and what is needed, is the scientific approach to fish farming. Research, and applied research, similar to what has been utilized in conventional animal agriculture.

In the last 10-15 years there has been great progress in the commercial cultivation of catfish, salmon, trout and shrimp. Research has illuminated much of the nutrient requirements, and with respect to catfish at least, great strides have been made in utilizing low cost feedstuffs such as soybean meal, cereal grains, synthetic amino acids etc., instead of high cost fish meals as the basis of the diet. As a result, catfish production has literally exploded, both in terms of tons of fish and number of farms entering the business.

Figure 28-2. As explained in the text, most current fish diets rely heavily upon fish meal and fish meal byproducts. Although these are very high cost ingredients, they are required since the definitive nutrient requirements for most species are not known. Photo is of "trash" fish discarded by the commercial fishing industry, in this case used as the sole diet for the Yellowtail culture of Japan.

Just recently, the reproduction of shrimp became commercially feasible. This has led to a number of new ventures in shrimp farming, which will undoubtedly become the nucleus of an even larger industry.

But there is an enormous amount of work to be done. As pointed out in the sidebar to this chapter, the ever increasing technology being applied to conventional fish capture, continues to decrease world marine fish populations. To offset this, increased commercial cultivation of fish will be required. In order for that to come about, research will be required to lower the costs and make the production economically feasible.

Probably the best example is the poultry industry. Forty years ago, poultry was one of the most expensive meats available. This was because of low, non-specialized, non-scientific production methods. But in the late 1950's and early 1960's, intensive research went into nutrition, reproduction, and genetics. As a result, the cost of raising poultry has been driven down to a fraction of what it once was. This greatly expanded the profitability, which has resulted in the development of an enormous specialized industry.

Commercial aquaculture/mariculture is at that crossroads today. Strides have been made in nutrition and reproduction, but an enormous amount remains to be done. Indeed, in the realm of research, the area of genetics and breeding has almost been untouched. Genetics, which has given us cows that can give 3 times more milk than the cows of 50 years ago, beef cattle that can gain weight 30-50% faster, and sheep that produce twins and triplets instead of single lambs, can undoubtedly be applied to greatly enhance the growth rate and other economically valuable traits of fish.

Science will provide the technology but profit will provide the development. Indeed, as discussed previously, for those who get in on the ground floor, the potential for profit is enormous.

Overview of the Chapter. The purpose of this chapter shall not be to provide technical detail on specific fishes. Indeed, trying to cover all of fish production would be like trying to cover all of animal agriculture in just one chapter. It cannot be done.

Rather, the purpose of this chapter will be to provide a basic overall view of fish farming. Generalities concerning problems and basic requirements will be discussed. The reader will be given a background of information that will be of use should a more detailed study of a particular species be desired.

MAJOR DIFFERENCES BETWEEN FISH FARMING AND CONVENTIONAL ANIMAL AGRICULTURE

Death Loss. Probably the most striking difference between fish farming and conventional animal agriculture is the potential for death loss. In conventional animal husbandry, as a matter of course, we expect some death loss. The actual rate, varying with the species, but from ½ of one percent to a high of 7% or 8% would be the range for most livestock or poultry operations.

In fish farming however, a 100% death loss is quite possible. In addition, death loss in fish farming can occur much more quickly. That is, whenever we have a serious problem with terrestrial livestock, we usually have some warning and a period of time to take corrective action. A few animals will appear ill, which gives us some lead time to diagnose the problem before the entire herd or flock incubates (in an infectious disease), or is otherwise affected. With fish, however, because of the aquatic environment, a rapid change in temperature, salinity, oxygen or ammonia content, can cause a virtual 100% death loss overnight.

Figure 28-3. A tank used in a striped bass research project being conducted in the U.S. (Hubbs-Sea World Research Institute).

This does not mean that such losses regularly occur, but the potential is there. Therefore management of aquatic farms must be much tighter than terrestrial farms in that the environment (the water) must be constantly monitored and analyzed.

Temperature control. As mentioned in the previous section, rapid temperature changes can cause death losses. Of course, this is true with respect to all livestock (especially poultry), but not to the extent it is with fish.

The reason is that fish are cold-blooded animals. Cold-blooded meaning that (as a general rule) they cannot control or regulate their body temperature. Terrestrial animals (mammals and avians), of course, can regulate their body temperature; therefore, within a reasonable range of temperatures, conventional livestock can adjust (and survive).

But, the body temperature of fish will be whatever the water temperature is. Whenever there are rapid temperature changes, fish are subject to what is known as "temperature shock".

The smaller the body of water, the greater the potential for temperature shock. This is because small bodies of water, such as ponds, are affected by air temperature to a greater degree. The catfish industry in the U.S., which is primarily located in the states of Mississippi, Louisiana, Alabama and Georgia, is probably more susceptible to temperature shock than most other types of aquaculture.

The reason is that while normally possessing warm, humid, almost semi-tropical climates, severe cold fronts are occasionally driven down from the north. The small ponds catfish are normally raised in are therefore susceptible to rapid temperature changes.

To guard against death losses due to temperature shock, ponds are normally dug relatively deep. This allows fish to move to the bottom of the pond where the water temperature will not change nearly so quickly (as the surface water). In large lakes, bays, or coastal ocean environments where areas can be screened or netted off, the large body of water acts as a buffer against rapid temperature changes. There can, however, be what are known as "up-wellings", where cold water from deeper areas can be brought up to the surface (usually due to prolonged winds across the water's surface). However, most mariculture projects are located in sheltered, shallow areas where the effect is minimized. Likewise, the temperature shifts are normally only in the area of about 10-15 degrees (Fahrenheit).

Indirect effect of temperature. Aside from direct effect upon the fish, water temperature has a number of indirect effects as well. Probably the most important of these is the solubility of oxygen. As the water temperature warms up, oxygen comes out of solution in the water, and is lost as it escapes from the water's surface to the air. This can significantly reduce the amount of oxygen available for fish. At the same time, warm water temperatures also speed up the metabolism of most fish (see next section), which creates a greater need for oxygen. The result can create a crisis in pond management. Indeed, occasionally hot weather can even cause fish kills in natural lakes with wild fish populations. Obviously then, commercial fish production with greater densities of fish, is even more vulnerable.

Temperature also affects the ability of algae to grow and reproduce. As water temperature increases, as a general rule, the rate of algae growth increases also. This can be either bad or good, depending upon the situation. In some cases, the growth of algae is required, or at least used, as a food source for brine shrimp, molluscs, and occasionally freshly hatched juvenile fish (fry). In that case the growth of algae is watched carefully and controlled.

The problem is that if algae growth is rapid and uncontrolled, the result can be oxygen depletion. Known as an "algae bloom", a rapid, uncontrolled growth overpopulates the environment and then dies back. The resulting degradation of the dead algae depletes the available oxygen and can result in a complete fish kill.

In pond environments, increased artificial aeration, fresh water exchange, and/or chemical treatment can be used to control or reduce algae related problems. In the marine environment, however, there can occasionally be what are known as plankton blooms which are totally beyond the control of the mariculture manager. Commonly known as "red tides" the plankton (microscopic marine animals) can totally deplete oxygen levels in large areas of seacoast resulting in catastrophic fish kills. Although very infrequent, (happening only every 10-20 years) "red tides" occasionally do occur and create a potential hazard for the open water mariculturist.

Temperature effects upon the growth of fish. As mentioned in the previous sections, fish are cold-blooded animals and their body temperature will be the same as their environment (the water). Temperature also has a direct effect upon fish metabolism. As temperature increases, so does the fish's metabolism. As also discussed previously, this increases the oxygen requirement of the fish, which can be detrimental, if steps have not been taken to provide increased oxygenation.

But the increased metabolism also has a very important benefit, increased growth. As a general rule, all fish will grow more rapidly in a warm environment. For that reason fish culturists generally try to keep their fish in water somewhat warmer than their indigenous environment.

For example, the warm water outflow from electric power plants is quite often used for fish culture. That is, it is common practice for coal, oil, and nuclear power plants to use water to cool their generators. In a great number of cases, fish culturists have been able to benefit by using the artificially warmed water for aquaculture/mariculture projects.

Obviously, there is a limit as to how warm of water can be used. If the water becomes too warm most fish will go off feed, and if temperature continues to rise, death may occur. But as a general rule, most fish can tolerate 5 to 10 degrees (F) higher temperatures than their native environment, and will respond with more rapid growth rates.

FISH NUTRITION AND FEEDING

As mentioned in the introduction to this chapter, except for a very few species, the nutrient requirements for fish are poorly defined. There have been great strides made in the area of a few individual species, but the majority of species (most marine species) remains to be illuminated.

Problems in nutritional research. Aside from the fact that conventional livestock research has been underway for nearly 50 years, the very nature of terrestrial animals makes research much easier. With terrestrial animals the exact amount and type of feed eaten can be accurately controlled, and the urine and feces can be collected for analysis.

Neither of these items are normally feasible with fish. That is, the feed must be put into the water, and certainly some will be dissolved in the water. Likewise, we cannot collect waste products directly from the fish, as they are excreted into the water.

Making things even more difficult is the fact that many fish can extract some nutrients back from the water for use in the body. Therefore, if we attempt to analyze the water (which is commonly done in fish research), it is sometimes difficult to separate food nutrients from waste products; and it is exceedingly difficult to know how many nutrients the fish recycled. It can be done, but it requires chemical labeling (such as

Figure 28-4 A&B. Above: A boat being loaded with "trash fish" for feeding of Yellowtail in Japan. Below: Similar boats with a small bay full of fish cages in the background.

the use of radioisotopes) which reduces the number of laboratories that can conduct this type of research. Therefore, the amount of research heretofore conducted has been limited.*

What is known about fish nutrition. It is known that the amino acid ratios required by fish are similar to terrestrial animals. The total protein requirement, however, is generally considered to be much higher. Conventional livestock are normally fed rations of 10% to a maximum of about 18% protein; fish are often fed rations of 30% to 50% protein.

What is poorly understood, however, is whether fish actually need that much protein since much of it is used for energy. The confounding factor is that fish have requirements for specific fats and oils.

Fish meal is high in the fats and oils required by many species (fish meal contains a number of fats and oils not ordinarily found in terrestrial animal or plant foods). When large amounts of fish meal are not included in the diet, growth is often retarded. The question becomes whether growth was retarded from reduced fish oils in the diet, or reduced protein?

Great progress in regards to this have been made with catfish, and much is being learned about salmonoids. In the case of catfish, enough has been learned to allow substitution of at least some vegetable proteins (soybean meal, cottonseed meal, etc.) for fish meal in practical diets. However, catfish are omnivorous fish (eating both vegetarian and carnivorous foods) and therefore little can be extrapolated to carnivorous species.

Therefore research continues into the fat requirements of carnivorous species. At this point one thing is clear, fish can utilize much higher fat levels in the feed than terrestrial livestock. With most farm animals, fat levels of only about 3 to about a maximum of 8 percent can usually be efficiently utilized. With fish, however, fat levels often run 20 to 30%.

In addition to high fat, some species require very high cholesterol and other cholesterol-like substances. Shrimp, lobster, and crabs have the highest requirement, but some fin fishes, such as salmon, also require substantial amounts of sterols in the diet. (Again, fish meal is usually a good source of these compounds.)

To date, indications are that most fish can utilize only limited amounts of carbohydrates. Carbohydrates, of course, in the form of feed grains, are normally the cheapest source of energy (for terrestrial livestock). Therefore it is very important to research and understand the maximum amount of carbohydrates that may be utilized for fish.

Much has been learned in regards to this with catfish, and many commercial catfish feeds utilize up to 40% carbohydrates. With carnivorous species, however, diets normally contain less than 20%; but as research continues, it may be possible to include more. For example, research with a terrestrial carnivore, dogs, has led to formulations that use carbohydrates as a major source of energy.

Practical feeding considerations. For the individual fish farmer who uses enclosed ponds, one of the most important considerations is estimating how much feed will be consumed. Unlike conventional livestock, we cannot overfeed to see how much the animals will eat and then adjust subsequent feedings.

That is, with terrestrial livestock if not all the feed is eaten, there is no big problem other than some wasted feed. With fish raised in ponds, however, excess feed will deteriorate water quality. Excess nutrients will allow growth of unwanted algae and/or other microbial populations. This can result in disease problems, depletion of oxygen, and ammonia buildup.

For the nutritionist formulating the feed, palatability is a foremost consideration for the same reason. If the fish are slow or reluctant to eat the feed, this will allow decomposition and subsequent water quality deterioration.

Other considerations, with respect to the feed itself, include integrity of the physical form of the feed. Unless we are feeding extremely small, juvenile fish, we do not want the feed to dissolve in the water.

Most commercial feeds are pelleted. To keep the pellet from dissolving, at least 20% carbohydrates must usually be included in the formulation. The formulation is then usually formed into pellets by the process known as "extrusion". In the extrusion process the feed is forced through tiny openings under immense force. This causes an enormous amount of heat and pressure which causes the starch in the carbohydrate to "gelatinize" and "glue" the formulation together.

Another consideration is whether the pellet should float or sink. For most species, it is better to have a pellet that floats. This allows the farmer to see the fish feeding. If the fish go off feed, the farmer will know it immediately.

Some species, such as shrimp, will not feed on the surface. In that case, a sinking pellet is required.

---

* **The answer for future research.** Basically there are two answers for fish nutrition research; both require increased funding. With increased funding for laboratories, more scientific basic metabolism studies can be conducted. In addition, with increased funding practical feeding trials under commercial conditions can be done.

This would be synonymous with the situation in conventional animal agriculture. That is, nearly all land-grant universities conduct basic nutrition and metabolism research. In addition, there are numerous experimental farms that are used to conduct practical research.

To date, the funding for fish nutrition research has been minimal. Given the reduction in coastal wild fish populations and the great potential for production, it would seem logical for fish research to receive the same amount of governmental support as conventional livestock research.

FISH REPRODUCTION

The knowledge of fish reproduction is probably more on a par with terrestrial livestock than any other aspect of aquaculture or mariculture. A great deal of this knowledge has been contributed by state and federal agencies involved in hatchery development and management.

For several decades game fish have been artificially reared for the "stocking" of lakes, streams, and less commonly, coastal waters. Nutrition research never really developed because there was no need. Empirical formulas of fish meal and other high quality proteins worked, and there was no economic need to change them. That is, the agencies were not constrained to produce fish for a profit, only produce them. Therefore, the bulk of the research went into reproduction. Trout and salmon have probably been the most thoroughly researched species; but a number of warm water fishes such as black and striped bass have been well researched and successfully reproduced.

More recently, the advent of commercial production brought about research into other species, such as shrimp and lobsters as well as catfish. As mentioned previously, the reproduction of shrimp has just recently become feasible.

At the time of writing, the state of Texas recently began operation of a redfish reproduction facility. The redfish facility is unique in that the purpose is to serve both the mariculture industry, as well as sport and commercial fisheries. And realistically, for aquaculture/mariculture to really develop this is the type of cooperation that is needed.

That is, fish reproduction is normally quite complicated and is beyond the scope of most small farmers. Therefore, either large commercial or government hatcheries are normally required as a source of juvenile fish.

Developing a practical system of reproduction for a species requires an enormous outlay in research. Normally only governmental agencies are equipped with the facilities and funding to engage in such activities. Once the technology is developed, however, large commercial concerns can go into production. These commercial breeders, and/or the governmental facilities then serve as a source of stockers for smaller farmers.

Practical limitations in fish reproduction. As discussed previously, more research effort has been focused on fish reproduction than any other phase of aqua/mariculture. Moreover great strides in the practical reproduction of many species have been made. Indeed, artificial insemination as well as artificial incubation of eggs in several species, has been accomplished.

But there are some practical problems that may be difficult to overcome. One is cannibalism. In many carnivorous species even very small juveniles will engage in cannibalism. Under natural conditions, dispersal of the population over wide areas reduces cannibalistic problems. Under intensive cultivation, however, where large numbers of hatchlings must be kept in small enclosures, cannibalism can be a serious problem.

Another practical limitation is that for some marine species, the young go through a larval form. After hatching, the young exist, in some cases for as long as a year, as a microscopic larvae. This is true of a number of economically important species such as abalone, spiny lobster, and some shrimp.

In the case of the spiny lobster, the female carries the eggs on her tail for up to one year before they hatch. After hatching, the young exist for another year as a larval form. Making things more complicated yet, is the fact that the larval form goes through several metamorphic changes.

Practical advantages to reproducing fish. While there are a number of problems associated with reproducing fish, it should be pointed out that there are also some enormous advantages. The first of which is economic value. For example, the spiny lobster may be technically very difficult to reproduce with the long incubation time and the subsequent larval forms, but once the technical problems are worked out, we are talking about a species with enormous commercial value.

The other aspect, which actually interrelates with the economic value, is the magnitude of what is feasible. We are not talking about an animal bearing a single or twin offspring every 6 to 9 months. Neither are we talking a bird that lays an egg every other day, or even every day. Rather, we are talking about the ability to produce hundreds of thousands and even millions of offspring.

In the case of the spiny lobster, quite possibly it may take two or more years before any juvenile "stockers" are ready from a laboratory hatching. But what we must keep in mind is that from one female, offspring worth ten of thousands of dollars can be produced.

FISH YOU DON'T FEED

The main thrust of this chapter has been to discuss the fish species that are of the most value in the U.S. and other developed nations. In many undeveloped nations, however, there is, and has been for many centuries, a detritus form of fish culture. That is, a form of fish culture in which no commercial or purchased feed is utilized. Most commonly, swine and occasionally poultry waste is used to fertilize ponds.

The animal waste allows aquatic plants, algae, and other microbial growth to take place. Fish, usually carp and tilapia, then feed on this growth. Practiced extensively in China and Southeast Asia, rice fields quite often serve a dual role as fish culture ponds.

From an aesthetic point of view, detritus fish culture will probably never become a major industry in developed nations. That is, once the origin of these

Figure 28-5. The crayfish is an example of fish production, in which commercial feed need not be provided. Raised in rice fields, crayfish are an added bonus in that they can forage on their own.

species of fish became known, the consumer will probably be "turned off", and marketing problems would ensue.

But other forms of detritus fish culture do exist in the U.S. and continue to grow. Probably the best example being crayfish production. In Louisiana and other deep southern states, crayfish production has increased several fold in the last few years.

Stocked in rice fields, in some cases crayfish are fed, but in many cases the crayfish are left to forage on their own. Natural occurring food, such as bacterial degeneration of rice plants can support a moderate population of crayfish.

Shrimp are also a candidate for this type of production. However, because their market value is so high, most operations utilize commercial feed.

But brine shrimp, an exceedingly small shrimp often used as feed for home aquariums, have been raised in this manner for decades. In this case, marine algae is allowed to grow in salt water ponds, which is in turn fed upon by the almost microscopic brine shrimp. (Brine shrimp also can and have served as a feed for developing juvenile fin fish used in mariculture.)

SUMMARY

Aquaculture and mariculture, currently diminutive industries, represent an exciting potential for growth and development. Exciting in terms of potential profit for individual producers, as well as increased protein production for humanity.

But for individual producers, there will be risks as well as rewards. As discussed in the chapter, the potential for death loss with fish is much higher than conventional livestock. Indeed, 100% death losses are quite possible. The potential causes have to do primarily with

Figure 28-6. A Portugese fishing vessel loaded with monofilament gill nets. The advent of this type of technology increased the take of local inshore fishes enormously. The end result has been depletion of near shore fish stocks, and is a primary reason why mariculture research is required.

water quality. Water quality must be checked and managed constantly as rapid, undesirable changes can result in complete fish kills within a matter of hours.

The potential rewards are tied to the theoretical efficiency of fish, as well as the high market value of many species. That is, most fish have the capability to convert feed to gain at ratios of 1.5 to 1 (or better); which is even better than poultry, (normally about 2:1), but many species of fish have market values several times higher than poultry.

What has been holding profits back has been lack of technology, in relation to nutrition and practical diets. Most fish diets in use today rely heavily on fish meal, a very expensive feed ingredient. Fish meal is used both as a source of energy as well as protein. Practical research will be required so as to allow the use of more carbohydrates and common fats and oils for energy, as well as vegetable protein and/or synthetic amino acids for protein.

THE AESTHETIC NEED FOR MARICULTURE

The main thrust of this chapter discusses production,

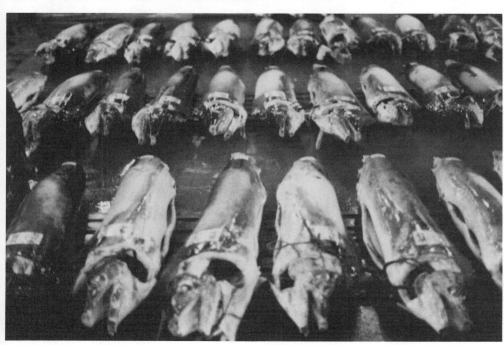

Figure 28-7 A&B. The Tokyo fish market. Hundreds of tons of tuna (above), and billfish such as marlin (below) are auctioned 3 timely weekly. Japan has been in the forefront of fish capture technology, and has effectively reduced fish stocks off her own shores. As a result Japan must send its fleet far offshore into the waters of other nations to maintain capture tonnages.

profits, and risks for the producer.

In the text of this chapter, the aspects of aqua/mariculture are discussed in relation to attributes one would normally associate with agriculture; that is, contribution to the food supply, and/or profitability to the producer. But for mariculture in particular, there are aesthetic needs for development as well.

Specifically, due to intense pressure by commercial fishing, the coastlines of many maritime nations are becoming severely depleted. Mariculture offers the potential of reducing some of that pressure by producing fish for consumption, as well as, hopefully, serving as a source of juvenile fish for replenishment of the marine coastal environment.

The basic problem is that, heretofore, the technology and research has been directed toward the capture of wild fish, rather than the culture and/or replenishment of fish. Some nations, such as Japan, Portugal and the Soviet Union, actually have fishing colleges where the student is taught the technology of sonar tracking, satellite navigation, current mapping, and other intricacies of hi-tech commercial fishing. This emphasis upon technology was necessary due to depletion of fish stocks along their own coastlines. Indeed, the coasts of Japan, Korea, Taiwan, and most European nations are almost devoid of fish in significant amounts. These nations have responded by traveling farther offshore and/or invading the territorial waters of other nations.

The U.S., Mexico and Canada have long felt the fishing pressure of the Japanese and Russian fishing fleets, and just recently, the Koreans. In addition, the U.S. and Canada also have their own rapidly expanding fishing fleets.

The U.S. and Canada are, therefore, beginning to undergo the coastal depletion that Japan and Europe experienced decades ago. That is, a growing flotilla of small coastal fishing boats is reducing inshore coastal fish populations, while larger, more sophisticated craft are working offshore.

In some cases, individual states have attempted to protect the inshore species within their own territorial limit (3 miles). However, this has not always been successful. In some cases lack of cooperation between states and/or the federal government has left large loopholes. The most common problem being boats fishing in the waters of restricted states and then simply unloading the catch in a neighboring state. Other problems include strong political lobbying by commercial fishing groups. For example, in California the state Fish and Game Department cannot regulate commercial fishing; only the state legislature may. Thus, commercial interests, through lobbying efforts, have been able to obtain fishing rights based on political rather than biological/scientific grounds.

Whether legislation, and subsequent regulation, can ever be effective is highly doubtful. Certainly it will not be effective with respect to the Japanese, the Soviets, Koreans, and/or the East Europeans. They most certainly are not going to restrict themselves, and/or submit to regulation by anyone else.

With respect to the U.S., conservation, ecology, and sport fishing groups may be able to bring about some control of inshore fish stocks. That is, in their view, fish populations are a public resource and should not be over-exploited by commercial interests. Still, due to heavy lobbying efforts by organized commercial interests, this will be difficult. With respect to Canada, provincial governments openly recognize commercial endeavors as having priority over public interests, and therefore little or no change can be expected.

Clearly, development of mariculture will be the only effective way of curtailing this problem internationally. The reason is due to economics.

Maritime nations are able to send their fleets around the world, due to the high market value of fish. Declining

Figure 28-8. Turtle eggs offered for sale in the Tokyo Fish Market. Considered an endangered species in most countries, in Japan there are no restrictions on sales of the turtle meat or eggs.

catches have been, at least partially, offset by increased prices. As the fishing industry puts more and more pressure on fish stocks, they do not necessarily damage themselves economically. The declining supply simply results in a higher price.

But if mariculture could be developed to the point of significantly increasing supply, this trend could possibly be reversed or at least moderated. If mariculture could significantly increase the supply of fish, the upward spiral of fish prices could, hopefully, be reduced. This would mean that over-fishing would have an economic impact on world fishing fleets. Hopefully, this would force them to engage in conservation practices, and/or become involved in restocking efforts. To date, of course, it has economically been more advantageous to put research and effort into the technology of capture, and therefore very little has been done in the area of restocking.

324

APPENDICES

Appendix Table 1 Nutrient Requirements of Poultry[1]

Type of Poultry	ME Mcal/ lb of feed	Crude pro- tein %	Lysine %	Cal- cium %	Phos- phorus %	Sod- ium %	Magne- sium %	Vit. A (IU/lb of feed)	Vit. D (IU/lb of feed)	Vit. E (IU/lb of feed)	Thia- mine (mg/lb of feed)	Nia- cin (mg/lb of feed)	Pano- thenic acid (mg/lb of feed)	Ribo- flavin (mg/lb of feed)	B12 micro- gram/ lb of feed
Broilers 0-3 weeks	1.45	23	1.2	1.0	.9*	.15	.06	700	91	4.5	.8	12.3	4.5	1.6	.004
Broilers 3-6 weeks	1.45	20	1.0	.9	.8*	.15	.06	700	91	4.5	.8	12.3	4.5	1.6	.004
Broilers 6-8 weeks	1.45	18	1.85	.8	.7*	.15	.06	700	91	4.5	.8	5.0	4.5	1.6	.002
Laying hens	1.32	14.5	.64	3.4	.7*	.15	.05	1820	227	2.3	.4	4.5	1.0	1.0	.002

* This is total phosphorus as apposed to "available" phosphorus. That is, for poultry, phosphorus from plant sources is only 30 to 40% available.

[1] Adapted from National Research Council.

Appendix Table 2. Nutrient Requirements of Dairy Cattle[1]

Type of animal	Crude protein (%)	ME (Mcal/lb)	TDN (%)	NE_L (Mcal/lb)	Calcium (%)	Phosphorus (%)	Vit. A (IU/lb)	Vit. D (IU/lb)
Mature, pregnant cow (last 2 mo. pregnancy)	10.0	.9	57	NA	.31*	.24	1700	
Mature bull	8.0	.9	55	NA	.23	.18	1400	
Growing replacement bulls (see beef cattle section)								
Growing heifer calves (100-220 lbs)	15.0	1.25	75	NA	.38	.30	1500	225
Growing heifer calves (220-440 lbs)	11.0	1.1	65	NA	.36	.29	1000	110
Growing heifer calves (440-660 lbs)	10.5	1.0	61	NA	.32	.25	800	
Growing heifer replacements (660-880 lbs)	10.0	.95	58	NA	.30	.23	800	
Growing heifer replacements (880-1100 lbs)	9.5	.90	55	NA	.28	.22	800	
Growing heifer replacements (1100-1300 lbs)	10.2	.92	56	NA	.29*	.22	1200	

[1] Adapted from National Research Council with some modification by the senior author.

* Just prior to calving, calcium levels should be kept at minimum levels to reduce the incidence of milk fever (see text).

Nutrients Required per Pound of Milk Production

Percent milk fat	grams of protein	Megacalories required	lbs of TDN	Megacalories of NE_L	grams of calcium	grams of phosphorus
2.5	30	.41	.12	.27	1.10	.77
3.0	32	.45	.13	.29	1.14	.81
3.5	34	.48	.14	.31	1.18	.86
4.0	35	.51	.15	.34	1.22	.90
4.5	37	.55	.16	.36	1.27	.95

Appendix Table 3. Nutrient Requirements of Beef Cattle[1]

Type of animal	Weight gain (lb)	ME (Mcal/lb)	TDN (%)	NE$_m$ (Mcal/lb)	NE$_g$ (Mcal/lb)	CP (%)	Calcium (%)	Phosphorus (%)	Sodium (%)	Magnesium (%)	Vit. A (IU/lb)
Pregnant yearling heifers	1.5	1.1	68.0	.65	.40	10.5	.35	.21	.08	.1	1300
Dry, pregnant cows first 6 mo. of pregnancy	.1	.8	50.0	.45	NA	7.5	.22	.18	.08	.1	1200
Dry, pregnant cows last 3 mo. of pregnancy	.8	.9	55.0	.52	NA	8.5	.26	.22	.08	.1	1200
First calf heifers nursing calves	.5	1.1	65.0	.68	.41	11.0	.38	.25	.10	.15	1800
Cow nursing calf (ave. milking ability)	0	.95	58.0	.58	NA	10.0	.32	.23	.10	.15	1800
Cow nursing calf (dual purpose breeds)	0	1.3	68.0	.70	NA	12.5	.40	.28	.15	.20	1800
Yearling bulls	1.5	1.0	60.0	.62	.36	10.0	.28	.22	.08	.1	1800
Two year old bulls	.5	.9	54.0	.50	.28	8.5	.24	.21	.08	.1	1800
Mature bulls	0	.8	50.0	.42	NA	7.0	.21	.20	.08	.1	1800
Growing calves 400 to 600 lbs	1.75	1.2	69.0	.72	.43	11.0	.30	.21	.08	.1	1100
Finishing yearlings 600 - 1100 lbs	3.0	1.4	78.0	.88	.52	11.5	.38	.28	.08	.1	1100

[1]Source: Adapted from NRC with modification by the senior author.

See also, Table   Net Energy Requirements of Growing and Finishing Cattle.

Net Energy Requirements of Growing and Finishing Steers

Daily gain lb	350	360	370	380	390	400	410	420	430	440	450	460	470	480	490	500	510	520	530	540
	\multicolumn{20}{c}{NE$_m$ required, mcal per day}																		
0	3.48	3.55	3.63	3.70	3.77	3.85	3.92	3.99	4.06	4.13	4.20	4.27	4.34	4.41	4.48	4.55	4.61	4.68	4.75	4.82
	\multicolumn{20}{c}{NE$_g$ required, mcal per day}																		
0.5	0.55	0.56	0.57	0.59	0.60	0.61	0.62	0.63	0.64	0.65	0.67	0.68	0.69	0.70	0.71	0.72	0.73	0.74	0.75	0.76
0.6	0.67	0.68	0.69	0.71	0.72	0.74	0.75	0.76	0.78	0.79	0.80	0.82	0.83	0.84	0.86	0.87	0.88	0.90	0.91	0.92
0.7	0.78	0.80	0.81	0.83	0.85	0.86	0.88	0.89	0.91	0.93	0.94	0.96	0.97	0.99	1.00	1.02	1.03	1.05	1.07	1.08
0.8	0.90	0.92	0.94	0.95	0.97	0.99	1.01	1.03	1.05	1.06	1.08	1.10	1.12	1.14	1.15	1.17	1.19	1.21	1.22	1.24
0.9	1.01	1.04	1.06	1.08	1.10	1.12	1.14	1.16	1.18	1.20	1.22	1.25	1.27	1.29	1.31	1.33	1.35	1.37	1.39	1.40
1.0	1.13	1.16	1.18	1.21	1.23	1.25	1.28	1.30	1.32	1.35	1.37	1.39	1.41	1.44	1.46	1.48	1.50	1.53	1.55	1.57
1.1	1.25	1.28	1.31	1.33	1.36	1.39	1.41	1.44	1.46	1.49	1.51	1.54	1.56	1.59	1.61	1.64	1.66	1.69	1.71	1.74
1.2	1.38	1.40	1.43	1.46	1.49	1.52	1.55	1.58	1.60	1.63	1.66	1.69	1.71	1.74	1.77	1.80	1.82	1.85	1.88	1.90
1.3	1.50	1.53	1.56	1.59	1.62	1.66	1.69	1.72	1.75	1.78	1.81	1.84	1.87	1.90	1.93	1.96	1.99	2.02	2.05	2.07
1.4	1.62	1.66	1.69	1.73	1.76	1.79	1.83	1.86	1.89	1.93	1.96	1.99	2.02	2.05	2.09	2.12	2.15	2.18	2.21	2.24
1.5	1.75	1.78	1.82	1.86	1.90	1.93	1.97	2.00	2.04	2.07	2.11	2.14	2.18	2.21	2.25	2.28	2.32	2.35	2.39	2.42
1.6	1.87	1.91	1.95	1.99	2.03	2.07	2.11	2.15	2.19	2.22	2.26	2.30	2.34	2.37	2.41	2.45	2.48	2.52	2.56	2.59
1.7	2.00	2.04	2.09	2.13	2.17	2.21	2.25	2.29	2.34	2.38	2.42	2.46	2.50	2.54	2.57	2.61	2.65	2.69	2.73	2.77
1.8	2.13	2.18	2.22	2.27	2.31	2.35	2.40	2.44	2.49	2.53	2.57	2.62	2.66	2.70	2.74	2.78	2.82	2.87	2.91	2.95
1.9	2.26	2.31	2.36	2.40	2.45	2.50	2.55	2.59	2.64	2.68	2.73	2.78	2.82	2.86	2.91	2.95	3.00	3.04	3.09	3.10
2.0	2.39	2.44	2.49	2.54	2.59	2.64	2.69	2.74	2.79	2.84	2.89	2.94	2.98	3.03	3.08	3.12	3.17	3.22	3.27	3.31
2.1	2.52	2.58	2.63	2.69	2.74	2.79	2.84	2.89	2.95	3.00	3.05	3.10	3.15	3.20	3.25	3.30	3.35	3.40	3.45	3.49
2.2	2.66	2.72	2.77	2.83	2.88	2.94	2.99	3.05	3.10	3.16	3.21	3.26	3.32	3.37	3.42	3.47	3.53	3.58	3.63	3.68
2.3	2.79	2.85	2.91	2.97	3.03	3.09	3.15	3.20	3.26	3.32	3.37	3.43	3.48	3.54	3.59	3.65	3.70	3.76	3.82	3.87
2.4	2.93	2.99	3.05	3.12	3.18	3.24	3.30	3.36	3.42	3.48	3.54	3.60	3.65	3.71	3.77	3.83	3.89	3.94	4.00	4.06
2.5	3.07	3.13	3.20	3.26	3.33	3.39	3.45	3.52	3.58	3.64	3.70	3.77	3.83	3.89	3.95	4.01	4.07	4.13	4.19	4.25
2.6	3.21	3.28	3.34	3.41	3.48	3.54	3.61	3.68	3.74	3.81	3.87	3.94	4.00	4.06	4.13	4.19	4.25	4.32	4.38	4.44
2.7	3.35	3.42	3.49	3.56	3.63	3.70	3.77	3.84	3.91	3.97	4.04	4.11	4.17	4.24	4.31	4.37	4.44	4.50	4.57	4.63
2.8	3.49	3.56	3.64	3.71	3.78	3.86	3.93	4.00	4.07	4.14	4.21	4.28	4.35	4.42	4.49	4.56	4.63	4.70	4.76	4.83
2.9	3.63	3.71	3.79	3.86	3.94	4.01	4.09	4.16	4.24	4.31	4.38	4.46	4.53	4.60	4.67	4.74	4.81	4.89	4.96	5.03
3.0	3.78	3.86	3.94	4.02	4.09	4.17	4.25	4.33	4.41	4.48	4.56	4.63	4.71	4.78	4.86	4.93	5.01	5.08	5.16	5.22
3.1	3.92	4.00	4.09	4.17	4.25	4.33	4.41	4.49	4.57	4.65	4.73	4.81	4.89	4.97	5.04	5.12	5.20	5.28	5.35	5.43
3.2	4.07	4.15	4.24	4.33	4.41	4.49	4.58	4.66	4.75	4.83	4.91	5.00	5.07	5.15	5.23	5.31	5.39	5.47	5.55	5.63
3.3	4.21	4.30	4.39	4.48	4.57	4.66	4.74	4.83	4.92	5.00	5.09	5.17	5.26	5.34	5.42	5.51	5.59	5.67	5.76	5.83
3.4	4.36	4.46	4.55	4.64	4.73	4.82	4.91	5.00	5.09	5.18	5.27	5.36	5.44	5.53	5.61	5.70	5.79	5.87	5.96	6.04

Net Energy Requirements of Growing and Finishing Steers (continued)

Daily gain lb	Body Weight lb																			
	550	560	570	580	590	600	610	620	630	640	650	660	670	680	690	700	710	720	730	740
	NE$_m$ required, mcgal per day																			
0	4.88	4.95	5.02	5.08	5.15	5.21	5.28	5.34	5.41	5.47	5.53	5.60	5.66	5.73	5.79	5.85	5.91	5.98	6.04	6.10
	NE$_g$ required, mcgal per day																			
0.5	0.77	0.78	0.80	0.81	0.82	0.83	0.84	0.85	0.86	0.87	0.88	0.89	0.90	0.91	0.92	0.93	0.94	0.95	0.96	0.97
0.6	0.93	0.95	0.96	0.97	0.98	1.00	1.01	1.02	1.03	1.05	1.06	1.07	1.08	1.09	1.11	1.12	1.13	1.14	1.15	1.17
0.7	1.10	1.11	1.13	1.14	1.15	1.17	1.18	1.20	1.21	1.23	1.24	1.26	1.27	1.28	1.30	1.31	1.33	1.34	1.35	1.37
0.8	1.26	1.28	1.29	1.31	1.33	1.34	1.36	1.38	1.39	1.41	1.43	1.44	1.46	1.48	1.49	1.51	1.52	1.54	1.56	1.57
0.9	1.42	1.44	1.46	1.48	1.50	1.52	1.54	1.56	1.58	1.60	1.61	1.63	1.65	1.67	1.69	1.71	1.72	1.74	1.76	1.78
1.0	1.59	1.61	1.64	1.66	1.68	1.70	1.72	1.74	1.76	1.78	1.80	1.82	1.85	1.87	1.89	1.91	1.93	1.95	1.97	1.99
1.1	1.76	1.79	1.81	1.83	1.85	1.88	1.90	1.92	1.95	1.97	1.99	2.02	2.04	2.06	2.09	2.11	2.13	2.15	2.18	2.20
1.2	1.93	1.96	1.98	2.01	2.03	2.06	2.09	2.11	2.14	2.16	2.19	2.21	2.24	2.26	2.29	2.31	2.34	2.36	2.39	2.41
1.3	2.10	2.13	2.16	2.19	2.22	2.24	2.27	2.30	2.33	2.36	2.38	2.41	2.44	2.47	2.49	2.52	2.55	2.57	2.60	2.63
1.4	2.28	2.31	2.34	2.37	2.40	2.43	2.46	2.49	2.52	2.55	2.58	2.61	2.64	2.67	2.70	2.73	2.76	2.79	2.81	2.84
1.5	2.45	2.49	2.52	2.55	2.58	2.62	2.65	2.68	2.72	2.75	2.78	2.81	2.84	2.88	2.91	2.94	2.97	3.00	3.03	3.06
1.6	2.63	2.67	2.70	2.74	2.77	2.81	2.84	2.88	2.91	2.95	2.98	3.01	3.05	3.08	3.12	3.15	3.18	3.22	3.25	3.29
1.7	2.81	2.85	2.89	2.92	2.96	3.00	3.03	3.07	3.11	3.15	3.18	3.22	3.26	3.29	3.33	3.37	3.40	3.44	3.47	3.51
1.8	2.99	3.03	3.07	3.11	3.15	3.19	3.23	3.27	3.31	3.35	3.39	3.43	3.47	3.51	3.54	3.58	3.62	3.66	3.70	3.74
1.9	3.17	3.22	3.26	3.30	3.34	3.39	3.43	3.47	3.51	3.56	3.60	3.64	3.68	3.72	3.76	3.80	3.84	3.88	3.92	3.96
2.0	3.36	3.41	3.45	3.49	3.54	3.58	3.63	3.67	3.72	3.76	3.80	3.85	3.89	3.94	3.98	4.02	4.06	4.11	4.15	4.19
2.1	3.54	3.59	3.64	3.69	3.73	3.78	3.83	3.88	3.93	3.97	4.02	4.06	4.11	4.16	4.20	4.25	4.29	4.34	4.38	4.43
2.2	3.73	3.78	3.83	3.88	3.93	3.98	4.03	4.08	4.13	4.18	4.23	4.28	4.33	4.38	4.42	4.47	4.52	4.57	4.61	4.66
2.3	3.92	3.98	4.03	4.08	4.13	4.18	4.24	4.29	4.34	4.40	4.44	4.49	4.55	4.60	4.65	4.70	4.75	4.80	4.85	4.90
2.4	4.11	4.17	4.23	4.28	4.33	4.39	4.44	4.50	4.56	4.61	4.66	4.71	4.77	4.82	4.87	4.93	4.98	5.03	5.08	5.14
2.5	4.31	4.37	4.43	4.48	4.54	4.59	4.65	4.71	4.77	4.83	4.88	4.93	4.99	5.05	5.10	5.16	5.21	5.27	5.32	5.38
2.6	4.50	4.57	4.63	4.68	4.74	4.80	4.86	4.92	4.99	5.04	5.10	5.16	5.22	5.28	5.33	5.39	5.45	5.51	5.56	5.62
2.7	4.70	4.77	4.83	4.89	4.95	5.01	5.08	5.14	5.20	5.27	5.32	5.38	5.45	5.51	5.57	5.63	5.60	5.75	5.81	5.87
2.8	4.90	4.97	5.03	5.10	5.16	5.22	5.29	5.35	5.42	5.49	5.55	5.61	5.68	5.74	5.80	5.87	5.93	5.99	6.05	6.12
2.9	5.10	5.17	5.24	5.30	5.37	5.44	5.41	5.57	5.65	5.71	5.78	5.84	5.91	5.98	6.04	6.11	6.17	6.24	6.30	6.37
3.0	5.30	5.37	5.45	5.51	5.58	5.65	5.72	5.79	5.87	5.94	6.00	6.07	6.14	6.21	6.28	6.35	6.41	6.49	6.55	6.62
3.1	5.50	5.58	5.65	5.73	5.80	5.87	5.94	6.02	6.09	6.17	6.24	6.30	6.38	6.45	6.52	6.59	6.66	6.73	6.80	6.87
3.2	5.71	5.79	5.87	5.94	6.02	6.09	6.17	6.24	6.32	6.40	6.47	6.54	6.62	6.69	6.76	6.84	6.91	6.99	7.06	7.13
3.3	5.92	6.00	6.08	6.16	6.23	6.31	6.39	6.47	6.55	6.63	6.70	6.78	6.86	6.94	7.01	7.09	7.16	7.24	7.31	7.39
3.4	6.13	6.21	6.29	6.37	6.45	6.54	6.62	6.70	6.78	6.86	6.94	7.02	7.10	7.18	7.26	7.34	7.41	7.50	7.57	7.65

Net Energy Requirements of Growing and Finishing Steers (continued)

Daily gain lb	Body Weight lb											Body Weight lb									
	750	760	770	780	790	800	810	820	830	840	850	860	870	880	890	900	910	920	930	940
	NE_m required, mcgal per day											NE_m required, mcgal per day									
0	6.16	6.23	6.29	6.35	6.41	6.47	6.53	6.59	6.65	6.71	6.77	6.83	6.89	6.95	7.00	7.06	7.13	7.18	7.24	7.30
	NE_g required, mcgal per day											NE_g required, mcgal per day									
0.5	0.98	0.99	1.00	1.01	1.02	1.02	1.03	1.04	1.05	1.06	1.07	1.08	1.09	1.10	1.11	1.12	1.13	1.14	1.15	1.16
0.6	1.18	1.19	1.20	1.21	1.22	1.24	1.25	1.26	1.27	1.28	1.29	1.31	1.32	1.33	1.34	1.35	1.36	1.37	1.38	1.40
0.7	1.38	1.40	1.41	1.42	1.44	1.45	1.46	1.48	1.49	1.50	1.52	1.53	1.55	1.56	1.57	1.58	1.60	1.61	1.62	1.64
0.8	1.59	1.61	1.62	1.64	1.65	1.67	1.68	1.70	1.71	1.73	1.75	1.76	1.78	1.79	1.81	1.82	1.84	1.85	1.87	1.88
0.9	1.80	1.82	1.83	1.85	1.87	1.89	1.90	1.92	1.94	1.96	1.97	1.99	2.01	2.03	2.04	2.06	2.08	2.09	2.11	2.13
1.0	2.01	2.03	2.05	2.07	2.09	2.11	2.13	21.5	2.17	2.19	2.21	2.22	2.24	2.26	2.28	2.30	2.32	2.34	2.36	2.38
1.1	2.22	2.24	2.27	2.29	2.31	2.33	2.35	2.37	2.40	2.42	2.44	2.46	2.48	2.50	2.52	2.55	2.57	2.59	2.61	2.63
1.2	2.44	2.46	2.48	2.52	2.53	2.56	2.58	2.60	2.63	2.65	2.68	2.70	2.72	2.75	2.77	2.79	2.82	2.84	2.86	2.89
1.3	2.65	2.68	2.71	2.73	2.76	2.78	2.81	2.84	2.86	2.89	2.92	2.94	2.97	2.99	3.02	3.04	3.07	3.09	3.12	3.14
1.4	2.87	2.90	2.93	2.96	2.99	3.01	3.04	3.07	3.10	3.13	3.16	3.18	3.21	3.24	3.27	3.29	3.32	3.35	3.38	3.40
1.5	3.09	3.13	3.16	3.19	3.22	3.25	3.28	3.31	3.34	3.37	3.40	3.43	3.46	3.49	3.52	3.55	3.58	3.61	3.64	3.67
1.6	3.32	3.35	3.39	3.42	3.45	3.48	3.51	3.55	3.58	3.61	3.65	3.68	3.71	3.74	3.77	3.80	3.84	3.87	3.90	3.93
1.7	3.54	3.58	3.62	3.65	3.69	3.72	3.75	3.79	3.82	3.86	3.90	3.93	3.96	4.00	4.03	4.06	4.10	4.13	4.17	4.20
1.8	3.77	3.81	3.85	3.89	3.92	3.96	4.00	4.03	4.07	4.11	4.15	4.18	4.22	4.25	4.29	4.33	4.36	4.40	4.43	4.47
1.9	4.00	4.05	4.08	4.12	4.16	4.20	4.24	4.28	4.32	4.36	4.40	4.44	4.48	4.51	4.55	4.59	4.63	4.67	4.70	4.74
2.0	4.23	4.28	4.32	4.36	4.40	4.45	4.49	4.53	4.57	4.61	4.66	4.69	4.74	4.78	4.82	4.86	4.90	4.94	4.98	5.02
2.1	4.47	4.52	4.56	4.61	4.65	4.69	4.74	4.78	4.82	4.87	4.91	4.95	5.00	5.04	5.08	5.13	5.17	5.21	5.25	5.30
2.2	4.71	4.76	4.80	4.85	4.90	4.94	4.99	5.03	5.08	5.13	5.17	5.22	5.26	5.31	5.35	5.40	5.44	5.49	5.53	5.58
2.3	4.95	5.00	5.05	5.10	5.14	5.19	1.24	5.29	5.34	5.39	5.44	5.48	5.53	5.58	5.62	5.67	5.72	5.77	5.81	5.86
2.4	5.19	5.24	5.29	5.35	5.40	5.45	5.50	5.55	5.60	5.65	5.70	5.75	5.80	5.85	5.90	5.95	6.00	6.05	6.10	6.15
2.5	5.43	5.49	5.54	5.60	5.65	5.70	5.76	5.81	5.86	5.92	5.97	6.02	6.07	6.13	6.18	6.23	6.28	6.33	6.38	6.44
2.6	5.68	5.74	5.79	5.85	5.91	5.96	6.02	6.07	6.13	6.18	6.24	6.29	6.35	6.40	6.46	6.51	6.57	6.62	6.67	6.73
2.7	5.93	5.99	6.05	6.11	6.16	6.22	6.28	6.34	6.39	6.45	6.51	6.57	6.63	6.68	6.74	6.80	6.85	6.91	6.97	7.02
2.8	6.18	6.24	6.30	6.36	6.42	6.48	6.54	6.60	6.67	6.73	6.79	6.85	6.91	6.97	7.02	7.08	7.14	7.20	7.26	7.32
2.9	6.43	6.50	6.56	6.62	6.69	6.75	6.81	6.87	6.94	7.00	7.07	7.13	7.19	7.25	7.31	7.37	7.44	7.49	7.56	7.62
3.0	6.69	6.76	6.82	6.89	6.95	7.02	7.08	7.15	7.21	7.28	7.35	7.41	7.47	7.54	7.60	7.67	7.73	7.79	7.86	7.92
3.1	6.94	7.02	7.08	7.15	7.22	7.29	7.35	7.42	7.49	7.56	7.63	7.69	7.76	7.83	7.89	7.67	7.73	7.79	7.86	7.92
3.2	7.20	7.28	7.35	7.42	7.49	7.56	7.63	7.70	7.77	7.64	7.92	7.98	8.05	8.12	8.19	8.26	8.33	8.39	8.46	8.53
3.3	7.46	7.54	7.61	7.69	7.76	7.83	7.91	7.98	8.05	8.13	8.20	8.27	8.34	8.42	8.48	8.56	8.63	8.70	8.77	8.84
3.4	7.73	7.81	7.88	7.96	8.04	8.11	8.19	8.26	8.34	8.41	8.49	8.56	8.64	8.71	8.78	8.86	8.93	9.00	9.08	9.16

Net Energy Requirements of Growing and Finishing Steers (continued)

Daily gain lb	Body Weight lb															Body Weight lb						
	950	960	970	980	990	1000	1010	1020	1030	1040		1050	1060	1070	1080	1090	1100
	NE_m required, megcal per day											NE_m required, megcal per day					
0	7.36	7.42	7.47	7.53	7.59	7.65	7.71	7.76	7.82	7.87		7.93	7.99	8.05	8.10	8.16	8.21
	NE_g required, megcal per day											NE_g required, megcal per day					
0.5	1.17	1.18	1.18	1.19	1.20	1.21	1.22	1.23	1.24	1.25		1.26	1.27	1.27	1.28	1.29	1.30
0.6	1.41	1.42	1.43	1.44	1.45	1.46	1.47	1.48	1.49	1.51		1.52	1.53	1.54	1.55	1.56	1.57
0.7	1.65	1.66	1.68	1.69	1.70	1.71	1.73	1.74	1.75	1.77		1.78	1.79	1.80	1.82	1.83	1.84
0.8	1.90	1.91	1.93	1.94	1.96	1.97	1.99	2.00	2.02	2.03		2.05	2.06	2.07	2.09	2.10	2.12
0.9	2.15	2.16	2.18	2.20	2.21	2.23	2.25	2.26	2.28	2.30		2.31	2.33	2.35	2.36	2.38	2.39
1.0	2.40	2.42	2.43	2.45	2.47	2.49	2.51	2.53	2.55	2.57		2.58	2.60	2.62	2.64	2.66	2.68
1.1	2.65	2.67	2.69	2.71	2.73	2.76	2.78	2.80	2.82	2.84		2.86	2.88	2.90	2.92	2.94	2.96
1.2	2.91	2.93	2.95	2.98	3.00	3.02	3.05	3.07	3.09	3.11		3.14	3.16	3.18	3.20	3.22	3.25
1.3	3.17	3.19	3.22	3.24	3.27	3.29	3.32	3.34	3.37	3.39		3.42	3.44	3.46	3.49	3.51	3.54
1.4	3.43	3.46	3.48	3.51	3.54	3.56	3.59	3.62	3.64	3.67		3.70	3.72	3.75	3.78	3.80	3.83
1.5	3.69	3.72	3.75	3.78	3.81	3.84	3.87	3.90	3.93	3.95		3.98	4.01	4.04	4.07	4.10	4.12
1.6	3.96	3.99	4.02	4.06	4.09	4.12	4.15	4.18	4.21	4.24		4.27	4.30	4.33	4.36	4.39	4.42
1.7	4.23	4.27	4.30	4.33	4.37	4.40	4.43	4.46	4.50	4.53		4.56	4.60	4.63	4.66	4.69	4.72
1.8	4.50	4.54	4.58	4.61	4.65	4.68	4.72	4.75	4.79	4.82		4.86	4.89	4.93	4.96	4.99	5.03
1.9	4.78	4.82	4.86	4.89	4.93	4.97	5.01	5.04	5.08	5.12		5.15	5.19	5.23	5.26	5.30	5.34
2.0	5.06	5.10	5.14	5.18	5.22	5.26	5.30	5.34	5.37	5.41		5.45	5.49	5.53	5.57	5.61	5.65
2.1	5.34	5.38	5.42	5.47	5.51	5.55	5.59	5.63	5.67	5.71		5.76	5.80	5.84	5.88	5.92	5.96
2.2	5.62	5.67	5.71	5.76	5.80	5.84	5.89	5.93	5.97	6.02		6.06	6.10	6.15	6.19	6.23	6.28
2.3	5.91	5.96	6.00	6.05	6.09	6.14	6.19	6.23	6.28	6.32		6.37	6.42	6.46	6.50	6.55	6.59
2.4	6.20	6.25	6.29	6.34	6.39	6.44	6.49	6.54	6.58	6.63		6.68	6.73	6.78	6.82	6.87	6.92
2.5	6.49	6.54	6.59	6.64	6.69	6.74	6.79	6.84	6.89	6.94		7.00	7.04	7.09	7.14	7.19	7.24
2.6	6.78	6.84	6.89	6.94	7.00	7.05	7.10	7.15	7.20	7.26		7.31	7.36	7.41	7.47	7.52	7.57
2.7	7.08	7.14	7.19	7.25	7.30	7.35	7.41	7.47	7.52	7.57		7.63	7.69	7.74	7.79	7.85	7.90
2.8	7.38	7.44	7.49	7.55	7.61	7.67	7.73	7.78	7.84	7.89		7.95	8.01	8.07	8.12	8.18	8.23
2.9	7.68	7.74	7.80	7.86	7.92	7.98	8.04	8.10	8.16	8.22		8.28	8.34	8.40	8.45	8.51	8.57
3.0	7.98	8.05	8.11	8.17	8.23	8.30	8.36	8.42	8.48	8.54		8.61	8.67	8.73	8.79	8.85	8.91
3.1	8.29	8.36	8.42	8.49	8.55	8.61	8.68	8.74	8.81	8.87		8.94	9.00	9.06	9.13	9.19	9.25
3.2	8.60	8.67	8.73	8.81	8.87	8.94	9.01	9.07	9.17	9.20		9.27	9.34	9.40	9.47	9.53	9.60
3.3	8.91	8.98	9.05	9.13	9.19	9.26	9.33	9.40	9.47	9.54		9.61	9.68	9.74	9.81	9.88	9.95
3.4	9.23	9.30	9.37	9.45	9.52	9.59	9.66	9.73	9.80	9.87		9.95	10.02	10.09	10.16	10.23	10.30

Net Energy Requirements of Growing and Finishing Heifers

Daily gain lb	Body Weight lb																			
	350	360	370	380	390	400	410	420	430	440	450	460	470	480	490	500	510	520	530	540
	NE$_m$ required, megcal per day																			
0	3.48	3.55	3.63	3.70	3.77	3.85	3.92	3.99	4.06	4.13	4.20	4.27	4.34	4.41	4.48	4.55	4.61	4.68	4.75	4.82
	NE$_g$ required, megcal per day																			
0.5	0.60	0.16	0.62	0.64	0.65	0.66	0.67	0.69	0.70	0.71	0.72	0.73	0.75	0.76	0.77	0.78	0.79	0.81	0.82	0.83
0.6	0.73	0.74	0.76	0.77	0.79	0.80	0.82	0.83	0.85	0.86	0.88	0.89	0.90	0.92	0.93	0.95	0.96	0.98	0.99	1.00
0.7	0.85	0.87	0.89	0.91	0.93	0.94	0.96	0.98	1.00	1.01	1.03	1.05	1.07	1.08	1.10	1.12	1.13	1.15	1.17	1.18
0.8	0.99	1.01	1.03	1.05	1.07	1.09	1.11	1.13	1.15	1.17	1.19	1.21	1.23	1.25	1.27	1.29	1.31	1.33	1.35	1.36
0.9	1.12	1.14	1.17	1.19	1.21	1.24	1.26	1.28	1.31	1.33	1.35	1.37	1.40	1.42	1.44	1.46	1.48	1.51	1.53	1.55
1.0	1.26	1.28	1.31	1.33	1.36	1.39	1.41	1.44	1.46	1.49	1.52	1.54	1.56	1.59	1.61	1.64	1.66	1.69	1.71	1.74
1.1	1.39	1.42	1.45	1.48	1.51	1.54	1.47	1.60	1.63	1.65	1.68	1.71	1.74	1.76	1.79	1.82	1.85	1.87	1.90	1.93
1.2	1.53	1.57	1.60	1.63	1.66	1.69	1.73	1.76	1.79	1.82	1.85	1.88	1.91	1.94	1.97	2.00	2.03	2.06	2.09	2.12
1.3	1.68	1.71	1.75	1.78	1.82	1.85	1.89	1.92	1.96	1.99	2.02	2.06	2.09	2.12	2.16	2.19	2.22	2.26	2.29	2.32
1.4	1.82	1.86	1.90	1.94	1.98	2.01	2.05	2.09	2.13	2.16	2.20	2.24	2.27	2.31	2.34	2.38	2.42	2.45	2.49	2.52
1.5	1.97	2.01	2.05	2.09	2.14	2.18	2.22	2.26	2.30	2.34	2.38	2.42	2.46	2.49	2.53	2.57	2.61	2.65	2.69	2.73
1.6	2.12	2.16	2.21	2.25	2.30	2.34	2.39	2.43	2.47	2.52	2.56	2.60	2.64	2.68	2.73	2.77	2.81	2.85	2.89	2.93
1.7	2.27	2.32	2.37	2.42	2.46	2.51	2.56	2.60	2.65	2.70	2.74	2.79	2.83	2.88	2.92	2.97	3.01	3.06	3.10	3.14
1.8	2.43	2.48	2.53	2.58	2.63	2.68	2.73	2.78	2.83	2.88	2.93	2.98	3.02	3.07	3.12	3.17	3.22	3.26	3.31	3.36
1.9	2.58	2.64	2.69	2.75	2.80	2.85	2.91	2.96	3.01	3.07	3.12	3.17	3.22	3.27	3.32	3.37	3.42	3.47	3.53	3.57
2.0	2.74	2.80	2.86	2.92	2.97	3.03	3.09	3.14	3.20	3.25	3.31	3.36	3.42	3.47	3.53	3.58	3.63	3.69	3.74	3.79
2.1	2.90	2.96	3.03	3.09	3.15	3.21	3.27	3.33	3.39	3.45	3.50	3.56	3.62	3.68	3.73	3.79	3.85	3.91	3.96	4.02
2.2	3.07	3.13	3.20	3.26	3.32	3.39	3.45	3.52	3.58	3.64	3.70	3.76	3.82	3.88	3.94	4.01	4.07	4.13	4.19	4.24
2.3	3.23	3.30	3.37	3.44	3.50	3.57	3.64	3.71	3.77	3.84	3.90	3.97	4.03	4.09	4.16	1.22	4.29	4.35	4.41	4.47
2.4	3.40	3.47	3.54	3.62	3.69	3.76	3.83	3.90	3.97	4.04	4.10	4.17	4.24	4.31	4.37	4.44	4.51	4.58	4.64	4.71
2.5	3.57	3.65	3.72	3.80	3.87	3.95	4.02	4.09	4.17	4.24	4.31	4.38	4.45	4.52	4.59	4.66	4.73	4.81	4.88	4.94
2.6	3.74	3.82	3.90	3.98	4.06	4.14	4.21	4.29	4.37	4.44	4.52	4.60	4.67	4.74	4.82	4.89	4.96	5.04	5.11	5.18
2.7	3.92	4.00	4.09	4.17	4.25	4.33	4.41	4.49	4.57	4.65	4.73	4.81	4.89	4.96	5.04	5.12	5.20	5.27	5.35	5.42
2.8	4.10	4.18	4.27	4.36	4.44	4.53	4.61	4.70	4.78	4.86	4.95	5.03	5.11	5.19	5.27	2.35	5.43	5.51	5.59	5.67
2.9	4.28	4.37	4.46	4.55	4.64	4.73	4.81	4.90	4.99	5.08	5.16	5.25	2.33	5.42	5.50	5.59	5.67	5.75	5.84	5.92
3.0	4.46	4.55	4.65	4.74	4.83	4.93	50.2	5.11	5.20	5.29	5.38	5.47	5.56	5.65	5.74	5.82	5.91	6.00	6.09	6.17
3.1	4.64	4.74	4.84	4.94	5.03	5.13	5.23	5.32	5.42	5.51	5.60	5.70	5.79	5.88	5.97	6.06	6.16	6.25	6.34	6.43
3.2	4.83	4.93	5.03	5.14	5.24	5.34	5.44	5.54	5.64	5.73	5.83	5.93	6.02	6.12	6.21	6.31	6.40	6.50	6.59	6.68
3.3	5.02	5.12	5.23	5.34	5.44	5.55	5.65	5.75	5.86	5.96	6.06	6.16	6.26	6.36	6.46	6.56	6.65	6.75	6.85	6.95
3.4	5.21	5.32	5.43	5.54	5.65	5.76	5.87	5.97	6.08	6.19	6.29	6.40	6.50	6.60	6.70	6.81	6.91	7.01	7.11	7.21

Net Energy Requirements of Growing and Finishing Heifers (continued)

Daily gain lb	Body Weight lb																			
	550	560	570	580	590	600	610	620	630	640	650	660	670	680	690	700	170	720	730	740
	NE$_m$ required, mcgal per day																			
0	4.88	4.95	5.02	5.08	5.15	5.21	5.28	5.34	5.41	5.47	5.53	5.60	5.66	5.73	5.79	5.85	5.91	5.98	6.04	6.10
	NE$_g$ required, mcgal per day																			
0.5	0.84	0.85	0.86	0.87	0.89	0.90	0.91	0.92	0.93	0.94	0.95	0.96	0.97	0.99	1.00	1.01	1.02	1.03	1.04	1.05
0.6	1.02	1.03	1.05	1.06	1.07	1.09	1.10	1.11	1.13	1.14	1.15	1.17	1.18	1.19	1.21	1.22	1.23	1.25	1.26	1.27
0.7	1.20	1.22	1.23	1.25	1.26	1.28	1.30	1.31	1.33	1.34	1.36	1.37	1.39	1.41	1.42	1.44	1.45	1.47	1.48	1.50
0.8	1.38	1.40	1.42	1.44	1.46	1.48	1.49	1.51	1.53	1.55	1.57	1.59	1.60	1.62	1.64	1.66	1.67	1.69	1.71	1.73
0.9	1.57	1.59	1.61	1.63	1.66	1.68	1.70	1.72	1.74	1.76	1.78	1.80	1.82	1.84	1.86	1.88	1.90	1.92	1.94	1.96
1.0	1.76	1.79	1.81	1.83	1.86	1.88	1.90	1.93	1.95	1.97	2.00	2.02	2.04	2.07	2.09	2.11	2.13	2.16	2.18	2.20
1.1	1.96	1.98	2.01	2.03	2.06	2.09	2.11	2.14	2.17	2.19	2.22	2.24	2.27	2.29	2.32	2.34	2.37	2.39	2.42	2.44
1.2	2.15	2.18	2.21	2.24	2.27	2.30	2.33	2.35	2.38	2.41	2.44	2.47	2.50	2.52	2.55	2.58	2.61	2.63	2.66	2.69
1.3	2.35	2.39	2.42	2.45	2.48	2.51	2.54	2.57	2.61	2.64	2.67	2.70	2.73	2.76	2.79	2.82	2.85	2.88	2.91	2.94
1.4	2.56	2.59	2.63	2.66	2.69	2.73	2.76	2.80	2.83	2.87	2.90	2.93	2.96	3.00	3.03	3.06	3.10	3.13	1.36	1.39
1.5	2.76	2.80	2.84	2.88	2.91	2.95	2.99	3.02	3.06	3.10	3.13	3.17	3.20	3.24	3.28	3.31	3.35	3.38	3.42	3.45
1.6	2.97	3.02	3.06	3.09	3.13	3.17	3.21	3.25	3.29	3.33	3.37	3.41	3.45	3.49	3.52	3.56	3.60	3.64	3.68	3.72
1.7	3.19	3.23	3.27	3.32	3.36	3.40	3.44	3.49	3.53	3.57	3.61	3.65	3.70	3.74	3.78	3.82	3.86	3.90	3.94	3.98
1.8	3.40	3.45	3.50	3.54	3.59	3.63	3.68	3.72	3.77	3.82	3.86	3.90	3.95	3.99	4.03	4.08	4.12	4.17	4.21	4.25
1.9	3.62	3.68	3.72	3.77	3.82	3.87	3.91	3.96	4.01	4.06	4.11	4.15	4.20	4.25	4.29	4.34	4.39	4.43	4.48	4.53
2.0	3.85	3.90	3.95	4.00	4.05	4.11	4.16	4.21	4.26	4.31	4.36	4.41	4.46	4.51	4.56	4.61	4.66	4.71	4.76	4.81
2.1	4.07	4.13	4.19	4.24	4.29	4.35	4.40	4.45	4.51	4.57	4.62	4.67	4.72	4.78	4.83	4.88	4.93	4.99	5.04	5.09
2.2	4.30	4.36	4.42	4.48	4.54	4.59	4.65	4.71	4.77	4.82	4.88	4.93	4.99	5.05	5.10	5.16	5.21	5.27	5.32	5.38
2.3	4.54	4.60	4.66	4.72	4.78	4.84	4.90	4.96	5.02	5.08	5.14	5.20	5.26	5.32	5.38	5.44	5.49	5.55	5.61	5.67
2.4	4.77	4.84	4.90	4.97	5.03	5.09	5.16	5.22	5.29	5.35	5.41	5.47	5.53	5.60	5.66	5.72	5.78	5.84	5.90	5.96
2.5	5.01	5.08	5.15	5.22	5.28	5.35	5.41	5.48	5.55	5.62	5.68	5.75	5.81	5.88	5.94	6.01	6.07	6.13	6.20	6.26
2.6	5.26	5.33	5.40	5.47	5.54	5.61	5.68	5.75	5.82	5.89	5.95	6.02	6.09	6.16	6.23	6.30	6.36	6.43	6.49	6.57
2.7	5.50	5.58	5.65	5.72	5.80	5.87	5.94	6.01	6.09	6.16	6.23	6.31	6.38	6.45	6.52	6.59	6.66	6.73	6.80	6.87
2.8	5.75	5.83	5.91	5.98	6.06	6.14	6.21	6.29	6.37	6.44	6.51	6.59	6.67	6.74	6.81	6.89	6.96	7.04	7.11	7.18
2.9	6.00	6.09	6.17	6.25	6.33	6.40	6.48	6.56	6.65	6.73	6.80	6.88	6.96	7.04	7.11	7.19	7.27	7.34	7.42	7.50
3.0	6.36	6.45	6.43	6.51	6.60	6.68	6.76	6.84	6.93	7.01	7.09	7.17	7.26	7.34	7.42	7.50	7.58	7.66	7.73	7.82
3.1	6.52	6.61	6.70	6.78	6.87	6.95	7.04	7.13	7.22	7.30	7.38	7.47	7.56	7.64	7.72	7.81	7.89	7.97	8.05	8.14
3.2	6.78	6.87	6.96	7.05	7.14	7.23	7.32	7.41	7.51	7.60	7.68	7.77	7.86	7.95	8.03	8.12	8.21	8.30	8.38	8.47
3.3	7.04	7.14	7.24	7.33	7.42	7.52	7.61	7.70	7.80	7.89	7.98	8.08	8.17	8.26	8.35	8.44	8.53	8.62	8.71	8.80
3.4	7.31	7.42	7.51	7.61	7.71	7.80	7.90	8.00	8.10	8.20	8.29	8.38	8.48	8.58	8.67	8.76	8.85	8.95	9.04	9.14

Net Energy Requirements of Growing and Finishing Heifers (continued)

Daily gain lb	\multicolumn{20}{c}{Body Weight lb}
	750	760	770	780	790	800	810	820	830	840	850	860	870	880	890	900	910	920	930	940
	\multicolumn{20}{c}{NE$_m$ required, megcal per day}
0	6.16	6.23	6.29	6.35	6.41	6.47	6.53	6.59	6.65	6.71	6.77	6.83	6.89	6.95	7.00	7.06	7.13	7.18	7.24	7.30
	\multicolumn{20}{c}{NE$_g$ required, megcal per day}
0.5	1.06	1.07	1.08	1.09	1.10	1.11	1.12	1.13	1.14	1.15	1.17	1.17	1.19	1.20	1.20	1.22	1.23	1.24	1.25	1.26
0.6	1.28	1.30	1.31	1.32	1.34	1.35	1.36	1.37	1.39	1.40	1.41	1.42	1.44	1.45	1.46	1.47	1.49	1.50	1.51	1.52
0.7	1.51	1.53	1.54	1.56	1.57	1.59	1.60	1.62	1.63	1.65	1.66	1.68	1.69	1.71	1.72	1.73	1.75	1.76	1.78	1.79
0.8	1.75	1.76	1.78	1.80	1.81	1.83	1.85	1.87	1.89	1.90	1.92	1.93	1.95	1.97	1.98	2.00	2.02	2.03	2.05	2.07
0.9	1.98	2.00	2.02	2.04	2.06	2.08	2.10	2.12	2.14	2.16	2.18	2.20	2.22	2.23	2.25	2.27	2.29	2.31	2.33	2.35
1.0	2.22	2.25	2.27	2.29	2.31	2.33	2.35	2.38	2.40	2.42	2.44	2.46	2.48	2.51	2.53	2.55	2.57	2.59	2.61	2.63
1.1	2.47	2.49	2.52	2.54	2.57	2.59	2.61	2.64	2.66	2.69	2.71	2.73	2.76	2.78	2.80	2.83	2.85	2.88	2.90	2.92
1.2	2.72	2.74	2.77	2.80	2.82	2.85	2.88	2.90	2.93	2.96	2.98	3.01	3.04	3.06	3.09	3.11	3.14	3.17	3.19	3.22
1.3	2.97	3.00	3.03	3.06	3.09	3.12	3.14	3.17	3.20	3.23	3.26	3.29	3.32	3.35	3.37	3.40	3.43	3.46	3.49	3.52
1.4	3.23	3.26	3.29	3.32	3.35	3.39	3.42	3.45	3.48	3.51	3.55	3.57	3.61	3.64	3.67	3.70	3.73	3.76	3.79	3.82
1.5	3.49	3.52	3.56	3.59	3.63	3.66	3.69	3.73	3.76	3.80	3.83	3.86	3.90	3.93	3.96	4.00	4.03	4.06	4.10	4.13
1.6	3.75	3.79	3.83	3.86	3.90	3.94	3.97	4.01	4.05	4.08	4.12	4.16	4.19	4.23	4.27	4.30	4.34	4.37	4.41	4.45
1.7	4.02	4.06	4.10	4.14	4.18	4.22	4.26	4.30	4.34	4.38	4.42	4.46	4.50	4.54	4.57	4.61	4.65	4.69	4.73	4.77
1.8	4.30	4.34	4.38	4.42	4.47	4.51	4.55	4.59	4.63	4.68	4.72	4.76	4.80	4.84	4.88	4.92	4.97	5.01	5.05	5.09
1.9	4.57	4.62	4.66	4.71	4.75	4.80	4.84	4.89	4.93	4.98	5.03	5.07	5.11	5.16	5.20	5.24	5.29	5.33	5.37	5.42
2.0	4.85	4.90	4.95	5.00	5.05	5.09	5.14	5.19	5.24	5.28	5.33	5.38	5.43	5.47	5.52	5.57	5.61	5.66	5.70	5.75
2.1	4.14	5.19	5.24	5.29	5.34	5.39	5.44	5.49	5.54	5.60	5.65	5.70	5.75	5.80	5.84	5.89	5.94	5.99	6.04	6.09
2.2	5.43	5.49	5.54	5.59	5.65	5.70	5.75	5.80	5.86	5.91	5.97	6.02	6.07	6.12	6.17	6.23	6.28	6.33	6.38	6.43
2.3	5.72	5.78	5.84	5.89	5.95	6.01	6.06	6.12	6.17	6.23	6.29	6.34	6.40	6.45	6.51	6.56	6.62	6.67	6.73	6.78
2.4	6.02	6.08	6.14	6.20	6.26	6.32	6.38	6.44	6.50	6.55	6.62	6.67	6.73	6.79	6.84	6.90	6.96	7.02	7.08	7.13
2.5	6.32	6.39	6.45	6.51	6.57	6.64	6.70	6.76	6.82	6.88	6.95	7.01	7.07	7.13	7.19	7.25	7.31	7.37	7.43	7.49
2.6	6.63	6.70	6.76	6.83	6.89	6.96	7.02	7.09	7.15	7.22	7.29	7.35	7.41	7.48	7.54	7.60	7.67	7.73	7.79	7.86
2.7	6.94	7.01	7.08	7.15	7.22	7.28	7.35	7.42	7.49	7.55	7.63	7.69	7.76	7.83	7.89	7.96	8.02	8.09	8.16	8.22
2.8	7.25	7.33	7.40	7.47	7.54	7.61	7.68	7.76	7.83	7.90	7.97	8.04	8.11	8.18	8.25	8.32	8.39	8.45	8.52	8.60
2.9	7.57	7.65	7.73	7.80	7.87	7.95	8.02	8.10	8.17	8.24	8.32	8.39	8.47	8.54	8.61	8.68	8.76	8.82	8.90	8.97
3.0	7.90	7.98	8.05	8.13	8.21	8.29	8.36	8.44	8.52	8.59	8.68	8.75	8.83	8.90	8.98	9.05	9.13	9.20	9.28	9.35
3.1	8.22	8.31	8.39	8.47	8.55	8.63	8.71	8.79	8.87	8.95	9.04	9.11	9.19	9.27	9.35	9.43	9.51	9.58	9.66	9.74
3.2	8.55	8.64	8.73	8.81	8.89	8.98	9.06	9.14	9.23	9.31	9.40	9.48	9.56	9.64	9.72	9.81	9.89	9.97	10.05	10.13
3.3	8.89	8.98	9.07	9.15	9.24	9.33	9.41	9.50	9.59	9.67	9.77	9.85	9.94	10.02	10.10	10.19	10.28	10.36	10.44	10.53
3.4	9.23	9.32	9.41	9.50	9.59	9.68	9.77	9.86	9.55	10.04	10.14	10.22	10.31	10.40	10.49	10.58	10.67	10.75	10.84	10.93

Net Energy Requirements of Growing and Finishing Heifers (continued)

Daily gain lb	Body Weight lb									
	950	960	970	980	990	1000	1010	1020	1030	1040
	NE_m required, megcal per day									
0	7.36	7.42	7.47	7.53	7.59	7.65	7.71	7.76	7.82	7.87
	NE_g required, megcal per day									
0.5	1.27	1.28	1.29	1.30	1.31	1.32	1.33	1.34	1.34	1.35
0.6	1.53	1.55	1.56	1.57	1.58	1.59	1.61	1.62	1.63	1.64
0.7	1.81	1.82	1.83	1.85	1.86	1.88	1.89	1.91	1.92	1.93
0.8	2.08	2.10	2.12	2.13	2.15	2.17	2.18	2.20	2.21	2.23
0.9	2.37	2.39	2.40	2.42	2.44	2.46	2.48	2.50	2.51	2.53
1.0	2.65	2.68	2.70	2.72	2.74	2.76	2.78	2.80	2.82	2.84
1.1	2.95	2.97	3.00	3.02	3.04	3.06	3.09	3.11	3.13	3.15
1.2	3.24	3.27	3.29	3.32	3.35	3.37	3.40	3.42	3.45	3.47
1.3	3.54	3.57	3.60	3.63	3.66	3.68	3.71	3.74	3.77	3.79
1.4	3.85	3.88	3.91	3.94	3.97	4.00	4.03	4.06	4.09	4.12
1.5	4.16	4.20	4.23	4.26	4.29	2.33	4.36	4.39	4.42	4.46
1.6	4.48	4.52	4.55	4.59	4.62	4.66	4.69	4.73	4.76	4.79
1.7	4.80	4.84	4.88	4.92	4.95	4.99	5.03	5.07	5.10	5.14
1.8	5.13	5.17	5.21	5.25	5.29	5.33	5.37	5.41	5.45	5.49
1.9	5.46	5.50	5.55	5.59	5.63	5.67	5.72	5.76	5.80	5.84
2.0	5.80	5.84	5.89	5.93	5.98	6.02	6.07	6.11	6.16	6.20
2.1	6.14	6.19	6.13	6.28	6.33	6.38	6.43	6.47	6.52	6.57
2.2	6.48	6.54	6.58	6.64	6.69	6.74	6.79	6.84	6.89	6.94
2.3	6.83	6.89	6.94	7.00	7.05	7.10	7.16	7.21	7.26	7.31
2.4	7.19	7.25	7.30	7.36	7.42	7.47	7.53	7.58	7.64	7.69
2.5	7.55	7.61	7.67	7.73	7.79	7.85	7.91	7.96	8.02	8.08
2.6	7.92	7.98	8.04	8.10	8.17	8.23	8.29	8.35	8.41	8.47
2.7	8.29	8.35	8.42	8.48	8.55	8.61	8.68	8.74	8.80	8.87
2.8	8.66	8.73	8.80	8.87	8.93	9.00	9.07	9.14	9.20	9.27
2.9	9.04	9.12	9.18	9.26	9.33	9.40	9.47	9.54	9.61	9.68
3.0	9.43	9.50	9.58	9.65	9.72	9.80	9.87	9.94	10.02	10.09
3.1	9.82	9.90	9.97	10.05	10.13	10.20	10.28	10.36	10.43	10.51
3.2	10.21	10.30	10.37	10.46	10.53	10.61	10.69	10.77	10.85	10.93
3.3	10.61	10.70	10.78	10.87	10.95	11.03	11.11	11.19	11.28	11.36
3.4	11.02	11.11	11.19	11.28	11.36	11.45	11.54	11.62	11.70	11.79

Appendix Table 4 Nutrient Requirements of Swine[1]

Nutrient Requirements of Growing Swine

Weight of swine (lbs)	Expected gain (lb)	ME Mcal/lb of feed	Crude protein %	Lysine %	Calcium %	Phosphorus %	Sodium %	Magnesium %	Vit. A (IU/lb of feed)	Vit. D (IU/lb of feed)	Vit. E (IU/lb of feed)	Thiamine (mg/lb of feed)	Niacin (mg/lb of feed)	Panothenic acid (mg/lb of feed)	Riboflavin (mg/lb of feed)	B_{12} microgram/lb of feed
2-11	.4	1.5	24	1.4	.9	.7	.1	.04	1000	100	7.2	.7	9.1	5.5	1.8	9.0
11-22	.5	1.5	20	1.2	.8	.65	.1	.04	1000	100	7.2	.45	6.8	4.5	1.6	8.0
22-44	1.0	1.5	18	.95	.7	.60	.1	.04	800	90	5.0	.45	5.6	4.1	1.4	6.8
44-110	1.5	1.5	15	.75	.6	.50	.1	.04	590	70	5.0	.45	4.5	3.6	1.2	4.5
110-220	1.8	1.5	13	.60	.5	.40	.1	.04	590	70	5.0	.45	3.2	3.2	.9	2.3
Amounts Required by Breeding Swine																
Bred gilts, sows & adult boars	N/A	1.45	12	.43	.75	.6	.15	.04	1800	90	10	.45	4.5	5.4	1.7	6.8
Lactating sows & gilts	N/A	1.45	13	.60	.75	.6	.20	.04	900	90	10	.45	4.5	5.4	1.7	6.8

[1] Adapted from National Research Council.

Appendix Table 5. Nutrient Requirements of Sheep.[1]

Type of animal	Weight gain (lb)	TDN (%)	ME (Mcal/lb)	Crude protein (%)	Calcium (%)	Phosphorus (%)	Vit. A (IU/lb)	Vit. E (IU/lb)
Dry, open ewe	0	55	.9	9.4	.20	.20	1300	7
Dry, pregnant ewe (first 15 weeks)	.1	58	.95	9.5	.25	.20	1300	7
Ewe, last 4 weeks pregnancy	.4	60	.98	10.7	.35	.23	1500	7
Ewe, last 4 weeks w/twins expected	.5	65	1.1	11.3	.40	.24	1500	7
Ewe, suckling lamb first 6-8 weeks	(-.1)	67	1.2	13.4	.35	.26	1500	7
Ewe, suckling twins first 6-8 weeks	(-.15)	68	1.2	15.0	.40	.29	1500	7
Pregnant ewe lamb (first 15 weeks)	.3	59	.95	10.6	.35	.22	1300	7
Ewe lamb, last mo. gestation	.35	63	1.05	11.8	.39	.22	1300	7
Ewe lamb, last mo. expecting twins	.5	68	1.1	12.8	.48	.25	1300	7
Lactating ewe lamb first 6-8 weeks	(-.1)	68	1.1	13.1	.35	.23	1500	7
Lactating ewe lamb w/twins	(-.2)	70	1.2	13.7	.37	.26	1500	7
65 lb replacement ewe lambs	.5	65	1.1	12.8	.53	.22	600	7
90 lb replacement ewe lambs	.4	65	1.1	10.2	.42	.18	700	7
130 lb replacement ewe lambs	.25	59	.95	9.1	.31	.17	800	7
90 lb replacement ram lambs	.75	63	1.0	13.5	.43	.21	600	7
130 lb replacement ram lambs	.7	63	1.0	11.0	.35	.18	700	7
Finishing lambs 65 lb	.65	72	1.1	14.7	.51	.24	500	9
Finishing lambs 90 lb	.6	76	1.2	11.6	.42	.21	600	9
Finishing lambs 110 lb	.45	77	1.3	10.0	.35	.19	800	7

[1] Source: NRC with some modification by the senior author.

FEED TABLES[1]
(All analysis are on a dry matter basis)

Feedstuff	% Dry Matter	% Crude Protien	% Crude Fiber	% Crude Fat	% Calcium	% Phosphorus	Carotene[1] (vit. A activity 1000 IU's/lb)	ME Mcal/lb	% TDN	NE_m Mcal/lb	NE_g Mcal/lb	NE_l Mcal/lb	% Potassium	% Magnesium
Alfalfa hay prebloom	85-90	18-20	23-25	2.8-3.0	1.8-2.0	.25-.33	25-35	.98	60-63	.59	.34	.66	2.2-2.6	.25-.35
Alfalfa hay mid-bloom	85-90	15-17	26-30	2.6-2.8	1.5-2.0	.18-.24	20-30	.95	56-58	.56	.30	.57	2.0-2.4	.20-.25
Alfalfa hay mature	88-92	12-14	31-37	1.2-1.4	1.2-1.4	.12-.17	2-10	.81	50-52	.45	.20	.53	1.7-1.9	.12-.15
Barley	88-90	12-14	5.5	2.0	.05	.30-.38	.2	1.4	80-84	.94	.64	.97	.4-.5	.12-.16
Bermuda-grass hay[2]	90-92	6-10	30-35	2.0	.3-.4	.10-.15	6-15	.7	42-46	.38	.13	.46	.6-.7	.1-.15
Brewers grains (wet)	14-22	22-24	15	6.5-7.0	.3	.5-.6	0	1.1	65-68	.68	.41	.70	.09	.16
Carrots	12	10	9-10	1.5	.4	.35	122	1.4	84	.94	.63	.96	2.8	.2
Cassava fresh	35-38	3.0	5.0	1.0	.28	.18	0	1.3	80	.88	.59	--	.2-.3	--
Clover hay immature	86-90	17-20	22-26	3.0-5.0	1.4-2.0	.2-.35	4-9	.9-1.0	58-60	.56	.32	.62	1.6-2.0	.5
Corn grain	85.5	10	2.5	4.0-4.2	.02	.28-.35	.3-.4	1.47	90	1.02	.68	1.1	.35-.4	.35
Corn distillers grains (dehydrated)	90-94	23.0	12.1	9.8	.11	.43	.6	1.41	86	.96	.66	.99	.18	.07
Corn gluten feed	90	25.6	9.7	2.4	.36	.82	1.1	.136	83	.92	.62	.96	.64	.36
Corn gluten meal	90-91	46.8	4.8	2.4	.16	.50	3.0	1.46	86	.96	.66	.99	.03	.06
Corn silage	32-38	7-8	25-35	3.0	.2-.35	.16-.21	1.4-6	1.01	60-68	.65	.41	.77	1.0-1.4	.18-.10
Cottonseed hulls	90-92	4.0	48	1.5-2.0	.15	.09	0	.69	42	.31	.07	.34	.87	.14
Cottonseed meal (solvent method)	91	41	14.1	1-1.5	.22	1.0-1.2	0	1.31	80	.88	.59	.86	1.2	.55
Cottonseeds whole	92	22-24	21	23	.16	1.16	0	1.57	96	1.09	.76	1.21	1.2	.35
Fats & Oils	96-99	0	0	99-100	0	--	0	2.9	177	2.15	1.59	.99	0	0
Fish meal (all)	90-93	59-72	1.0	4-10	5-8.9	2.7-3.8	--	--	--	.75	.49	--	3.0	--
Grass hay[2]	88-90	6-10	30-38	2.0	.18-.35	.08-.15	1.5-7	.6-.75	42-46	.38	.12-.15	.42-.46	.6-.8	.1-.2
Grape pumace	90	.7	32	7.9	.6	.06	0	.44	27	.04	0	--	.6	--
Molasses beet	78	8.5	--	.2	.17	.03	0	1.3	79	.86	.58	.86	6.0	.29
Molasses cane	75	4-6	--	.1	1.0	.11	0	1.2	72	.77	.49	1.1	3.8	.43
Oats	89-92	11-13	12	5.4	.07	.3-.38	0	1.2	77	.84	.55	.86	.44	.14
Rapeseed meal (solvent method)	90	43.6	15.3	1.8	.67	1.0	0	1.1	69	.72	.45	.75	1.3	.60
Sorghum grain	88-90	9-10.5	2.5	2.4	.04	.24-.35	0	1.35	81-83	.91	.61	.93	.3-.35	.18
Sorghum flaked	81-86	9-10.5	2.5	2.4	.04	.24-.35	0	1.51	84-86	.97	.64	--	.3-.35	.18
Sorghum silage	24-28	6-7	28-34	2.6	.28-.35	.15-.17	2-6	.95	54-58	.56	.31	.58	1-1.2	.25-.3
Soybeans (whole)	92	42	5.8	18.8	.27	.65	.1-.2	1.49	91	1.03	.71	1.15	1.8	.28
Soybean hulls	91	12	40.1	2.1	.49	.21	0	1.05	64	.65	.39	.89	1.3	--
Soybean meal	89	49	7.0	1.5	.33	.71	0	1.38	84	.93	.63	1.00	2.1	.3

FEED TABLES[1]
(All analysis are on a dry matter basis)

Feedstuff	% Dry Matter	% Crude Protien	% Crude Fiber	% Crude Fat	% Calcium	% Phosphorus	Carotene (vit. A activity 1000 IU's/lb)	ME Mcal/lb	% TDN	NE_m Mcal/lb	NE_g Mcal/lb	NE_l Mcal/lb	% Potassium	% Magnesium
Alfalfa hay prebloom	85-90	18-20	23-25	2.8-3.0	1.8-2.0	.25-.33	25-35	.98	60-63	.59	.34	.66	2.2-2.6	.25-.35
Sunflower meal w/hulls	90	25.9	35	1.2	.23	1.03	0	.72	44	.34	.10	--	1.1	.75
Sunflower meal w/o hulls	93	49.8	12.2	3.1	.44	.98	0	1.06	65	.67	.40	.69	1.1	.77
Urea (40% N)	99	287	0	0	0	0	0	0	0	0	0	0	0	0
Wheat grain	88	11-14	2.4	1.8	.05	.35-.43	0	1.44	88	1.00	.68	.70	.4	.15-.18
Wheat bran	89	16-17	11.3	4.4	.13	1.38	6	1.15	70	.74	.47	.77	1.5	.6
Wheat midds	89	17-18	8.2	4.9	.13	.9-.99	.65	1.13	69	.72	.45	.97	1.1	.4

Special notes on use of the Feed Tables:

[1] In many cases ranges are given for feeds instead of absolute values. This makes use more confusing, but absolute values as printed in most feed tables can be very misleading. The use of ranges is more realistic.

[2] Many tables give individual values for various types of grasses and grass hay. This is misleading since the variations between most species of grasses is not nearly as important as the maturity of the grass (hay). The ranges given will represent the variation for most grasses. If green and immature the higher figures will be applicable and if yellow and immature, the lower figures will be more applicable.

[3] Excerpted and condensed from Modern Practical Feeds, Feeding and Animal Nutrition by D.P. Price.

Major Common Minerals

Compound	Crude protein equivalency (NPN) %	Calcium %	Phosphorus %	Sodium %	Chlorine %	Potassium %	Magnesium %	Sulfur %	Iron ppm	Copper ppm	Zinc ppm	Cobalt ppm	Manganese ppm
Ammonium monophosphate	70.9	.28	24.7	.06			.46	1.5	17,400	10	100	10	400
Ammonium phosphate dibasic (Diammonium phosphate)	115.9	.50	20.0	.05			.46	2.2	12,400	10	100		400
Bone meal steamed	8.4	24.0	14.0	.06									
Calcium carbonate	0	39.4	.04	.06		.06	.05		300				300
Calcium sulfate (gypsum)		22.0	.01				2.6	19.9	2,010				
Deflourinated phosphate rock	0	32.0	18.0	4.0		.09	.6	1.1	10,000	80	220		300
Dicalcium phosphate	0	20.0	18.5	.08			.6	1.1	14,400		100		
Sodium Chloride (salt)				40	60								
Magnesium oxide							50.0						
Magnesium sulfate							10-20.0	13-26					

Common Trace Minerals

Compound	Iron %	Copper %	Zinc %	Cobalt %	Manganese %
Cobalt carbonate				45	
Cobalt sulfate				21-33	
Copper carbonate		53			
Copper oxide		75			
Copper sulfate		25			
Iron carbonate	43				
Iron sulfate	21				
Manganous oxide	3.4				60
Manganous sulfate					25
Zinc oxide			73		
Zinc sulfate			36		

GLOSSARY

abomasum - the "true" stomach of the ruminant animal. Located after the rumen, feeds are digested essentially the same as in monogastric animals (after they have passed through the rumen).
accrual accounting - a form of accounting in which inventories must be taken into consideration when figuring profit and loss (as opposed to cash accounting).
acetic acid - a volatile fatty acid produced in the rumen in greater quantities than other volatile fatty acids. Used as a source of energy by ruminant animals. Of great importance in dairying since acetic acid is the precursor for synthesis of butterfat.
acidosis - a metabolic disturbance caused by giving ruminant animals grain or other concentrates too rapidly.
ad libitum - in animal feeding, giving as much as the animal will eat.
additive effect - in genetics, refers to genes that complement one another in a predictable arithmetic manner. Genes that affect quanitative traits. This is apposed to dominant or recessive genes that affect traits in an "all or nothing" manner (qualitative traits).
aerobic - relates to bacteria and other microorganisms that require oxygen to survive.
air dry - feeds at a normal dry matter of 90%.
alkali - soil containing sodium salts in amounts that damage plants.
alkaloids - toxic compounds found in some undesirable range plants.
alternate feeding (skip feeding) - generally refers to the feeding of range supplements every other day or occasionally every third day.
amino acid - organic molecules containing nitrogen necessary for the formation of protein; e.g. the building blocks of protein. A feed is said to be "balanced" for amino acids if it contains all the amino acids the animal requires for the formation of body tissue.
amylolytic bacteria - rumen bacteria that digest starch.
anaerobic bacteria - bacteria that do not require oxygen as a gas to survive.
anhydrous ammonia - a form of ammonia salts used as fertilizer.
anions - a negatively charged ion.
annuals - pasture plants or crops propagated by seeds, and must be planted or regenerated each year.
antibodies - circulating protein molecules that help neutralize disease organisms.
apparent digestible protein - a laboratory analysis used to estimate the digestibility of protein in forages.
application efficiency - percent of irrigation water that is stored in soil for crop use.
appraised values - values which are supposed to represent market value (as apposed to book value).
appreciate - in finance refers to assets that increase in value over time.
aquifer - underground water-bearing rock, sand or gravel.
as fed - refers to feeds analyzed, formulated or reported on the amount of moisture normally contained within them. (As apposed to a dry matter basis.)
asexual reproduction - reproduction such as cellular reproduction in which the nucleus divides on its own, and two cells appear out of one.
ash - in feeds refers to what is left after all organic matter is burned off. Represents the total amount of mineral contained in the feed.
assets - in finance refers to any item which has tangible value to the firm.
associative effects of feedstuffs - a concept which deals with the manner in which the values of various feedstuffs change depending upon how they are combined in a ration.
available moisture - moisture in soil which is available for plant growth (not bound to soil particles).
awn - the bristle extending from the tip of a grass (including small grain) seed.

balance sheet - the basic accounting analysis of a business. Outlines the assets, liabilities and owner's equity.
balled tree - a nursery tree that is dug from the nursery row and transported to the orchard site with soil around the roots remaining intact. (see bare root)
bare root - a nursery tree dug from the nursery row without soil around the roots.
base - in relation to soils and animal nutrition, a positively charged mineral such as Ca, K or Na.
basin irrigation - irrigation by flooding levelled fields with borders built up around to hold the water.
basis - in futures trading the difference between the future's price and the actual cash price for a particular area.
beard - the awn of grasses.
bin - a container (usually holding about one-thousand pounds) used by growers or pickers to transport fruit from the orchard to the packinghouse.
black face breeds - meat breeds of sheep.

black fleece - fleece containing so many black fibers that white or light colored cloth cannot be made from it; thus, the wool's value is reduced.

black layer - an abscission (separation) layer found at the base of corn kernels after all grain filling has been completed.

book values - accounting values which reflect purchase price of an asset less any calculated depreciation (rather than true market value).

booster vaccination - a second or multiple vaccination given to increase an animal's resistance to a specific disease.

boot stage - the initiation of the formation of seed (grain) in cereal crops.

border irrigation - irrigation water is applied at the upper end of a strip with earth borders to confine the water to the strip.

breech birth - a birth in which the hind feet of the young are presented first.

broiler - male chicken raised for meat.

broken-mouth - a sheep that has lost part of its permanent incisors, usually at 5 or more years of age.

Browning Reaction - the name of the chemical reaction that occurs when feeds (usually forages) become wet and overheat (destroying the nutritional value).

buffer - a compound used to minimize the acidity or alkalinity of a fluid.

bulk density - weight of dry soil per unit volume, normally expressed as gram per cubic centimeter.

bummer (orphan) - a lamb that is not raised by its mother; usually raised on a bottle.

calcareous - containing lime (calcium carbonate).

call (option) - an option which gives the purchaser the right to "call" the issuer at any time (during the life of the option) and demand that a future's contract be sold to him for a set price.

candled - egg examined for the presence of a developing embryo.

capillary water - water which remains in the pore spaces of the soil after all gravitational water is removed.

carbohydrate - a combination of carbon, hydrogen and oxygen which animals use as a source of energy. Starch (from grains) is normally thought of as carbohydrate, but cellulose (fiber) is also a carbohydrate.

carotene - compounds found in plants that can be used by animals such as ruminants, for synthesis of vit. A.

carrier - in feed formulation, carrier usually refers to a bulky feed product used to dilute more concentrated ingredients such as vitamins, minerals, medications, etc.

cash accounting - a form of income accounting which does not take inventory into consideration. Usually used only for tax purposes.

cash flow - an analysis of the movement of cash in and out of a business for a given time period.

cash market - refers to the actual physical market price(s) for agricultural commodities (as apposed to future's prices).

cation - a positively charged ion.

cellulolytic bacteria and microorganisms - refers to rumen bacteria and protozoa that digest cellulose or fiber.

cellulose - the structural carbohydrates in plants. Usually referred to as fiber.

challenge feeding - feeding a dairy cow for a high production until she proves incapable of producing that much.

chelate - a mineral bound to an organic molecule.

chemical thinner - a spray material applied at bloom or to young fruit to remove part of the crop.

chilling requirement - a dormant rest period required by plants before they bloom and grow normally the following spring (usually measured in hours below 45 degrees F).

chlorosis - when diseased plants lose their green color and become yellow.

chromosomes - the "strands" of genetic material on which the genes are located.

clay - (1) soil particles less than 0.002 mm in diameter or (2) a soil containing more than 40% clay, less than 45% sand, and less than 40% silt.

clinical - as a term used to describe a disease, the word clinical means the animal is noticeably ill. This is opposed to the term subclinical, which means the animal contracts the disease, but outwardly appears to be healthy.

clip - wool from a given flock; also, total yearly production.

cloaca - the combined external genitalia and waste vent in birds.

coccidosis - a parasitic disease of livestock that involves bloody scouring and occasionally central nervous disorders.

collateral - in finance, an asset that is used to secure a loan. Upon default the lender takes possession of the asset.

colostrum - first milk given after birth. High in antibodies, this milk protects the newborn against diseases.

commodity exchange - the institution in which futures contracts are bought and sold.

compensatory gain - a phenomenon whereby livestock which have been previously deprived of nutrients, gain at a more rapid rate when realimented.

component pricing - a method of pricing milk by which the amount of milk solids determine the price rather than simply the volume of milk.
concentrates - a feedstuff high in energy and/or protein; e.g. grains and oilmeals.
condition - amount of fat and muscle tissue on an animal's body.
confinement operations - livestock operations which are totally enclosed and environmentally controlled.
consistency - the ability of soil to resist crushing and change in shape. Terms used to describe consistency in soils are plastic, sticky, friable, loose, and firm.
container grown tree - nursery trees grown in plastic or metal sleeves or cans for transporting to permanent sites.
continuous cropping (mono-culture) - planting the same crop year after year.
continuous grazing - leaving grazing animals in one pasture year 'round.
corrugation - irrigation water applied to small, closely spaced furrows, frequently in grain and forage crops, to confine the flow of irrigation water to one direction.
cotyledons - primary leaves of the embryo that furnish carbohydrates for growth of emerging plants.
craw - the food storage area of a bird or insect.
creep - feeding area available to young animals but mothers are excluded.
crimp (wool) - natural waviness of wool fibers.
crimping - in feeds a form of steam rolling grain for a desired texture. Also, having equipment that uses a corrugated roller to speed up the drying process.
crop rotation - a definite sequence of crops grown on a particular field over a 2-6 year period and that is repeated when the cycle is completed.
crossbreeding - usually refers to the breeding of two different breeds of livestock.
crude fiber - a chemical analysis which theoretically includes the materials in feeds that have a low digestibility.
crude protein - a chemical approximation of the total amount of protein in a feed. Specifically, the crude protein analysis measures the nitrogen in a feed (nitrogen is associated with the protein fraction of a feed), and then the percent nitrogen is multiplied times 6.25.
crustaceans - an eight legged arthropod.
crutching (noun) - wool removed from sheep during the crutching or tagging process. This wool is usually free of manure as opposed to tags - which contain a lot of manure.
crutching (tagging) - removing wool from the inside of a sheep's back legs and belly.
cubed - a form of hay packaging. Hay is pressed into small 3 or 4 inch "cubes".
cull - an animal removed from a herd or flock for non or low productivity.
cultivar (cultivated variety) - any variety of a species maintained true to type under cultivation.

dam - a female parent.
deciduous - plants that lose their leaves in the fall.
deep-water rice (floating) - rice produced in a system with standing water during most of the growing season.
delivery point - a specified city (usually there are several possible cities) commodities may be delivered to when physical delivery consumates a futures contract.
demand - in economics, the sum of the amount of a particular good or commodity consumers will purchase at various prices.
dental pad - an extension of the gums on the front part of the upper jaw in ruminants, it is a substitute for top front teeth.
dentrification - a microbial process that reduces nitrogen (NO_3^-, NO_2^-) to gaseous compounds (N_2, N_2O, NO) that may be lost from the soil. Particularly a problem in waterlogged soils.
dicotyledon - a plant with two seed leaves.
dicumerol - a toxic compound produced by mold in sweet clover.
diffusion - the movement of individual gases across a gradient; from areas of high concentration to those of low concentration. Nutrients move through the water during diffusion.
digestibility - percentage of a feedstuff digested and absorbed in digestive tract of animal.
docking - to remove the sheep's tail.
dolomite - a combination of limestone, calcium carbonate, and magnesium carbonate.
dominant - in genetics, single genes that control a particular trait.
dough stage - a stage in grain maturation just before maturity.
drip (trickle) irrigation - irrigation applied slowly and frequently, often to individual plants through emitters in tubing.
dry matter - the amount of feed by weight left when all the water is removed.
dystocia - difficult birth.

ecology - a branch of science concerned with patterns of relationship among organisms and their environment.
economic injury level - the lowest pest population density that will cause economic damage.
economic threshold - the density at which control measures should be applied to prevent an increasing pest population from reaching the economic injury level.
eluviation - process of removing soil material in suspension from a layer or layers of soil.
emaciation - loss of flesh resulting in extreme leaness.
empirically derived - a formula or conclusion developed by trial and error or other non-scientific means.
endosperm - pertains to the starchy inner core of grains.
ensiling - the process of preserving wet, moist feeds by excluding all air, and allowing acid producing bacteria to lower the pH.
entomology - the scientific study of insects and their relatives.
enzyme - compounds that are necessary or otherwise facilitate biochemical reactions; e.g. proteolytic enzymes cause the breakdown of dietary proteins into their constituent amino acids for absorption.
epistasis - interaction between genes which otherwise have no relation to one another, to produce unusual or unexpected traits.
epithelial - pertaining to the skin.
equilibrium price - where demand and supply curves intersect.
equity - commonly referred to as the amount of money an investor has placed in an investment that has come out of his own pocket; i.e. contrasted to the amount of borrowed money; e.g. a 30% equity investment means the investor has put up 30% of the money and borrowed 70%.
estrogen - the primary female hormone causing estrus to occur.
estrus - an animal is receptive to mating and can conceive.
estrus synchronization - synchronizing female farm animals to come into heat at the same time.
evapotranspiration - total water loss from field to air; includes both evaporation from soil surface and transpiration.
evergreen - plants whose leaves remain all year.

fallow - land left idle in order to restore productivity, mainly through accumulation of water, nutrients or both.
fallow efficiency - precentage of the total precipitation received during the fallow period that is stored in the soil.
farrow - giving birth to a litter of pigs.
fat-soluble vitamins - vitamins stored in fat (A, D, E and K).
fermentative heat loss - heat generated by microorganisms during fermentation which constitutes an energy loss in feeds.
fiber - the cellulose and lignified portions of feeds.
fibrilate - a term describing a dysfunction of the heart. A fibrilating heart is one that does not beat with its normal rhythm, but rather, spasmotically contracts and relaxes at an increased, but irregular and disorganized pace. The end result is that blood is not pumped through the circulatory system, and if not corrected, will usually result in death.
field capacity - the greatest amount of water that the soil will hold under conditions of free drainage, usually expressed as a percent of oven dry weight of soil (also called water holding capacity).
financial statement - another name for the balance sheet.
finishing - the act of feeding an animal to produce a desirable carcass for market.
fixed cost - a cost that is incurred by a business whether anything is produced or not (e.g. real estate taxes, etc.).
fleece - wool as it is shorn from sheep; the fleece should remain in one piece.
flooding - irrigation water released from field ditches and allowed to flood over the land.
flushing - increasing the plane of nutrition of an animal before and during the breeding season.
fly strike - when green and blue blowflies lay eggs in wet and stained wool and maggots develop.
fob - free on board. Price quote at shipping point, not including freight and other expenses.
foliar - pertaining to the leaves of a plant.
follicle - a growth on the ovary which produces estrogen and eventually releases ova.
forage - generally refers to succulent plant growth used for feeds.
founder - in animal nutrition, refers to chronic acidosis in cattle and horses, which results in abnormal hoof growth and lameness.
friction head - pressure decreases (head loss) due to water flowing along rough walls of pipe and through fittings.
frost - in horticulture, generally meaning temperatures below freezing.
fruitwood - vegetative portion of a tree upon which fruit is produced.
full bearing - an orchard old enough to be in maximum production.

fungicide - a chemical or other substance (natural or man made) that kills fungi.
furrow irrigation - water applied between row crops in ditches made by tillage implements.

gametes - reproductive cells used in sexual reproduction (sperm cells and ova).
ganglion - a mass of nerve cells that serves as a center of nervous influence (in lieu of a true brain).
gene frequency - the percentage of a particular gene that appears within a population.
genes - the chemical codes that carry inheritance.
genotype - the genetic makeup of an animal or plant (as apposed to the phenotype).
germ - refers to the genetic material in a seed.
germination - the sprouting of a seed.
gilt - a young female pig.
gonadotrophic hormones - hormones used to stimulate or control the reproductive functions and the expression of sexual characteristics.
graft - inserting part of one tree or plant into another to grow.
graft (animal) - using some sort of trick to make a cow or ewe believe an orphan calf or lamb is her own.
grass tetany - a phenomenon in grazing cattle in which massive muscle spasms cause death which is related to magnesium and calcium content of the grass (also called magnesium tetany). Most common in small grain pastures (also called wheat pasture poisoning).
gravitational water - water in excess of the soil water storage capacity that moves down through the soil profile due to the force of gravity.
gravity flow water - water from a surface source (river, stream, canal) that does not require pumping (as in wells).
grease wool - wool shorn from the sheep before it has been cleaned.
green chop - forage crops harvested and delivered to livestock on a daily basis.
groundwater - water found underground in porous rock and soils.
gypsum - calcium sulfate.

hammer mill - a piece of machinery used to grind feeds by means of a corrugated screen through which the feed is forced by means of a series of rotary hammers.
hardpan - a hardened layer of soil resulting from cementation of soil particles that restricts root growth and water and nutrient movement, occurring only in certain soil types.
hay conditioners - commonly refers to corrugated rollers over which hay is squeezed, which allows juice to physically escape, thereby speeding up the drying process. However, can also refer to chemical agents which retard mold growth and allow hay to be harvested at a higher moisture level.
head - pressure measured by means of a column of water where the pressure at any point consists of the weight of water above that point and expressed in feet.
heat capacity - the heat required to raise the temperature of one square centimeter of soil one degree Celsius.
hectare - 10,000 square meters of land, or 2.47 acres.
hedging - commonly referred to as using the futures market to guarantee a sale price and thereby reduce market risk.
heeled in - to temporarily cover the roots of nursery trees so they will not dry out.
herbicide - a chemical used to kill weeds.
heretability - the mathmatical calculation of the ability of an animal to pass on quantitative traits to its offspring.
heterosis - the phenomenon associated with crossbreeding which concerns the yet unexplained improvement in traits that are otherwise of low heretibility.
homozygous - contains only dominant or recessive genes for a particular trait.
horizon - a layer of soil with distinct properties such as color, pH, and structure.
hormone - a chemical compound that has profound effect on specific target tissues.
hybrid - a plant or animal derived from two genetically different parents. A hybrid will not breed true (produce plants or offspring genetically identical to itself).
hybrid vigor - common term for heterosis.
hydraulic conductivity - flow rate of water through soil.
hydrologic cycle - the process that collectively describes the movement of water through the soil-plant system. Evaporation, transpiration, precipitation, runoff, and infiltration are all involved in the cycle.
hygroscopic water - water that is held tightly by the soil particles and is not available to plants.

immobilization - the conversion of inorganic N (NO_3^-, NH_4^+) to organic N, a form unavailable for plant uptake.
improved pastures - pastures planted with commercial grass or forage.

inbreeding - breeding animals that are related to one another.
indigenous - native
infiltration rate - rate of flow of water into the soil.
insecticide - a chemical or other substance (natural or man made) that kills insects.
integrated pest management (IPM) - the evaluation and consolidation of all available techniques into a unified program to manage pest populations so that economic damage is avoided and adverse side effects on the environment minimized.
integument - the enveloping layers which cover an animal's body (primarily pertains to insects).
intercropping - two or more crops grown simultaneously in the same, alternate, or paired rows in the same area.
ionophores - antibiotics that are used as coccidiostats in poultry, and as methane reducers in ruminants.
ions - mineral atoms that carry a plus or minus charge.

jug - a small pen where a ewe and her lambs are put for the first 24 hours after birth.

ketones - byproducts of fat metabolism in mammalian physiology.
ketosis - a build up of ketones in the bloodstream that can bring about tetany and death.

lactation - when an animal is giving milk.
lactose - the form of sugar found in milk.
legume - a type of plant that is capable of utilizing nitrogen from the air; i.e. does not require nitrogen fertilization. Indeed, legumes typically deposit nitrogen in the soil which can be used by other plants.
leverage - in finance refers to borrowed capital.
liabilities - in accounting refers to expenses or debts that must be paid.
lignin - extremely coarse fiber, indigestible to ruminants.
lime - calcium carbonate.
linebreeding - a form of inbreeding. Refers to breeding to one exceptional but related individual.
loam - a soil containing moderate amounts of clay (7 to 27%), sand (less than 52%), and silt (28 to 50%).
lodge - in agronomy refers to grain crops that lay over (due to wind, rain, etc.) making combining difficult.

macro-minerals - minerals required by animals in relatively large percentages; e.g. Ca, P, Na, Mg, K and S.
maintenance requirement - the amount of energy required by an animal to maintain its bodyweight.
mass flow - the flow of gases or nutrients in one direction, such as water containing nutrients from the soil to the roots.
mastitis - inflammation in the udder.
megacalorie - one thousand kilo calories.
metamorphic - refers to total and complete change (as in pupae to butterfly).
metamorphosis - the marked changes in form during the postembryonic development of insects.
methane - a carbon gas formed in the decomposition of vegetative matter, as in a rumen.
micro-minerals - minerals required in very small quantities; e.g. Fe, Cu, Co, Zn, Mn, and Se.
mineralization - the process of converting an element such as organic matter or soil to an inorganic form that is available to plants,
moldboard plow - a tillage tool which cuts 6 by 12 inch (usually) furrows of soil and turns them over.
molt - when birds shed their feathers.
monocotyledon - plants having a single seed leaf.
monogastric - an animal with only one stomach.
morphology - a branch of biology that deals with the form and structure of organisms.
multiple cropping - growing more than one crop on the same land in one year. Within this concept there are many possible patterns of crop arrangement in space and time.
mycotoxins - powerful poisons toxic in extremely minute quantities, produced by molds.

necrosis - a decaying of plant or animal tissue.
net worth - assets minus liabilities.
nicking - in animal breeding refers to matings between specific lines that produce exceptional individuals (for unknown genetic reasons).
nitrates - the form of nitrogen absorbed by plants (NO_3) which can be toxic to animals.
nitrification - the conversion of ammonia (NH_4+) to nitrite (NO_2^-) to nitrate (NO_3^-) by oxidation.
nitrogen fixation - bacteria associated with legumes that convert gaseous N from air into a form usable by plants.

nominal interest rate - the interest rate printed on a contract which may or may not represent the true, effective interest rate.
non-protein nitrogen (NPN) - nitrogen ruminants can synthesize into protein; e.g. urea.

omasum - the "second stomach" of a ruminant animal
organic matter - the end product of accumulation and decay of residues, mainly plants and roots, in the soil.
osmotic pressure - pressure difference of two liquids of different concentrations through a semipermeable membrane.
osteomalacia - a physiological condition whereby Ca, P and Mg are pulled out of the bones, thereby weakening the bones.
osteoporosis - porous bone tissue resulting from reabsoption of minerals.
outbreeding - breeding unrelated individuals (usually refers to within a breed).
ovarian cycle - the regular cycle whereby non-pregnant animals will go into estrus (heat), ovulate (release an egg cell), and then become sexually unreceptive until the next onset of estrus.
overshot (parrot mouth) - when the lower jaw is shorter than the upper.
oviposition - the act of laying eggs.
ovulation - egg released from the ovary.

paddy (lowland) rice - rice produced with wetland preparation of fields and grown with 2-20 inches of standing water.
parakaratosis - in swine, a severe condition of skin lesions brought about by a lack of zinc. In cattle, a sloughing of the rumen papillae, usually caused by an acid pH in the rumen.
parasitoid - an insect whose larvae feed upon the living tissues of a host insect in such a way that the host is not killed until larval development is completed.
parthenocarpy - fruit development without pollination, fertilization or seed.
parturition - act of birth.
pathogen - an organism which causes disease in plants or animals.
ped - an aggregate that is formed by natural processes; eventually becomes a specific soil structure, such as platy, granular, or block structure.
perennials - refers to plants that endure from season to season (do not need reseeding or regeneration).
pericarp (ovary) - lower part of the pistil enclosing the ovules (young seeds).
peristaltic action - waves of non-voluntary muscle contraction along a muscular tube.
pH - a measure of the acidity or alkalinity of a liquid. Measured on a logarithmic scale of 1 to 14, 7 is neutral, 1 is the most acid and 14 is the most alkaline.
phenotype - the visually apparent traits of an animal. (As apposed to genotype, which is the true genetic makeup.)
pheromone - an odor given off by an animal that stimulates other animals into activity.
photosynthesis - the process by which plants take sunlight and convert it into organic energy.
pica - depraved appetite as in the chewing of inedible objects.
pistillate flower - a flower containing only female parts (see staminate flower).
plankton bloom - a rapid growth of marine micro-organisms.
plow pan - a compacted soil layer formed immediately below the plowing depth.
plumule - the bud of a plant while still in the embryo.
polled - the absence of horns.
pollen - sporelike grains or particle originating in the anther and containing the male gametophyte.
polyestrus - multiple ovulations.
pome fruit - fruit type derived from the fusion of the ovaries, calyx cup and floral tube (apple, quince, pear).
porosity - the total volume of pore space in a soil.
power charge - the monetary cost of power used to pump a given amount of water from underground to the surface.
pregnancy toxemia (twin lamb disease) - ketosis.
pressure head - operating pressure of system converted to feet of head.
primitively wingless - insects lacking wings in the adult stage, and whose ancestors also lacked wings.
profile (soil) - the various horizons or distinctive layers of soil down to the bedrock.
progesterone - a hormone produced in great quantities during pregnancy which supresses estrus and causes the cervix to seal.
prolapse - when the rectum or vagina are turned inside-out and forced out of the body.
prostaglandins - extremely powerful hormone-like substance.
pterosaur - a member of the now-extinct group of flying reptiles.

puddled soil - dense, massive soil artificially compacted by destroying soil structure.
pullets - a hen less than one year old.
pump horsepower - actual power required to perform a task, also called brake horse power.
purebred animal - an animal of a recognized breed kept pure for many generations. A purebred animal may or may not be registered, but all registered animals are purebred.

qualitative traits - genetic traits that can be seen; horns, coat color, etc.
quantitative traits - genetic traits that must be measured; weight gain, milk production, etc.

radicle - lower part of plant embryo which forms the roots.
ram or buck - male sheep of any age that has not been castrated.
ratoon cropping - shoots that come from plant roots; e.g. sugarcane.
real interest rates - interest rate paid minus inflation.
recessive - in genetics refers to genes that express themselves only when homozygous for a particular loci.
rectal prolapse - a portion of the rectum protrudes past the anus.
registered animal - a purebred animal that has a registration certificate and number issued by a breed association.
relay cropping - planting the second crop into the first crop prior to harvesting the first crop.
reticulum - the forward, ventral portion of the rumen.
rhizobium - bacteria that attach themselves to the roots of legumes (e.g. soybeans, alfalfa, clovers) and take nitrogen from the air and change it into a form that the plants can use.
rhizomes - the vegetative growth (runners) used in seedless reproduction by some grasses.
rod weeder - a weeder designed to minimize surface soil disruption.
root interception - direct contact of a root with a nutrient in the soil that allows for uptake by the plant.
rootstock - a plant having certain desirable root characteristics to which other cultivars are grafted or budded to produce a superior performing compound tree.
roughage - feeds high in fiber.
rumen - the first "stomach" in ruminant animals used for fermenative digestion.
ruminants - animals with "rumens", which are capable of digesting fibrous feeds and synthesizing protein from non-protein nitrogenous compounds.

salmonids - fish of the salmon family, including trout.
sand - soil particle between 0.05 and 2.0 mm in diameter.
scarification - scratching the pericarp or outer coat of seeds to facilitate moisture absorption.
scion - part of a plant with certain desirable fruit characteristics that is grafted to a rootstock (see rootstock).
sclerotized - hardened and darkened, as with insect cuticle.
scoured wool - wool that has been cleaned or scoured.
scouring - watery feeces (diarrhea).
secondarily wingless - insects lacking wings in the adult stage, but whose ancestors possessed wings.
selection differential - the measured difference in measured traits between new animals introduced into the breeding herd and the average of the breeding herd.
semi-arid - area of little rainfall, usually 10 to 20 inches per year.
sequential cropping - one crop planted after harvest of the first. (Sometimes called relay planting in West Africa.)
shifting cultivation - several crop years are followed by several fallow years with the land not under management during the fallow. The shifting cultivation may involve shifts around a permanent homestead or village site, or the entire living area may shift location as the fields for cultivation are moved.
silage - generally refers to moist forages preserved through fermentation.
silt - soil consisting of particles between 0.05 and 0.002 mm in diameter.
sodic - containing large amounts of sodium (salt).
soil structure - the manner in which soil particles have joined or arranged themselves to form a distinct aggregate with unique size and shape. Platy, prismlike, crumb and angular blocky are all terms used to describe soil structure.
soil texture - the relative proportions of various soil components (sand, silt, clay, and organic matter).
soil type - the definite sequence of soil texture layers of a soil.
solid-non-fat (SNF) - the protein, mineral and lactose content of milk.
solum - the part of the soil profile containing A and B horizons.
somatic cell - a body cell as opposed to a germ cell. When found in milk can indicate an infection.

spermatheca - a sac-like organ in the reproductive system of female insects which receives and retains sperm cells until time for fertilizing the eggs.
staminate flower - a flower containing only male parts (see pistillate flower).
staple - common reference to length of wool fibers.
static head - the vertical distance water is raised.
stolons - a stem just below the soil surface which produces new plants.
stone fruit - fruit (Prunus species) containing a hard pit (peach, plum, cherry).
subirrigation - irrigation applied slowly through tubes below ground level.
subsistence agriculture - agriculture in which land tenants produce only for themselves.
super-phosphate - a phosphoric acid compound used as fertilizer.
supplementary irrigation - water in excess of rainfall necessary to produce a crop and sustain growth.
supply - In economics refers to the amounts of a good suppliers are willing to sell over a range of prices.
sweep plow - a tillage tool with horizontal blades which cut plant roots below ground level, leaving the surface undisturbed.

tagging - see crutching.
taxonomy - the systematic classification and naming of organisms.
temperature shock - a cause of death loss in fish caused by rapid changes in water temperature (since fish cannot regulate their body temperature).
tempering - adding water to soften grain.
tender wool - wool that has a weak or tender area in it. The tender area is called a break. Wool fiber that breaks at this point reduces wool value.
terminal cross - crossbred animals used only for slaughter.
tetany - a condition involving powerful muscle spasms.
thermal conductivity - the ability of a soil to transmit heat down through the soil.
top soil - surface soil.
topping - a practice of removing growth in the upper part of a tree (by hand or machine).
total digestible nutrients (TDN) - a measure of the digestibility of a feed.
total dynamic head - the total force at which a system is operating. For water at rest in a container, the pressure at any point consists of the weight of water above that point. The column of water, real or imaginary, above the point of interest is referred to as head and is expressed as feet. Pressure (in pounds per square inch (psi) and head are different ways of expressing the same thing. For water, psi x 2.32 = head in feet.
total mixed rations (TMR) - complete rations containing roughages, concentrates, minerals and vitamins.
trace-minerals - see micro-minerals.
transitional horizon - a layer of soil that contains properties of two different horizons. Designated as AB, EB, BC, or E/B, B/C, etc.
transpiration - the process by which water vapor is released to the air mainly from plant leaves.
tubers - a thickened underground plant stem.
turgor pressure - the pressure inside plant cells exerted by their contents.

unsaturated oils - oils that contain less hydrogen due to double bonding, which makes them soft or liquid at room temperature.
upland rice - rice produced in a system with dryland preparation of fields and grown with no standing water.

variable cost - costs the farmer or business has direct control over, and will vary with production. As apposed to fixed costs, which occur regardless of whether anything is produced or not.
variety - (cultivar) genetically identical plants that produce plants identical to the parents.
vector - an insect capable of transmitting a pathogen from one organism to another.
velocity head - pressure (head) necessary to maintain the velocity (speed) of water flowing in a pipe.
vernalization - exposure to cold necessary for flowering and seed (grain) production in some grasses, including some, but not all varieties of small grains.
vertical integration - in business refers to entering into a different stage of production or marketing of a product.
volatile fatty acids (VFA) - the breakdown products in the rumen used for energy.
volatilization - the loss of gaseous nitrogen to the atmosphere. Ammonia (NH_3) losses can occur from surface-applied urea and other fertilizers containing ammonia.

water holding capacity - the greatest amount of water that the soil will hold under conditions of free drainage, usually expressed as a percent of oven dry weight of soil (also called field capacity).
water soluble vitamins - vitamins soluble in water, and not stored in the body (all the B vitamins plus C).
water use efficiency - crop production per unit of water used.
wether - a male sheep castrated before the development of secondary sex characteristics.
wheat pasture poisoning - common term for magnesium tetany (an induced magnesium deficiency).
white face breeds (sheep) - wool breeds of sheep.
white muscle disease - muscular distropy in young animals caused by selenium deficiency.
wilting point - (or permanent wilting point) the water content of the soil on oven dry basis at which plants wilt and fail to recover when rewatered. It is the soil moisture content at approximately 15 bars suction.
windrow - a row of hay or other crops that have been swept together to be picked up by machine.

xanthophylls - yellow pigments, related to carotene, found in plants.

INDEX

accounting 23-27

accrual accounting 24

acid soils 69-70, 138

acidosis 179-180, 261

anhydrous ammonia 59

aquaculture
  death loss 315
  ecomonics 314
  nutrition 313-314, 317-318, 319
  reproduction 319
  temperature 316-317

artificial insemination 201, 202-205

available water 84, 106

balance sheet 23-24

barley production 115-117

basis (futures market) 30-32, 33, 34

beef cattle
  breeding cows 249, 250
  cow-calf operations 250, 254-257
  feedlot operations 260-261
  growing 260
  ranching 245-255
  stocker operations 249, 257-260

bloat 258-259

budding 128, 135

bulbs 121

calcium
  deficiencies 162
  in feeds 162-163, 166
  in plants 65-66
  in soil 71

California Net Energy System 195

calving, spring vs fall 250

capillary water 42

carbohydrates 159, 167
  cellulose 159
  digestibility 157
  fiber 160

cash accounting 24

cash flow 5, 6, 25-27

cattle breeds
  British 261-262
  continental 262-264
  dairy 241, 242, 243
  dairy beef 264-265
  Zebu 264

chromosomes 209, 213-214

clay 42, 44-45, 59, 77, 84

climate zones 99, 127, 137

cobalt 164

collateral 6

colostrum
  dairy cattle 240
  sheep 297
  swine 282

concentrates 157
  dairy cattle 236, 238-239

copper 163, 164

corn
  production 113
  moisture 114, 160
  silage 114

costs
  fixed 19-20
  opportunity 20
  total 19-20
  variable 19-20

cotton 118-119

cow-calf operations 250, 254-257
  breeding 253, 256
  confined 254-257
  feeding 250, 251-254
  on pasture 250-251
  supplements 251-254

crop production
  dry matter 86
  yield 100-101, 105, 110, 140

crop residue 47-50, 59, 75, 104, 110
  grazing 250, 251, 258

crop spacing 78, 113, 114, 116-117, 118, 119

crop water requirement 85

cropping systems 48, 102-105, 108-109, 111, 138
  burning 138, 143
  inter-cropping 109, 139, 140

multiple cropping 139-140
pasture 141-143
plantation 143
ratoon cropping 109
relay cropping 109
sequential cropping 139-140
shifting 138-139, 144

crossbreeding 211, 270-271, 302-303

cultivars 127-128, 135

cultivation 76

dairy cattle 231-244
challenge feeding 238
feeding 235-243
genetic improvement 243-244
housing 234
milking 234
production grouping 237-238
production schedule 241, 244
raising replacements 239-241
total mixed rations 236-238

demand 15, 21
elasticity of 15, 21
shift 18

depreciation 24-25

dicotyledon seed 121

diffusion 43, 55-56, 63

digestion 175-180
enzymes 175-176
monogastric 175-176, 180
poultry 216
ruminant 176-180, 237

discing 75-76, 77

diseconomies of scale 20

dry matter composition of feeds
air dry 159, 189
as fed 160, 189, 191

dryland agriculture 99-112
zones 99

ducks 210

E. coli 255

economies of scale 20

eggs
formation 215
incubation 218
processing 225-226

eluviation 40

embryo transplant 201-202, 205

endosperm 121

enterotoxemia 303

epistasis 213

equilibrium (price) 18

equity 10

estrus 200
synchronization 205, 279

evaporation 106

evapotranspiration 85

fallowing 102-103, 104, 105, 138

fats 159

feeds
dehydrating 173
dry matter 159, 160
moisture 157-158, 189
processing 169-174

fertilizer 44-50, 57, 59, 62, 66-67, 69, 71-73, 78, 83, 110
nitrogen 59-60, 63, 143
phosphorus 63
tree 132
urea ammonium nitrate 59

fiber 160, 161
cellulose 159, 160, 179
digestion 169, 179-180
lignin 160

field capacity (water) 42, 84

financial risk 3, 6, 9-14, 29

financial statement 23

financing (loans) 4, 5

forward contracting 3, 9, 13
marketing 10
pricing 10-11, 14
selling 3, 10-12

founder 179-180, 261

fruit
harvesting 132
processing 133
production 130-133
storage 132

fungicides 123, 132

futures market 11, 14, 29-36
delivery 30, 34
options 32-33

genes 209-211

genotype 210

grafting 128, 135-136

grain
  in dairy rations 236, 238-239, 242
  feeding 261
  grinding 169
  high moisture 171
  in swine rations 274
  popping 171
  pressure flaking 171
  processing 169-173
  reconstituted 172
  roasting 173
  rolling 169, 170-171
  steam flaking 170-171
  tempering 173

gravitational water 42

grazing
  continuous 248
  deferred 249
  deferred-rotational 249
  high intensity, low frequency 249
  multiple pasture systems 248-249
  residues 251, 258
  rest-rotational 249
  rotational 249
  seasonal 248
  short-duration 249

handmating 202

hay 181-185
  conditioners 183
  moisture 158, 160, 182, 183
  preservatives 183
  quality 185
  rained-on 184
  small grain 115
  tropical 143

heat capacity (soil) 43

hedging 29-32, 33, 34

herbicides 78, 83, 105

heritability 211, 214

heterosis 211, 214

heterozygous 210, 214

homozygous 210, 214

hybrid vigor 211

hybrids
  plants 113
  seeds 122

hydraulic conductivity 84

hygroscopic water 42

income statement 24-26

inflation 21-22

insect control
  biological 152
  chemical 152
  cultural 152
  genetic 152
  mechanical 152
  regulatory 152

insecticides 83, 132, 148, 152
  resistance 148

insects 145-154
  control 83, 132, 147-148, 149
  economic injury level 149
  identificiation 146, 147
  metamorphosis 146
  reproduction 145-146
  sampling 148-149

insurance
  crop 11-12, 32
  deposit 21

integrated pest management 146, 148-149, 153

interest rates 5
  nominal 21
  real 21-22

iodine 164

iron 164, 282

irrigation 81-89, 91-98
  application efficiency 89
  basin 81
  canals 93, 95
  drip 82
  furrow 81
  ground water 88
  pumping plant 88
  sediment 94
  sprinkler 83
  sub-irrigation 82
  surface water 87, 91-92
  system maintenance 93-95, 97
  timing 86-87
  tree 132
  uniformity of application 88-89

ketosis 242, 294-295

leaching 59, 64, 66-67, 70-71, 78, 138

levered capital 4

liming 69-70

linebreeding 212

liquidity 23-24, 26
  long term capital 5

magnesium
  in feeds 163
  plant 65-66
  tetany 163. 251

manganese 164

manure
  poultry 222
  swine 287-288

margin calls 11, 14, 31, 35

mass flow 43, 55, 57

mastitis 234-235, 304

micronutrients in plants 67-69

milk 231-233
  fever 241
  pricing 231-233
  productihn 233-234

mineralization 47, 48, 59

minerals
  dairy cattle 236, 241
  in nutrition 161, 167
  macro 162
  micro or trace 162, 163-165

monocotyledon seed 121

mycotoxin 286

net energy 161, 195-196

nicking 213

nitrate toxicity 259-260

nitrification 59

nitrogen
  dentrification 59
  immobilization 59
  in feeds 160
  non-protein 161
  plant 57-58, 59-60, 62, 138
  volatilization 59

non-protein nitrogen 161, 167

nuts
  harvesting 134
  production 134

oat production 115-117

operating capital 4, 7, 9

options call 34
  put 32-33
  trading 32-33

organic matter 43-45, 47, 49, 63, 69, 84, 102, 107-108

osteomalacia 162

outbreeding 212

ovarian cycle 199-200
  swine 278

pasture mating 202

pasture systems
  grazing 248-249, 250, 257-258
  legumes 143
  tropical 141-143

Pearson's square 189-191

pelleting 169, 274, 318

pheasants 210

phenotype 210

phosphorus
  deficiencies 162
  in feed 162-163, 166
  in plants 57, 62

pica 162

plant deficiencies 56
  nutrients 55, 138

plowing 75, 76-77

plumule 121

pollination 129-130

porcine stress syndrome 286

potassium
  in feeds 163
  in plants 57, 63-65

poultry
  beak trimming 220-221
  breeders 226
  broiler production 227-228
  digestion 216
  diseases 216-218
  hatchling management 219
  housing 222
  layer nutrition 224-225
  laying hens 223
  laying house management 225
  litter 220
  manure management 222
  molting 222-223
  pullet nutrition 224
  toe clipping 221-222

precipitation 41-42, 99-100, 102, 106, 108

pregnancy determination
  swine 279
  sheep 295

price competition 12-13

product differentiation 12-13

prolapsed uterus 304

protein 159
　amino acids 159
　apparent digestible 161
　crude 160
　digestible 160
　NPN 161

pruning 128-129

radicle 121

range management 245-249
　grazing systems 247-249
　overgrazing 245-247
　plant succession 246
　reseeding 247

ration formulation
　adjusting moisture 189
　balancing two nutrients 190
　computers 191, 193
　estimating consumption 194
　feedlot 260-261
　for milk production 196
　growing 260
　least cost 192-193
　meeting maintenance requirements 194-195
　predicting performance 193-194, 195
　swine 276-278

reproduction 199-207
　and nutrition 205-207
　beef cattle 252-253
　dairy cattle 244
　fish 319
　sheep 292
　swine 278

reproductive anatomy
　chicken 215
　mammalian female 199
　mammalian male 200-201

rhizomes 121

rice 140-141, 144

rickets 162

root interception 55, 56

roughage 157, 167
　dairy cattle 236, 238
　digestion 178
　pelleting 169
　processing 169, 181-188

rumen
　microorganisms 177-179
　pH 179, 237

ruminant
　concentrate feeding 179, 237
　energy digestion 177
　fat digestion 177-178
　protein digestion 177
　roughage utilization 178, 237

saline soil 70-71

salt 163, 242

salty water irrigation 83, 87, 93

sand 42-43, 84, 126

scours
　calf 255
　sheep 304

scutellum 121

seeding 125
　depth 126
　rate 126
　sodding 125
　sprigging 125
　timing 126

seeds
　breeder 124
　certification 124
　dormancy 124
　foundation 124
　germination 122-124
　laws 125
　quality 122, 125
　registered 124
　size 123-124
　sources 124-125
　storage 124
　varieties 122

selenium 164, 166, 304

sheep
　breeding 292, 293-294, 300-301, 302
　breeds 303, 306-311
　creep feeding 299, 300
　crutching 295
　diseases 303-304
　docking 298
　extensive production 292
　gestation supplementation 292
　intensive production 292
　lambing 295-298
　nutrition 299, 300-301
　parasites 304
　pregnancy determination 295
　pregnancy toxemia 294
　ram management 301
　reproduction 292-293
　stock selection 301-303
　weaning 299
　wool 304-306

silage 185-188
　additives 188

corn 114
digestibility 187, 188
moisture 185
pH 186, 187
small grains 115
sorghum 115
tropical grasses 143

small grains
grazing 115, 257-258
silage 115
grains hay 115
production 115-117

sodic soil 70-71

sodium 163

soil
aeration 43, 45
aggregate 44-45
buffering capacity 46
consistency 45
depth 46
development 39-40
erosion 50, 78, 83, 106-107, 139, 247
fertility 55-73
horizons 40
infiltration rate 84, 106
organisms 47
pH 46, 47, 62, 69-70, 71, 102, 138
porosity 45
structure 44-45
temperature 43
testing 71
texture 43-44
water 41-42, 106
water capacity 42-43, 84
water measurement 86-87
weathering 40, 46, 64-65

sorghum
moisture 115
production 114-115

soybeans
production 117-118
toasting 173

speculation 32, 35-36

stocker cattle 249, 257-260
health problems 258-260
on crop residues 258
on improved pastures 257-258
on range 257

stolons 121

subsistence agriculture 3

sulfates 163

sulfur
in feeds 163
in plants 65, 66

supply 16-18, 21
change 18
elasticity of 17-18, 21
shift 18

swine
boar management 280-281
breeding systems 270-271, 279-281, 283, 287
breeds 268-270
disease 284-285, 286, 287
estrus detection 278
estrus synchronization 279
farrowing 281-282
feeding programs 276-278, 287
gilts 278
growing-finishing 284-285, 287
housing 268, 281-283
marketing 287
mycotoxin 286
nutrition 273-278, 284, 287
operatiion layout 271-272
operation location 271
parasites 285-286
produciition systems 268, 271-272, 285
records 288
reproduction 278-279
stock selection 272-273, 283, 287
waste managment 287-288
weanling pigs 283-284

thermal conductivity (soil) 43

tillage 48, 51, 75-79, 103-105, 110

total digestible nutrients (TDN) 157, 161, 195, 252
pricing 192-193

transplanting
seedlings 125
trees 128

tropical agriculture 127, 137-144

tubers 121

turkeys 228-229

urea
as a feed 161 (see also non-protein nitrogen)
as fertilizer 59

velocity of money 21

vertical integration 12

vitamins 165- 168
A 165
B1-B12 166-167
C 166
D 166
dairy cattle 236
E 166, 304
K 166

water
as a feed 157-159

waterbelly 260

weaning
 calves 256-257
 sheep 299
 swine 283-284

wheat production 115-117

white muscle disease 304

wilting 42, 86

zinc 165

Order Form

SWI Publishing
P.O. Drawer 3 A&M
University Park, NM 88003
U.S.A.
505-525-1370

_____ copies **Modern Agriculture** @ $34.50 _____
science, finance, production & economics

_____ copies **Beef Production** @ $24.95 _____
science & economics; application & reality
hardcover @ $32.00 _____

_____ copies **Intelligent Dieting** @ $13.95 _____
hardcover @ $17.95 _____
(A review of animal food products in human nutrition.)

No shipping or handling charges.
20% discount on multiple books.
30 day unconditional guarantee on all books —
return for a full refund.

Name _____
Address _____
City _____
State, Zip _____

checks accepted or
VISA / MasterCard
Card No. _____
Expiration date _____
Signature _____